ADVANCES IN THE CRYSTALLOGRAPHIC AND MICROSTRUCTURAL
ANALYSIS OF CHARGE DENSITY WAVE MODULATED CRYSTALS

Physics and Chemistry of Materials with Low-Dimensional Structures

VOLUME 22

ADVANCES IN THE CRYSTALLOGRAPHIC AND MICROSTRUCTURAL ANALYSIS OF CHARGE DENSITY WAVE MODULATED CRYSTALS

Edited by

Frank W. Boswell

Department of Physics, Faculty of Sciences,
University of Waterloo, Ontario, Canada

and

J. Craig Bennett

Department of Physics, Faculty of Pure and Applied Sciences,
Acadia University, Wolfville, Nova Scotia, Canada

SPRINGER-SCIENCE+BUSINESS MEDIA, B.V.

A C.I.P. Catalogue record for this book is available from the Library of Congress.

ISBN 978-94-010-5945-9   ISBN 978-94-011-4603-6 (eBook)
DOI 10.1007/978-94-011-4603-6

Printed on acid-free paper

TABLE OF CONTENTS

PREFACE

Charge density wave modulated materials have been studied intensively for the past twenty years and a number of books and reviews have been published dealing with the physics and chemistry of these materials as well as the underlying theory of charge density waves. Particular emphasis, both experimental and theoretical, has been devoted to the unusual electrical properties of these materials. Although many aspects of the unusual physical and structural properties have now been explained, numerous subtleties of their behaviour remain unclear.

Since charge density wave modulations have periodicities which are generally incommensurate with the parent lattice, these materials inherently possess unique microstructures associated with the discommensurations, dislocations and antiphase boundaries required to accommodate the modulations and their phase transitions. An understanding of these microstructural defects is essential for any microscopic theory of charge density wave dynamics and phase changes. Experimental data on these microstructures is sparse for two main reasons: many of the low-dimensional charge density wave materials contain large numbers of conventional lattice defects which mask the modulation microstructures, and in most cases the charge density wave transitions occur below room temperature, imposing severe constraints on the usual techniques of analytical and high resolution microscopy. In spite of these difficulties, detailed work has been carried out on a very restricted set of charge density wave modulated crystals and in this book we have brought together the evidence gathered so far. The most direct evidence has been acquired by the methods of electron microscopy, however the interpretation of this data is highly dependent on related crystallographic and x-ray diffraction analyses. Chapters providing this essential background are therefore included. Closely related to the subject of modulation microstructures is that of the interaction of impurities and defects with charge density waves on the microscopic scale and in one chapter the current state of knowledge is presented. The final chapters deal with experimental and theoretical aspects of the powerful new methods of scanning tunneling and atomic force microscopy as applied to the study of charge density wave systems. The atomic resolution that can be achieved by these methods has already revealed significant new data and promises to play an increasingly prominent role in the investigation of charge density wave systems.

This book is intended as both an introduction for graduate students as well as a reference for researchers working in the field. It is written as the science of charge density waves stands poised to enter a new era of technological development. Recent advances in the synthesis of thin films and composites of these materials have brought a myriad of possible applications in electronic devices within reach. As experience with silicon based devices has abundantly shown, detailed microstructural characterization over a wide range of resolutions is vital if this potential is to be fully realized. We have found that no modern survey of the microstructural phenomena associated with charge density waves exists and present this volume with the objective of stimulating new interest in this fascinating aspect of the behaviour of a truly remarkable class of materials.

Frank Boswell and Craig Bennett

July 1998

ALTERNATIVE APPROACHES TO THE CRYSTALLOGRAPHIC DESCRIPTION OF CHARGE DENSITY WAVE MODULATED SYSTEMS

ALBERT PRODAN[1] and ANDRZEJ BUDKOWSKI[2]

[1]*"J.Stefan" Institute, Jamova 39, Ljubljana, Slovenia*
[2]*Institute of Physics, Jagellonian University, Reymonta 4, Cracow, Poland*

1. Introduction

It has become increasingly evident that many compounds do not easily conform to the standard definition of a crystalline solid, i.e. that of a lattice of atoms repeating periodically in three dimensions. As early as the beginning of this century, it was already known that in some minerals (e.g. calaverite $Au_{1-x}Ag_xTe_2$) the crystal faces violate the standard rule of rational indices [1,2]. However, it was not until after the discovery of X-rays and their subsequent use in structure determination that the origin of phenomena like crystal structure modulation and its incommensurability with the undistorted lattice could be investigated [3,4,5]. However, since modulation effects as a rule represent small perturbations of the basic structure, the corresponding X-ray diffraction effects are also weak. In addition, these effects are often restricted to small domains, which are below the resolution of X-ray analysis. The development of transmission electron microscopy and diffraction, with a wide variety of diffraction and microanalytical techniques, opened up new avenues of investigation. Structures modulated commensurately or incommensurately in one, two and three dimensions were found in metals, semiconductors and insulators [1,6]. These modulations can be generally classified as displacive or compositional in origin. Whereas in cases of compositional ordering and intergrowth phases [7,8,9], the very concept of modulation can often be questioned (since it is not always clear which of the sublattices should be considered the basic one), a displacive modulation as a rule exhibits weak distortions which follow a charge density [10] or magnetic ordering [11].

Charge density waves (CDW) and the accompanying periodic structural distortions (PSD) were intensively studied over the last decade, since they are the origin of structural phase transitions and temperature-dependent anomalies in physical properties [12,13,14]. There are two important conditions for the formation of a CDW/PSD [15]. First, a "nesting" instability in electronic susceptibility must take place and second, strong electron-phonon interactions are needed to trigger the structural distortion. The largest families of compounds, where these two

1

F. W. Boswell and J.C. Bennett (eds.),
Advances in the Crystallographic and Microstructural Analysis of Charge Density Wave Modulated Crystals, 1–39.

conditions are fulfilled, are the transition-metal di-, tri- and tetra-chalcogenides. These compounds undergo numerous phase transitions, in which high-temperature, often diffuse, scattering transforms into sharp satellites at low-temperatures. Although these effects can be observed by a range of techniques, transmission electron microscopy and diffraction [16,17,18] are the most suitable. Recently, scanning tunneling microscopy [19] and synchrotron radiation diffraction [20] have also been used.

Since the CDW is driven by electronic considerations, the PSD is often incommensurate with the underlying lattice. With the growing number of known incommensurate phases a need for appropriate crystallographic descriptions has also developed. A (3+d)-dimensional crystallography with corresponding superspace groups [21–24], the analogous dualistic interpretation with separate basic structure and modulation descriptions [25] and normal mode symmetry analysis [26,27,28] based on the Landau theory of second-order phase transitions, are today the most frequently used approaches to this problem.

In the first part of this article an overview of the various approaches used for the description of modulated structures is given. These include those mentioned above as well as the long-period superstructure description and some further methods, which are less commonly used but are included here for completeness. Our intention is by no means to offer a detailed description of the various approaches, but rather to provide an overview of the basic concepts and characteristics, so that their application, given in the second part of this article, can be readily followed. For those interested in more detail, a number of excellent review articles and books are available [1 and the references therein, 11–15]. The applications have been chosen to illustrate the utility of the different approaches in the description of CDW phases. These include two polytypes of one-dimensionally modulated NbS_3 and the $NbTe_4/TaTe_4$ system, where a series of temperature- and composition-dependent phase transitions have been reported.

2. The Symmetry Description of Modulated Structures

2.1. THE SUPERSPACE GROUP (SSG) APPROACH

In an incommensurately modulated structure (MS), ordinary symmetry properties are apparently lost but can be restored by considering the structure to be a three-dimensional section through a hypothetical (3+d)-dimensional entity, called a supercrystal [21]. The symmetry is then described in the corresponding superspace by means of superspace operators, whose rotational and translational parts are defined by the symmetry of the diffraction pattern and by the corresponding extinction rules.

All reflections of a MS are characterized in reciprocal space by a quasi-lattice M^* of rank (3+d), whose nodes k are given by:

$$k = \sum_{i=1}^{3} h_i a_i^* + \sum_{j=1}^{3} m_j q_j , \tag{1}$$

where a_i^*'s belong to the lattice Λ^* of the main spots, which corresponds to a hypothetical average structure (AS) as well as to the undistorted basic structure (BS) [29]. The satellites q_j are related to Λ^* via a (d x 3)-dimensional matrix σ_{ji}:

$$q_j = \sum_{i=1}^{3} \sigma_{ji} a_i^* \tag{2}$$

M^* is a three-dimensional projection of the (3+d)-dimensional reciprocal lattice Σ^*, spanned in the reciprocal superspace (i.e. a direct sum of the reciprocal external (V_E^*) space and internal (V_I^*) space) by the vectors:

$$b_i^* = (a_i^*, 0) \quad i = 1, 2, 3 \quad \text{and} \quad b_{j+3}^* = (q_j, e_j^*) \quad j = 1, ..., d. \tag{3}$$

The direct space is accordingly divided into V_E and V_I and the corresponding direct lattice Σ is defined by:

$$b_i = (a_i, -\Delta a_i) \quad i = 1, 2, 3 \quad \text{and} \quad b_{j+3} = (0, e_j) \quad j = 1, ..., d \tag{4}$$

where the internal components $-\Delta a_i$ of the external basic vectors a_i are given by the matrix σ_{ji}:

$$\Delta a_i = \sum_{j=1}^{d} \sigma_{ji} e_j = \sum_{j=1}^{d} (q_j \cdot a_i) e_j . \tag{5}$$

Thus, by introducing Σ and Σ^* any atom μ, located in the external space at:

$$X^\mu = \sum_{i=1}^{3} x_i^\mu a_i , \tag{6}$$

can be given (a) new coordinate(s) x_{3+j}^μ :

$$x_{3+j}^\mu = x_1^\mu q_j^1 + x_2^\mu q_j^2 + x_3^\mu q_j^3 + t_j , \tag{7}$$

where t_j specifies the three-dimensional section through a supercrystal. In this way the translational invariance is recovered [21,23,30]. In Figure 1, it is shown how the main reflections Λ^* and the satellites, creating together the quasi-lattice M^*, can be considered a projection of a four-dimensional lattice Σ^* in the corresponding space $R_4 = V_E^* \oplus V_I^*$ onto a three dimensional space $R_3 = V_E^*$. Unlike the three-dimensional quasi-lattice M^*, the corresponding four-dimensional lattice Σ^* exhibits full translational symmetry. Vectors b_1^*, b_2^* and b_3^* determine the lattice of the main reflections, which correspond to the dominant features of the structure, b_4^* determines the position of the satellites in R_3 and the unit vector e_1^* is perpendicular to R_3. In R_4 each atom is represented by a string in the b_4 direction, which again appears as an atom for each three-dimensional section t=constant. Cases for a normal crystal, a substitutional modulation and a displacive modulation are shown in Figure 2. In the first case the density ρ^s

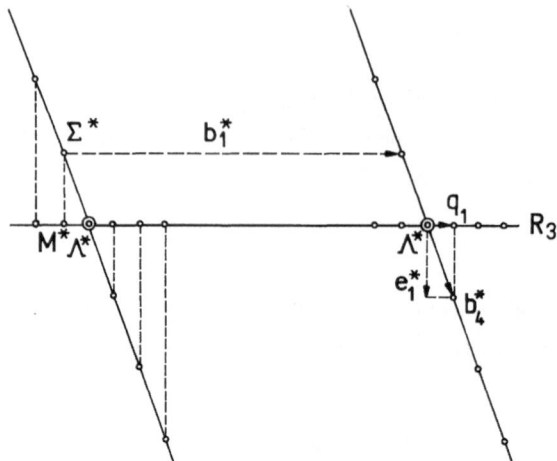

Fig.1 Main reflections Λ^* and satellites of quasi-lattice M^* in the three-dimensional space R_3 as a projection of points Σ^* in the corresponding four-dimensional space R_4 [21].

does not depend on x_4^μ and any section t=constant (note that t=0 in Figure 2) through the string-like "atoms" is perfectly periodic with a period a_1. Conversely, for substitutional and displacive modulations the periodicities are lost for t=constant, since the atoms S and S' differ in their occupational probabilities and external coordinates x_i^μ (i=1,2,3), respectively. The corresponding modulation functions, regardless of their origin, will be functions of internal coordinates x_{3+j}^μ of the undistorted structure [22].

The MS point symmetry is determined by the orthogonal transformations R [22], represented by ((3+d) x (3+d)) matrices $\Gamma(R)$ [24]:

$$\Gamma(R) = \begin{bmatrix} \Gamma_E(R) & 0 \\ \Gamma_M(R) & \Gamma_I(R) \end{bmatrix}. \tag{8}$$

From the four blocks, the (3 x 3)-dimensional $\Gamma_E(R)$ and the (d x d)-dimensional $\Gamma_I(R)$ matrices are active only in the external and internal spaces respectively, while the (d x 3)-dimensional integral matrix $\Gamma_M(R)$ can always be set to zero by a proper choice of the BS unit cell [24]. In that case each R is reduced to a pair (R_E, R_I) whose components act in V_E and V_I, respectively. The forms of $\Gamma(R)$ and σ_{ji} (Equation 2) are correlated, thus imposing symmetry restrictions onto the wave-vectors and vice versa [24]. Each matrix σ_{ji} can be expressed as a sum of two parts σ^i and σ^r, related to $\Gamma(R)$ as:

$$\sigma^i \Gamma_E(R) - \Gamma_I(R)\sigma^i = 0 \quad \text{and} \quad \sigma^r \Gamma_E(R) - \Gamma_I(R)\sigma^r = \Gamma_M(R) \tag{9}$$

where σ^r has only rational entries and is set to zero for $\Gamma_M(R) = 0$, while there are no restrictions on σ^i.

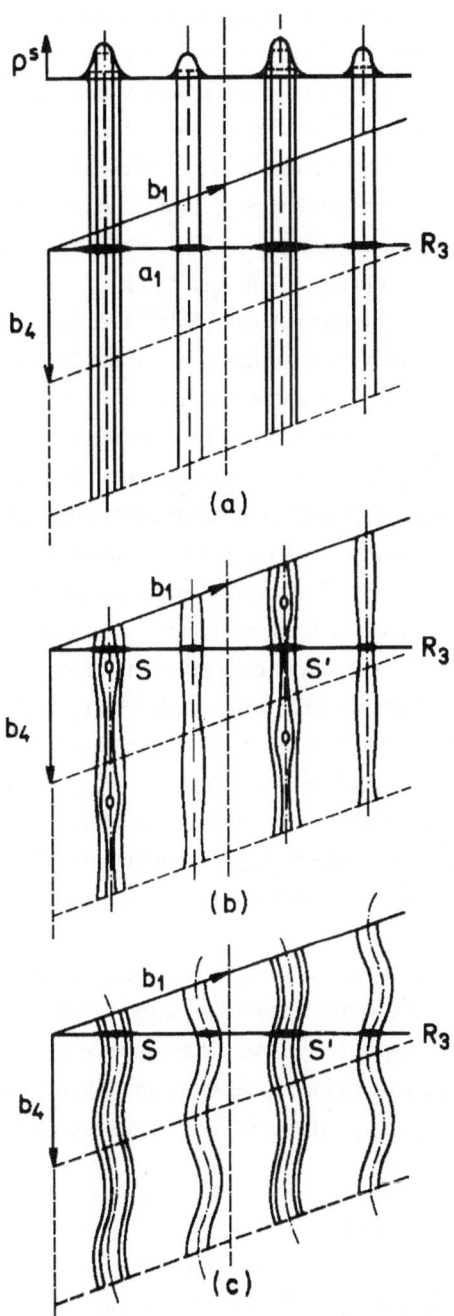

Fig. 2 The density function ρ^s in the four-dimensional space for (a) a normal crystal, (b) a structure with substitutional modulation and (c) a structure with displacive modulation [21]. The real structures are three-dimensional sections (here t=0) through ρ^s.

In analogy to three-dimensional crystallography, Bravais classes and corresponding SSG may be defined for supercrystals. A Bravais class is represented by an equivalence class of matrices $\Gamma(K)$, where K is the holohedral point group, which leaves M^* and Σ^* invariant. Bravais classes for incommensurate structures with d = 1,2,3 [24] as well as the additional ones for the commensurate cases are tabulated [31]. These tables are particularly useful in cases with d ≥ 2, for which the corresponding SSG's are yet to be worked out explicitly.

A full SSG, which determines the symmetry of a supercrystal $\rho^s(x_1,...,x_{3+d})$ in the super-space $V_E \oplus V_I$, is a group of operators $\left(R_E R_I | \tau' \tau_4...\tau_{3+d}\right)$, under action of which the supercrystal remains invariant. The components $\left(R_E | \tau'\right)$ belong to the space group G_B [23], referred to as the basic space group (SG), which is one of the 230 possible three-dimensional SG's [32]. A complete list of all 775 non-equivalent (3+1)-dimensional SSG's (excluding those applicable to commensurate cases only [31]) was published by de Wolff et al. [23] and completed by Yamamoto et al. [30]. Two notations are used. In the two-line notation, a SSG is represented by a capital latter, followed by a superscript and a subscript line. The capital letter indicates the form of σ^r (for $\sigma^r = 0$ the letter P is used for any $d \le 3$, whereas in the case of internal centering with $\sigma^r \neq 0$, different letters are used for d = 1 and for d > 1 [23,24]). The superscript line gives the basic SG and the subscript line the internal part of the SSG $\left(R_I | \tau_4...\tau_{3+d}\right)$. For d > 1, the subscript line is given in accordance with the notation of the International Tables of Crystallography [32], while for d = 1 a special notation [23] is used: for $\left(+1 | \tau_4\right)$ the intrinsic rational increment Δt of the internal parameter t is given (i.e. 1,s,t,q,h for 0,1/2,±1/3,±1/4,±1/6, respectively), while in the case of $\left(-1 | \tau_4\right)$ only $\bar{1}$ is possible, since an origin can always be found for which $\Delta t = 0$. In the analogous one-line symbol, the basic SG and the internal part of the SSG are separated by a set of vectors Q_i^*, which under action of the point group operators R and together with a_i^* ($i = 1,2,3$), generate all points of the quasi-lattice M^*.

Finally, the structure factors F_k and F_{k^s}, which determine the intensities at nodes k of M^* and k^s of Σ^*, are the Fourier transforms of the MS electron density $\rho(x_1,x_2,x_3)$ and that of the superstructure $\rho^s(x_1,...,x_{3+d})$, respectively. As a consequence of the one-to-one relationship between M^* and Σ^*, in the case of an incommensurate modulation these are identical and the non-lattice translations $(\tau',\tau_4,...\tau_{3+d})$ of the SSG operators are determined by the extinction rules [22,33]:

$$\sum_{j=1}^{3+d} H_j \tau_j \neq 0 \quad (\text{mod } 1) \tag{10}$$

for any reflection $\left(H_1,..,H_{3+d}\right)$, which fulfills the requirement:

$$H_i = \sum_{j=1}^{3+d} \Gamma_{ij}^T(R)H_j \quad (i = 1,...,3+d). \tag{11}$$

Additional extinction rules are obtained as a result of internal centering. It should be noted, however, that violation of the systematic extinctions is possible in the case of commensurate modulations, where the higher-order satellites coincide. The reflection conditions are published together with the corresponding SSG [23].

2.2. THE DUALISTIC INTERPRETATION (DI)

Since any modulated structure (MS) is obtained by multiplying the atomic density functions of its basic structure (BS) with the modulation pattern (MP), the diffraction pattern is a convolution of the individual Fourier transforms of BS and MP. Accordingly, the symmetry of an (in general incommensurate) MS can be described by the corresponding space groups S_B and S_M [25,34]. However, there are restrictions imposed on these two SG's by the virtual invariance of the MS under any translation t_i of the MP along the modulation direction(s). Thus, S_B and S_M are replaced by the appropriate subgroups $G_B \subset S_B$ and $G_M \subset S_B$.

The two dualistic principles, which formulate the required symmetry restrictions, are expressed through introduction of the modulation pattern elements (MPE), determined in those BS sub-spaces in which the translational symmetry is lost. A MPE, which depends on the type of modulation and on the BS, is represented by a one-, two- or three-dimensional space with a defined modulation function. An example of a planar MPE is shown in Figure 3, where a one-dimensional modulation in a three-dimensional monoclinic structure occurs in a plane perpendicular to the unique axis c. Translational symmetry exists along the c axis but is lost in the plane perpendicular to it. Such parallel planes form the two-dimensional MP with its two orthogonal basic vectors pointing along the modulation direction and along the unique axis. The MP unit cell constants, whose directions are contained in the MPE, are irrational and defined by the modulation wave-vectors $q_j = a^*_{Mj}$, while the additional rational constants coincide with those of the BS, including a possible centering. The two dualistic principles can be formulated as follows [25]:

(i) each atom of the BS belongs to only one MPE.

(ii) any MS symmetry operation, which is also a symmetry operation of the MP $(g_M \in S_M)$, can be described as a symmetry operation of the BS $(g_B \in S_B)$ (applied to the actual MS) supplemented by a shift t_i of the corresponding MP:

$$g_M = g_B t_i . \tag{12}$$

The first principle implies that no MP unit cell can be determined if an atom is a part of several MP's, while a direct consequence of the second principle is that point group components of g_M and g_B form identical point groups $K_M = K_B = K$ and that projections of lattices Γ_B of BS and Γ_M of MS coincide along the invariance translation t_i.

In analogy to the SSG description a special DI notation is also used. The MS is characterized by a two-line symbol [25]. In the top line, a small letter determining the crystal family of G_B, is written in front of G_B itself. The bottom line gives information on the orientation and dimensionality of the modulation, followed by the corresponding G_M. A one-line symbol can

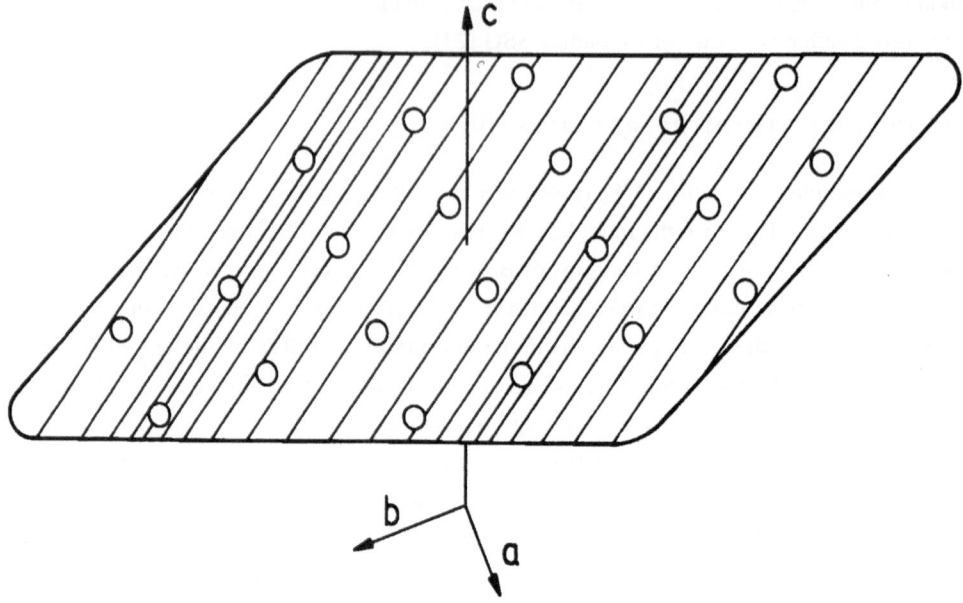

Fig.3 A planar MPE for a mono-atomic monoclinic structure, with the modulation (d=1) wave-vector perpendicular to the unique c-axis [25].

also be used if the described top and bottom lines of the standard notation are written sequentially and separated by a slash (/). If obvious, the data preceding both G_B and G_M are often omitted.

For cases with d = 1 there is a complete equivalence between the DI and the SSG approach. However, for d > 1, structures may exist where the DI is not applicable, since the corresponding MPE cannot be defined. In all other cases it is easy to establish a full equivalence between both descriptions since their BS's and the corresponding space groups G_B's are equivalent. For the simplest case with d = 1, the internal parameter $t = x_4 - q \cdot r$ in the SSG description represents the phase of modulation. In the DI, t defines the origin of the MP unit cell with regard to that of the BS. In the SSG description, any change of the internal parameter Δt corresponds to a different three-dimensional section through the supercrystal. Translated into the language of the DI, this is equivalent to a shift of $t / |q|$ of the MP with respect to BS. Thus, for (3+1)-dimensional MS's, the DI can be derived from the corresponding SSG and vice versa [25].

2.3 THE LONG-PERIOD SUPERSTRUCTURE (LPS) APPROXIMATION

There are two criteria for a modulated structure to be considered incommensurate (IC). The first requires that the modulation parameter (the ratio between the periodicities of the basic structure (BS) and the modulation pattern (MP) along the modulation direction(s)) be an irrational number. Secondly, the variation of the modulation wave-vector must be a continuous function of temperature or pressure (if such variation is present) [34]. Since both criteria depend on the accuracy of the experimental measurements and thus on finite statistical errors (see e.g. [35,36]), it is always possible to approximate an IC modulation parameter by a ratio of two integers u/v. Similarly, a continuous change of the modulation wave-vector can be regarded as series of infinitesimal rational steps, which may or may not be separated by continuous subregions [1,37,38]. As a consequence, under certain conditions, it is justified to treat an incommensurately modulated structure (MS) as a LPS, where the accuracy of the approximation depends on the choice of the unit cell. Its symmetry depends not only on the modulation period, but also on v, the corresponding multiplication of the unit cell. Since symmetry operations must be common to both BS and MP, an improperly chosen v will reduce the resulting symmetry and increase the number of structural parameters needed for a complete description of the modulated structure [34]. Such a symmetry dependence on the parity combination of u and v was studied for thiourea and Rb_2ZnX_4 (X=Br, Cl) [39], where it was shown that only one parity combination yielded glide-planes common to both the BS and MP.

The possibility of transforming an incommensurately MS into a long-period commensurate (COM) superstructure is based on an universal property of dynamical or structurally modulated systems. For a continuously changed (external) parameter, the ratio of competing frequencies or spatial periodicities locks to rational numbers, represented on the corresponding plot by an infinite series of steps. The result is a so-called "devil's staircase", where between any two steps there is an infinite number of new steps. If any region of this plot is magnified, the resulting curve will look like the original one and the process can be proceeded infinitely [37]. Depending on the nature of the parameters involved, an infinite number of high-order locked COM phases may or may not be separated by an infinite number of truly IC phases [38].

There has been a large number of structures reported where competition between spatial periodicities results in "devil's staircases". Many inter-metallic compounds, such as Au_3X (X=Mn,Cd,Zn) exhibit LPS's based on a periodic introduction of stacking faults [40]. A spectacular example of this kind of behaviour was found in the Ti-Al system, where more than 20 different LPS's were observed with temperature variation [40]. Another example is the temperature dependent magnetic structure of erbium [41] which can adopt various sequences of up-spin and down-spin ordered layers. Also, "staging" (i.e. periodic filling of the gaps between Van der Waals bonded layers by guest species) has been reported for intercalated graphite [42] and alternate ordering of cubic and hexagonal close-packed stacks form the superstructures in the ternary alloys $MgZn_2$-$MgAg_2$, known as Friauf-Lavais phases [43]. On the other hand, experimentally observed phenomena like phasons [44,45] and "satellite" single-crystal faces [1] leave no doubt that truly IC MS's indeed represent a novel class of materials. In the limit however, there is no clear distinction between a purely IC modulation phase and a long-range-ordered "devil's staircase" and it is thus justified to describe the symmetry of MS's, regardless of whether COM or IC, by either a SSG or by an appropriate LPS approximation. For example, a superstructure approximation was applied in case of the incommensurately modulated room-temperature $NbTe_4$ [46,47] and the SSG approach was used for the description of the ten-fold LPS of CuAu II [48].

The conditional equivalence of these two apparently diverse approaches requires some further consideration. The periodic perturbation of any MS is described by a modulation function, whose argument $q \cdot r$ (mod 1) can assume an infinite and continuous set of values in the interval (0,1) for a truly IC structure, whereas in the case of a COM modulation, only a finite set of v values exist. In the IC case, any phase shift of the modulation function results in an identical structure accordingly displaced, while in the COM case, equivalent structures result only for phase shifts which are multiples of $1/v$. All other shifts generate different structures [31]. Thus, for a v-fold superstructure, a phase t must be specified for the modulation function. However, a LPS in its "devil's staircase" limit (i.e. v infinite) becomes equivalent to a truly IC structure. A criterion for the IC/COM distinction was formulated by Pérez-Mato et al. [33]. The structure factor F for a reflection $H=(hkl)+mq=(hklm)$ with $q=(u/v)c*$ is given for a COM structure as:

$$F(hklm) = \sum_n F(hk, l - nu, m + nv).$$ (13)

If, for each reflection, this summation has only one non-negligible term, the structure is considered IC. This is what occurs for large LPS's.

A general one-dimensional Ising model [49,50] also provides a theoretical formalism for the consideration of IC structures. Various ferroelectric, magnetic and crystal LPS's are described by the following Hamiltonian:

$$E = \sum_{t=1}^{N} HS_t + \frac{1}{2} \sum_{i,j=1}^{N} J(i-j)(S_i + 1)(S_j + 1)$$ (14)

where $S_i = \pm 1$, H and $J(i-j)$ have different meanings for different types of ordering. In the case of ferromagnetic ordering, S_i, $J(i-j)$ and H represent the up- and down-spins, the antiferromagnetic interaction between the up-spins at sites i and j and the magnetic field, respectively. In the case of long-range periodic crystals, like $Ti_{1+x}Al_{3-x}$ [40], these quantities represent the presence or absence of defects, the interaction between them and a temperature and composition dependent chemical potential. Similarly, in the case of polytypism, as present in the Friauf-Lavais phases [43], an analogy is made between the up-spin or down-spin states and the cubic or hexagonal stacking of layers, with the interaction becoming one between regions of cubic stacking. It was shown [49,50] that for a given magnetization $q=m/n$ (fraction of up-spins), the energy is minimized for a configuration where the position x_p of the p-th up-spin is given by:

$$x_p = \text{integer } (pn/m)$$ (15)

Thus, for example if $q = 2/9$ a stable configuration is '+---+----'. For irrational values of q, Equation (15) leads to "quasi-crystalline" structures with lost periodicities. It was further shown that for each rational value of q the corresponding structure is stable in a finite field interval $\Delta H(q)$. These COM intervals fill the entire H-axis resulting in a *complete devil's staircase*. This is shown in Figure 4 where, in a certain interval ΔH, q passes through an infinity of rational values between 0 and 1. The most stable part of the interval is the one with $q = 1/2$. More complicated models, leading to *incomplete* staircases with IC phases also present, were also considered. Among them. "devil's flowers" were obtained by the ANNNI (axial next-nearest-neighbour Ising) model [51,52], a three dimensional thermodynamic model in which intra-layer and axial nearest-neighbour ferromagnetic interactions compete with antiferromagnetic axial next-nearest-neighbour interactions.

The LPS approximation, when compared to the SSG approach (or the essentially equivalent DI) [22], has a number of disadvantages. First, in principle $3v$ positional parameters of an atom in a v-fold superstructure correspond to the same number of parameters in the SSG approach, describing the average structure and its displacement field [33].

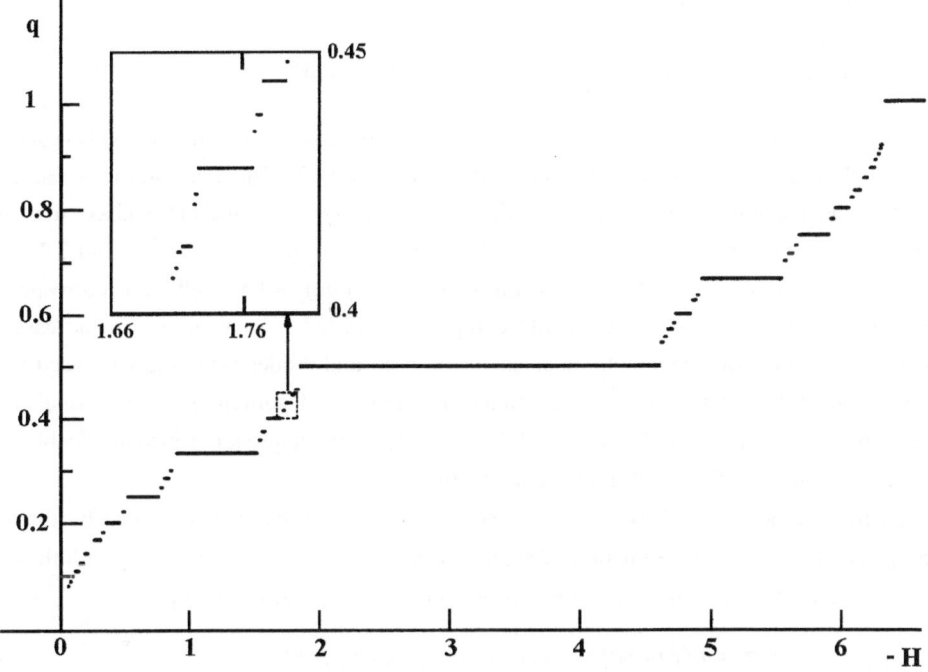

Fig.4 The ratio of up-spins over the total number of spins q as a function of the applied field H for the interaction $J(n) = 1/n^2$. The enlarged area shows self-similarity of the "devil's staircase" under magnification [50].

However, in cases of displacive modulations, with higher-order satellites neglected, the number of SSG parameters can often be appreciably reduced. For example, $Au_{2+x}Cd_{1-x}$ is described within the SSG approach with only as few as 1/8th of parameters required in the corresponding SG description [53]. Second, the SSG approach is not restricted to IC MS's only, but was shown to be very useful in many COM cases as well, as for example in the case of the room-temperature MS of $TaTe_4$ [54]. Third, SSG symmetry allows comparison of different COM and IC phases, which appear in a certain system under different thermodynamical conditions [55]. Finally, a SSG contains more symmetry information than the corresponding SG, since the superstructure represents only a section t through the supercrystal and is thus dependent on the parity of u and v, whose ratio determines the modulation wave-vector [56]. Conversely, the LPS approximation does have several advantageous features. First, it enables one to consider the complete modulated structure as a whole, regardless of whether the modulation pattern is COM or IC with the underlying BS lattice. Second, LPS derived superstructure models of IC MS's can more easily provide the most probable atomic positions, bond-lengths and bond-angles, and thus allow easier crystal-chemical interpretations as compared to the corresponding SSG [47] where these parameters are somewhat indirectly expressed as a continuous function of $q \cdot r$.

2.4 THE NORMAL MODE SYMMETRY (NMS) ANALYSIS

As a result of a periodic distortion, the symmetry of a modulated structure (MS) is reduced as compared to the one of the corresponding basic structure (BS). The distortion is characterized by the order parameter (OP), defined in the Landau theory of second-order phase transitions and related to the irreducible representation of the BS space group [57,58]. Thus, the latter determines the symmetry of the (in general incommensurate) MS as well. The corresponding distortion field can be decomposed into components, called normal modes in the case of a displacive modulation. Symmetry analysis of these normal modes was originally used for the description of crystal vibrations [28]. Since, in the case of softening, these normal modes represent time-independent distortions of the BS [59,60], the approach is also applicable to the crystallographic description of modulated structures.

At the second order phase transition the high-symmetry phase, determined by the space group G and with a density function $\rho(r)$, transforms to a low-symmetry phase with the corresponding density $\rho'(r)$. In terms of the irreducible representations of G [1]:

$$\rho'(r) = \rho(r) + \delta\rho(r) = \rho(r) + \sum_{q\mu\lambda} c_{\mu\lambda}(q)\phi_{q\mu\lambda}(r) \tag{15a}$$

where the components $c_{\mu\lambda}(q)$ are multiplied by the basis functions of these representations $\phi_{q\mu\lambda}$. The summation is taken over all wave-vectors q, over all irreducible multiplier representations $\hat{\tau}^\mu(q)$ of the μ-th point group $\hat{G}_q$ and over all eigenvectors (labeled by λ) of the same μ-th representation [1]. The Landau theory is usually restricted to a single representation $\hat{\tau}^{\mu}(q)$ of a primary displacive modulation [1,61]. It corresponds to a MS generated by a softening of a single vibrational mode [26]. An example of such a softening for the Σ_2 mode in K_2SeO_4 is shown in Figure 5. In this case the position $r(l,k)$ of the k-th atom in the l-th unit cell is given by:

$$r(l,k) = r_l + r_k + u_k(l) \ , \tag{16}$$

with the displacement field $u_k(l)$ derived from the theory of normal vibrations in crystals [28,62] with an assumption of time-independent atomic displacements. It is decomposed according to the μ-th representation [58]:

$$u_k(l) = \sum_{q\lambda} c_{\mu\lambda}(q)\varepsilon_k(q,\mu,\lambda)exp(iq \cdot r_\lambda) \ , \tag{17}$$

where $\varepsilon_k(q,\mu,\lambda)$ and $c_{\mu\lambda}(q)$ are a symmetry-adopted eigenvector of the normal mode and a normal (mode) coordinate, respectively [63].

In the theory of phase transitions an OP $\eta_{\mu\lambda}(q)$ is introduced, which equals zero in the high-symmetry phase, while condensation of a symmetry-breaking normal mode μ leads to the

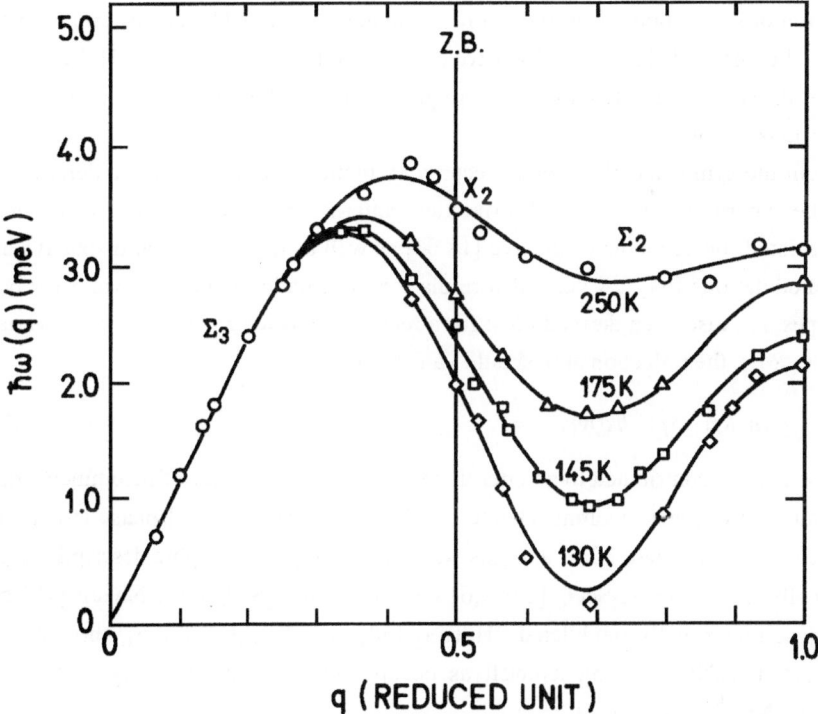

Fig.5 Softening of the Σ_2 vibrational mode in K_2SeO_4 at 130K as a precursor effect to the onset of a primary modulation with $q = (1- \delta)a^*/3$ [59].

low-symmetry modulated phase with the OP defined as a normal coordinate of this mode μ:
$\eta_{\mu\lambda}(q) = c_{\mu\lambda}(q)$.   The normal coordinate of mode μ is transformed according to the
irreducible multiplier representation $\hat{\tau}^{\mu}(q)$. For a one-dimensional representation, this matrix
becomes a character $\chi^{\mu}(q)$ and the normal coordinate $c_{\mu}(q)$ is transformed under action of the
space group operator $(R|v(R))$ [1,26] as:

$$c'_{\mu}(q) = (R|v(R))c_{\mu}(q) = c_{\mu}(q)\chi^{\mu}(q, R)exp(-iq \cdot v(R)) .$$ (18)

For a modulation, which is occupational instead of displacive, the transforming properties are
introduced in a similar way [26,64,65].

Although the soft mode μ makes the main contribution to the distortion, other secondary
modes are often superimposed [58].   The symmetry of these secondary modes must be
compatible with the distortion of the soft mode [58,61]. This requirement is a consequence of
the form of the Landau free-energy expansion:

$$F = F_0 + \sum_{l\mu_l} c_{\mu_l}(q_l)...c_{\mu_m}(q_m)\Delta(q_l+...+q_m)$$ (19)

where a symmetry invariance is required for all terms, including those which couple normal
coordinates of the primary (soft) mode to those of a secondary mode [58].   As a result,
symmetry restrictions are imposed on the entire modulated phase [61].   For example, in the
case of the two-dimensional charge density wave (CDW) modulated compound 2H-TaSe$_2$ with
$q = (\frac{1}{3} - \delta)a$*, the same Σ_1 mode symmetry is required for both the primary q_l and the secondary $2q_l$ distortions [59,60].

Finally, from the symmetry of the eigenvectors, extinction conditions can be derived and
compared to the experimentally observed diffraction patterns. Structure factors $F(hkl + mq)$
have been calculated for harmonic displacive [18,66] as well as non-sinusoidal displacive and
occupational modulations [64]. A general formula for systematic extinctions, valid for both
modulation types, has also been derived [26]: for a reciprocal vector $k = (hkl) + mq$ invariant
under R of $(R|v(R))$), the reflection at k should vanish unless:

$$[\chi^{\mu}(q, R)]^m exp(i(hkl)v(R)) = 1.$$ (20)

NMS analyses have contributed to a better understanding of a number of incommensurate-
to-commensurate [38,67] and incommensurate-to-incommensurate [68–72] phase transitions.
Representative examples, where NMS analysis was used for crystallographic descriptions, are
one-dimensionally modulated K$_2$SeO$_4$ [27,59,61,63,73], α'_1-Sr$_2$SiO$_4$ [66], NiGeP [18] and
NbS$_3$ [74], two-dimensionally modulated 2H-TaS$_2$ [60] and Au$_{2+x}$Cd$_{1-x}$ [75], three-dimensionally modulated V$_6$Ni$_{16}$Si$_7$ [76], as well as occupationally modulated C$_{12}$H$_{16}$Br$_2$ [65],
Yb$_{3-\delta}$ S$_4$ [64] and Nb$_2$Zr$_{x-2}$O$_{2x+1}$ [77].

Although the NMS analysis emerges from a purely physical property — that of an (at least
hypothetical) second-order phase transition [1,26] — while the SSG description is a standard

crystallographic procedure [23,30,78], both descriptions are to a large extent compatible. As shown by Pérez-Mato et al. [58], the abstract superspace concept can be replaced by the invariance properties of the free-energy expansion (Equation (19)). The corresponding invariance group consists of those operators which transform the normal coordinates of the soft mode according to the irreducible representation with an additional phase shift. The SSG is then defined as a sub-group, which also leaves the MS invariant. The way by which the SSG is derived from the representation and vice versa [26,73] is simple in the case of a one-dimensional displacive modulation and the corresponding one-dimensional irreducible representation of $\hat{G}_q$ [1,26,27]. The transformation properties of the soft mode normal coordinate $c_\mu(q)$ $[= \eta_\mu(q)]$ are given in the SSG description [27] by Equation (18), but with the right side expressed in terms of the aforementioned phase shift Δt, given by the SSG operator $\left(R, \varepsilon = +1 \middle| v(R), v_J\right)$:

$$\chi^\mu(q, R)exp(-iq \cdot v(R)) = exp(i\Delta t) \qquad (21)$$

where $\Delta t = v_J - q \cdot v$. The internal translation $v_J(R)$, applied in the SSG to the additional variable ϕ of the OP $\eta(q)exp(i\phi)$, corresponds in the NMS approach to the character $\chi^\mu(q, R)$, applied to the "bare" OP $\eta(q)$ [1].

There are two main differences between the two methods. First, unlike the SSG description, the NMS analysis is based upon inherent physical arguments, related to the Landau theory, which limit possible choices of the corresponding SSG's. The SSG approach is a purely crystallographic description which, without application of the Landau symmetry theorem [79] can lead to a physically incorrect SSG. It is, however, less straightforward in the Landau theory analysis to identify the secondary modes once the soft mode is known [26,58,61,63]. The symmetry of the OP is equivalent to a SSG assignment, which automatically generates restrictions on the form of the modulation function, including its higher harmonics [61] (see [77] for application). Such restrictions can be obtained within the NMS analysis only after a full consideration of all the possible coupled modes in the Landau potential. The second main difference between the two methods originates from a different definition of equivalence for SSG's and for representations [26,34]. When, due to temperature or pressure variation within the same crystal phase, the wave-vector q leaves the first Brillouin zone, the irreducible representation of the OP (for q folded into the first Brillouin zone [80]) is changed. However, the SSG stays the same (according to the q-equivalence [30]). Thus, compared to the SSG approach the NMS analysis seems to be too finely detailed for a general symmetry description [26,34].

2.5 OTHER SYMMETRY DESCRIPTIONS APPLIED TO MODULATED STRUCTURES

In addition to the commonly used methods described in the previous sections, some further approaches have also been used.

2.5(a) *Reciprocal Space Crystallography*

Reciprocal space crystallography [81–84] can be applied to both periodic as well as quasi-periodic structures. Both possess the common feature of being mesoscopically homogenous, i.e. homogenous on scales larger than atoms but smaller than typical crystals [82]. In quasi-crystals and incommensurately modulated structures, the microscopic structure of any region of size D is reproduced in other regions at distances of the order D without, however, exhibiting a perfect crystal periodicity. In this case the point group operators relate only *indistinguishable* densities, i.e. densities with an identical distribution of all mesoscopic substructures [82,83], which is less restrictive compared to the usual requirement of truly identical crystal densities. The crystal symmetry is considered in the Fourier space [85], where both (quasi)-periodicity and indistinguishability of densities can easily be defined. The first is determined by the position of nodes in the reciprocal space, while the second requires the Fourier coefficients to be of the same magnitude [82]. A symmetry classification has been introduced, based on the point group symmetry and on the phase-relationship between density Fourier coefficients at nodes related by this point-group [81,82,83]. In contrast to the superspace approach, the reciprocal space crystallography does not separate the main reflections from the satellites [23]. As a result, some Bravais classes, non-equivalent in the SSG approach, are identical in the reciprocal-space description, reducing the total number from 24 into 16 for d=1 [83,84].

2.5(b) *Geometrical Approach*

To simplify the application of the four-dimensional point and space groups, [86,87,88] a geometrical approach to crystallography in high-dimensional space has been introduced by Weigel, Phan and Veysseyre [89,90,91], with a corresponding WPV-notation. In analogy to the superspace approach a one-dimensionally modulated structure is considered as a three-dimensional section through a four-dimensional supercrystal. However, this mathematical object is described by four-dimensional point and space groups, without separating the coordinates into those of external and internal spaces [92]. Consequently, in this approach the one-dimensionally modulated structures are "mono-incommensurate" structures with no separate description of the three-dimensional symmetry of the average structure [92,93,94]. The WPV notation is more general as compared to the SSG one and there is no one-to-one relationship between the two. Some Bravais classes and SSG's, distinguished in the superspace classification, are identical within the WPV description. Again, only 16 crystallographic four-dimensional classes out of the 24 SSG Bravais classes are established and these are identical to those obtained within the reciprocal space crystallography [84]. A correspondence between the SSG [23,24,78,95] and WPV notations has also been established [92] and the geometrical approach was recently extended to the five-dimensional space [96,97,98] and the corresponding "di-incommensurate" structures.

2.5(c) *Alternative Superspace Methods*

There are also two alternative presentations to the superspace approach [27,33,58] which endeavor to avoid the abstract mathematical construction of the supercrystal. In the first [58] SSG are introduced through the invariance properties of the Landau free energy expansion as described in section 2.4., while in the second an atomic modulation function (AMF) is defined, which describes the displacement field $u_k(qT)$ of the k-th atom in the unit cell T with respect to its average structure position. The argument of the AMF, considered in the SSG approach as the internal coordinate, is interpreted here as a continuous label of the undistorted structure unit-cells. The SSG symmetry is replaced by an appropriate relationship between the AMF's. This relationship is finally transformed into a general formula, which describes the symmetry and the extinction rules of the corresponding diffraction pattern.

2.5(d) *Component Structures*

A further description by means of component structures was introduced by McConnel and Heine [99], where the atomic position in a one-dimensionally modulated structure was given by a superposition of three parts:

$$r + u(x_4) = r + u_0 + Re\, u_1 exp(ix_4)$$
$$= average\ structure + C_1 cos(x_4) + C_2 sin(x_4) \qquad (22)$$

where $x_4 = q \cdot r$. The real C_1 and imaginary C_2 parts of the modulation amplitude describe, in conjunction with the average structure (AS), two "component structures", whose symmetries are described by ordinary three-dimensional SG's G_1 and G_2, directly related to a corresponding SSG [100]. The method cannot be directly extended to two-dimensionally modulated structures and it is not convenient for the description of non-sinusoidal modulations [33].

2.5(e) *Wreath Groups*

Finally, a new class of symmetry groups was introduced. These so-called wreath groups [99] are similar to colour groups [102] and are used to describe the symmetry of the displacement field $u(r)$. Together with the basic structure space group G they describe the complete modulated structure.

3. Applications

3.1.   THE NbS_3-I AND NbS_3-II POLYTYPES

NbS_3 belongs to the family of transition-metal tri-chalcogenides MX_3 [103], for which CDW-modulated structures are commonly observed phenomena. All these compounds have a similar structure, based on X_6 trigonal prisms of chalcogen atoms with the metal atoms M at or near their centers. A base-to-base stacking of the prisms results in linear chains of metal atoms and in a strong one-dimensional character for these compounds, while the M-X inter-column bonds result in a half-prism height displacement of adjacent columns and in a corresponding ribbon-like crystal growth. All members of this family of compounds can be derived from the structure of undistorted $ZrSe_3$ [104] (Figure 6). Isosceles- or equilateral-like cross-sections of the chalcogen cages [105] result in columns with an oxidation state $(X_2)^2 X^{2-}$ or $3X^{2-}$ [105,106,107]. Dependent on their number and stacking the unit cells contain between 2 and 24 columns of different types [108], which are subject to Peierls distortions.

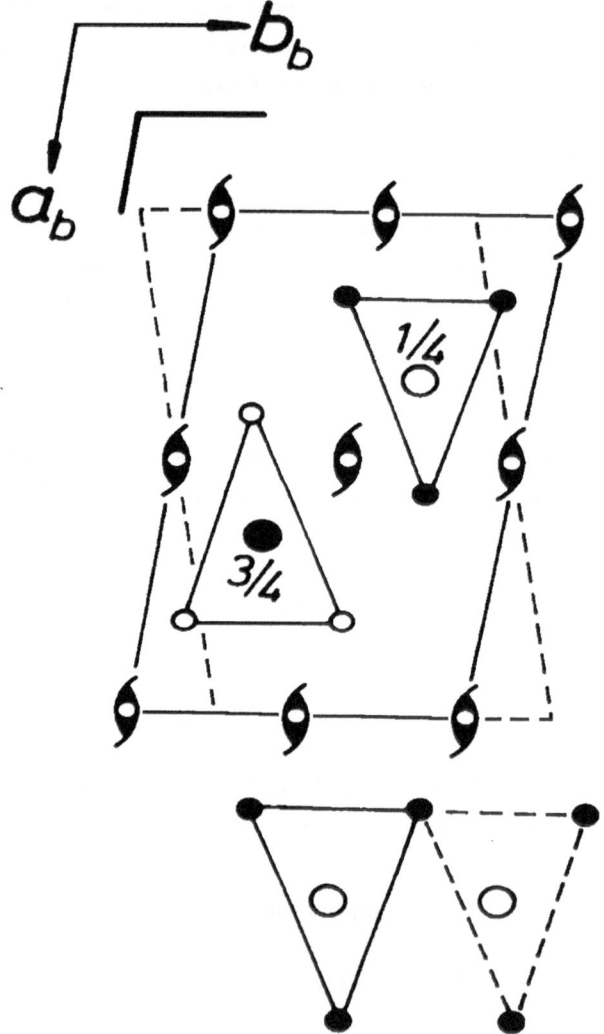

Fig.6 The projection of the basic ($ZrSe_3$-type) structure of NbS_3-I along the c -axis. Mid-sized circles represent S atoms and large circles Nb atoms. Symmetry elements and possible stacking disorder (broken lines) are shown [74].

Various NbS_3 polytypes have been reported [103], but only NbS_3-I and NbS_3-II were confirmed and investigated in detail. When grown from the elements in evacuated quartz tubes [109], both polytypes usually appear together, but can be separated under an optical microscope since NbS_3-I appears in the form of somewhat larger and straight crystals while NbS_3-II has a fine cotton-like appearance. Both polytypes are characterized by numerous stacking faults of the two-dimensional layers [110,111], which reduce the quality of single crystals, especially those of NbS_3-II.

A brief summary of various crystallographic descriptions for both polytypes follows.

3.1(a) *NbS$_3$-I*

The single crystal X-ray refinement of NbS3-I [112] revealed that this phase is a two-fold superstructure of the undistorted $ZrSe_3$ structure (Figure 6), where Nb and S atoms are displaced from the 2(e) positions of the space group (SG) $P2_1/m$ (Z=2) into the 2(i) positions of the SG $P\overline{1}$, (Z=4). This results in a doubling of the unit cell (a_0=9.14 Å, b_0=4.96 Å, $2c_0$=6.73 Å, $\alpha = \beta$ =90^0 , γ =97.2^0). The superstructure is a result of a CDW formation, stabilized by a Peierls distortion with $q = 2k_F = \frac{1}{2} c^*$ [113,114]. The half-filled metallic d_{z^2} energy band is thus split by a 0.44 eV energy gap at the Fermi level [108,115].

The reconstructed reciprocal space of NbS3-I, based on X-ray and electron diffraction data is shown in Figure 7. The lattice of main spots with forbidden reflections (00L): L=2n+1 [116] is described by the SG G_B=$P2_1/m$, while the superlattice reflections require a $(a_0 \times b_0 \times 2c_0)$ modulation pattern (MP) unit cell and obey the symmetry of G_M=P2/m. These two SG's, combined into a dualistic symbol $^{m\ P2_1\ m(a_0,b_0,c_0)}_{z\ P2\ m\ (a_0,b_0,2c_0)}$, can be replaced by the corresponding superspace group (SSG) $P^{P2_1\ m}_{1\ 1}$, where the $\left(\frac{2}{1}\right)$ operator generates in-phase modulation along adjacent columns and allows those Σ^* lattice spots, which obey the condition (00lm): l=2n (see Equation (11)). In the case of a commensurate modulation, the $\left(\frac{2}{1}\right)$ operator may become a three-dimensional SG element for $q=(2n/2m+1)c^*$, however this does not apply for NbS3-I where $q = \frac{1}{2} c^*$. A reduction of other SSG elements into SG operators depends on the internal parameter t (sections 2.1. and 2.3.) and the superstructure SG $P\overline{1}$ [112] is obtained for supercrystal cross-sections with t=0,1/2 and 1/4,3/4 [74,116]. Since NbS3-I is commensurately modulated, the forbidden Σ^* lattice spots are projected onto the same quasilattice M^* positions as other allowed ones (e.g. the extinct (0,0,2n+1,1) reflection coincides with the allowed $(0,0,2n + 2,-1)$ one along the same row of satellites). As a result some systematic absences in $M^*$are violated [116].

The symmetry properties of the modulated structure with G_B=$P2_1/m$ and $q = \gamma c^* (|\gamma| \langle 1/2)$ are given by the irreducible multiplier representations Λ_1 and Λ_2. These two modes act as SSG elements $\left(\frac{2}{1}\right)$ and $\left(\frac{2}{s}\right)$, since for the non-trivial element 2_z of $\ddot{G}_q$ $\chi(\Lambda_1,2_z)=1=exp(i0)$ and $\chi(\Lambda_2,2_z)=-1=exp(i/2)$; (see Equation (21)). On the other

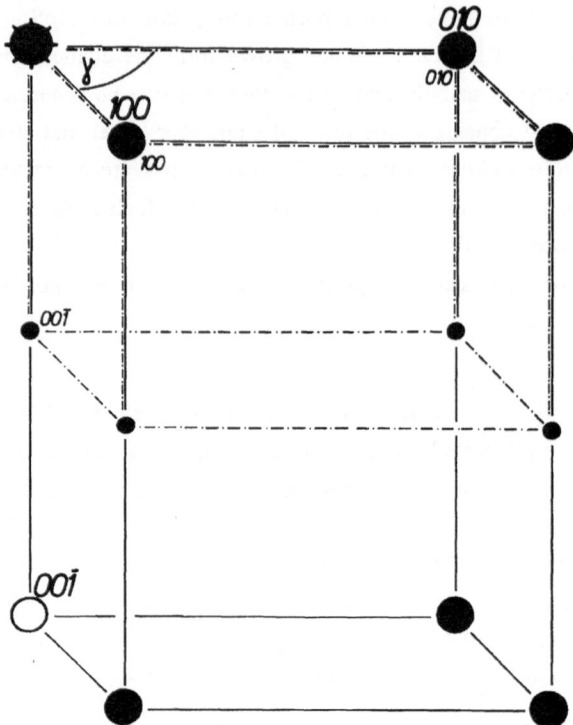

Fig.7 The reconstructed reciprocal lattice of NbS₃-I. G_B and G_M unit cells are given by full lines (large numbers) and broken lines (small numbers), respectively. Empty circles indicate forbidden reflections [74].

hand the equivalence rule for the normal coordinate $c_\mu^*(q) = c_\mu(-q)$ requires the same restrictions on the form of the modulation functions as the second generating SSG operator $\left(\frac{m}{\bar{1}}\right)$. Hence, for an incommensurate q, only Λ_1 corresponds to the SSG $P^{P2_1/m}_{1\ \bar{1}}$, while in case of a commensurate modulation with $q = \frac{1}{2} c^*$ the two modes "stick" together and yield a doubly degenerate Z_1 mode with no extinction conditions on satellite reflections.

3.1(b) *NbS₃-II*

Due to small crystal sizes and the presence of a large number of stacking faults, the incommensurate structure of NbS₃-II remains unknown. From a preliminary X-ray analysis [117], the unit cell parameters of the basic structure (BS) were found to be approximately a≈ a_0=9.6 Å, b≈4b_0=18.7 Å, c≈c_0=3.36 Å and γ=98°, indicating a four-fold enlargement along the *b*-axis as compared to the NbS₃-I BS unit cell. Electron diffraction patterns at room temperature [74,109,110,118] reveal doublets of incommensurate satellites which usually show pronounced streaking perpendicular to the c^* direction. These satellites are described by two wave-vectors, q_1=(0.5,0,0.353) and q_2=(0.5,0,0.297), as shown in the reconstructed reciprocal space (Figure 8). Although the two wave-vectors may be related as first- and second-order satellites $\left(q_{2z} = 1 - 2q_{1z}\right)$, their independent nature was confirmed by the disappearance of the q_2 component above $T_c \approx 350$ K [74,118]. In agreement with Wilson's considerations [113,114] a four-fold BS unit cell as compared to the one for NbS_3-I must contain eight trigonal-prismatic columns, which appear in symmetry related pairs. In analogy to the situation in $NbSe_3$ [20,113,114], it is proposed that three out of four column-pairs have chalcogen cages with isosceles cross-sections, while those of the remaining pair are equilateral. The corresponding oxidation states $((S_2)^{2-}$-S^{2-} and $3S^{2-}$, respectively) have been associated with a Peierls distortion with $q = 2k_F \approx c^* / 3$ along the three pairs with isosceles bases, where each Nb atom shares about 2/3 of an electron, while the columns with equilateral bases are fully stripped of the d_{z^2} electrons and do not participate in the modulation [74].

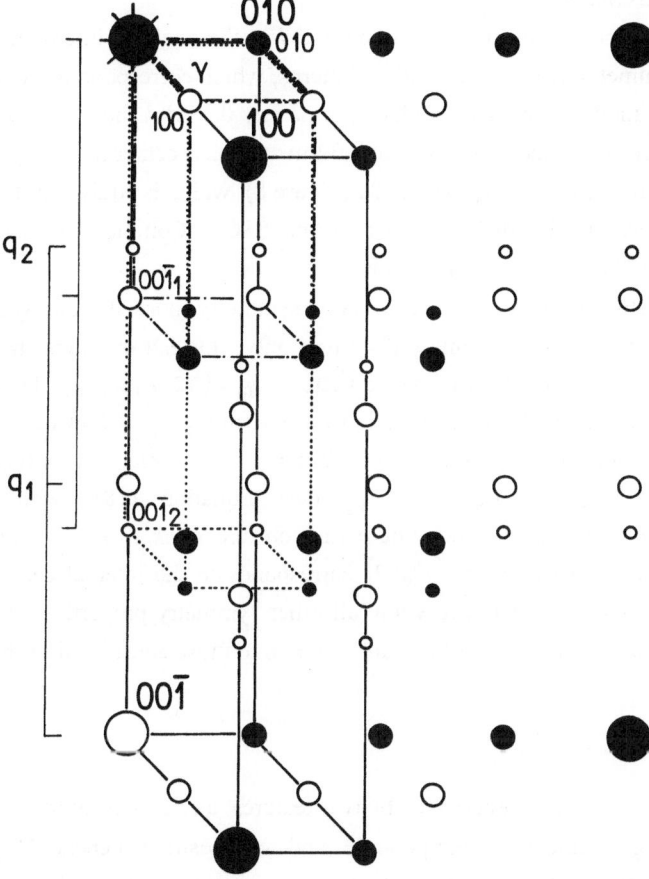

Fig.8 The reconstructed reciprocal lattice of NbS_3-II. G_B and G_{M1} (corresponding to q_1) and G_{M2} (corresponding to q_2) unit cells are shown by solid (large numbers), broken (small numbers with subscript 1) and dotted (small numbers with subscript 2) lines, respectively. Open circles indicate forbidden satellites [74]. Note that for clarity only first-order satellites are shown.

Based on structural considerations with regard to of NbS_3-I as well as on electron diffraction experiments, the basic SG G_B of NbS_3-II is best described by $G_B=P2_1/m$. The situation however is not straightforward regarding the symmetry of the modulation. From the geometrical point of view all nodes fit to a quasi-lattice M^* with *(hklm)=(hkl)+mq_1*, based on a (2a x b x c) unit cell and with $q_1=0.352c^*$, but there is no single (3+1)-dimensional SSG [23], which would account for the observed extinctions. Consequently, both modulations q_1 and q_2 must be considered as genuine distortions and the complete modulated structure symmetry described formally by one (3+2)-dimensional [24] SSG (the plausible one is $C_2 \, {}^{P2_1}_{p1} \, {}^{m}_{2}$). Both distortions have the same symmetry and are described by the same (3+1)-dimensional SSG $A^{P2_1}_{1} \, {}^{m}_{1}$. The SSG operators are the same as those of the SSG $P^{P2_1}_{1} \, {}^{m}_{1}$ (describing NbS_3-I), but with an additional internal centering translation $\left(E1|1/2\,0\,0\,1/2\right)$, which yields the condition for allowed reflections *(hklm)*: *h+m=2n*.

In the case of separate descriptions of both CDW's, the dualistic interpretation can also be applied. The symmetry of both modulation patterns, which correspond to the q_1- and q_2-type satellites with modulation unit cells $\left(2a\,x\,b\,x\,c\,/\,0.352\right)$ and $\left(2a\,x\,b\,x\,c/(2\,x\,0.352)\right)$, respectively, is given by the same SG G_M=B2/m. B-face centering of G_M limits the MP reciprocal lattice nodes to $(HKL)_M$: *H+L=2n* (Figure 8), which is equivalent to the restrictions imposed by the A-type internal centering of the SSG. Complete dualistic symbols are ${}^{m\,P2_1\,m(a,b,c)}_{z\,B2\,m\,(2a,b,2.84c)}$ and ${}^{m\,P2_1\,m(a,b,c)}_{z\,B2\cdot m\,(2a,b,l\,42c)}$, respectively.

Since both distortion waves exhibit the same symmetry, normal mode symmetry analysis can also be carried out for one primary distortion only. Instead of taking the small unit cell $\left(a\,x\,b\,x\,c\right)$ with a modulation wave-vector $q_V=(1/2,0,\gamma)$, a larger unit cell $\left(a'\,x\,b'\,x\,c'\right)=\left(2a\,x\,b\,x\,c\right)$ with a corresponding wave-vector $q_A=(0,0,\gamma)$ was considered. The resulting SG operator $\left(E|a'/2\right)$ transforms a normal coordinate c_μ into $c'_\mu=c_\mu exp\left(-iq_V\cdot a\right)=c_\mu(-1)exp\left(iq_A a'/2\right)=-c_\mu$ (see Equation (18)) with an associated character $\chi^\mu(E)=-1$ and a corresponding extinction rule $F(hkl)+mq_A=0$, unless *h+m=2n* (see Equation (20)). The operator $\left(E|a'/2\right)$ corresponds to the internal centering translation $\left(E1|1/2\,0\,0\,1/2\right)$ in the SSG approach, while all other symmetry properties, given by the irreducible representation point group of q_A, are identical to those considered for NbS_3-I.

3.2 THE $Nb_xTa_{1-x}Te_4$ SYSTEM

The $Nb_xTa_{1-x}Te_4$ system is characterized by two features, a relatively simple average structure (AS) and a large number of temperature and composition dependent phases, whose periodicities are either commensurate (COM) or incommensurate (IC) with the basic structure (BS) lattice. This section is mainly devoted to: (i) various descriptions of the room temperature (RT) charge-density wave (CDW) modulation phases of the end members $NbTe_4$ and $TaTe_4$, (ii) to two low-temperature (LT) transitional $NbTe_4$ phases studied primarily by means of electron diffraction and (iii) to a series of $Nb_xTa_{1-x}Te_4$ "mixed" compounds with

$0 \le x \le 1$. The latter two are considered in the long-period superstructure (LPS) approximation and studied by means of computer simulation of the corresponding electron diffraction data. This description is by no means complete. Large areas, e.g. the role of defects in the vicinity of phase transitions, the detailed crystallographic analyses and the high-resolution electron microscopy (HREM) of some of the phases, are omitted and the present chapter is to be considered supplementary to those in this volume which treat these subjects in detail.

The various $Nb_xTa_{1-x}Te_4$ phases represent CDW-driven perturbations of the AS. The latter is built of continuous Te anti-prismatic columnar cages, occupied by Nb or Ta chains [119,120]. The structure belongs to the space group (SG) P4/mcc with the Nb occupying 2(a) positions and the Te lying at 8(m) positions with unit cell parameters $a_0 \cong 6.50$Å and $c_0 \cong 6.84$Å. The Te sub-lattice forms a well defined three-dimensional framework while the metallic chains are isolated, such that the structure is generally referred to as "quasi-one-dimensional".

3.2(a) *The RT NbTe$_4$ and TaTe$_4$ Phases*

The RT modulated phases of $NbTe_4$ and $TaTe_4$ were the first to be detected [121-124] and, as a result, are the most thoroughly studied. Electron diffraction patterns obtained with the beam parallel to four low-index zone-axes are shown in Figures 9 and 10. They reveal two major differences as shown in the respective reconstructed reciprocal spaces (Figures 11 and 12). First, the satellites are IC with the periodicity of the main reflections in case of $NbTe_4$ but COM for $TaTe_4$ and second, there is a difference in the modulation unit cell base, presuming the weak diffuse spots in $NbTe_4$, which develop on cooling into streaks and further into narrow-spaced satellites, are ignored at RT.

Theoretical studies, based on the Landau theory of phase transitions, were carried out by Walker [126] and revealed that a single wave-vector $q = (0,0,0.691)$, is sufficient for a description of all satellites in RT $NbTe_4$. These originate from antiphased CDW's along neighbouring columns, which results in extinction rules for the corresponding diffraction patterns. Similar studies, based on free-energy minimization for a transverse-wave model, suggested structural solutions for the other phases of the system, known at that time [127].

A single crystal X-ray refinement of the RT $NbTe_4$ structure was carried out by Böhm and Von Schnering [46] and treated in the LPS approximation. Using two models (the first based on a COM three-fold superstructure and the second on a discommensuration model, where six Nb-triplets were followed by a doublet, with the doublets out of phase between adjacent columns) they were able to determine all important characteristics of the modulation. Nb atoms are shifted from their average positions along the metallic chain direction by as much as ± 0.25Å, resulting in the formation at Nb_3 and Nb_2 groups with larger metal atom spacings occurring between these groups. In addition, Te_4-squares are modulated by three modes, a breathing mode, a weak libration mode about [001] and a longitudinal mode along this direction, while the inter-column Te_2-pairs are kept at nearly constant distances of approximately 2.91 Å.

The RT NbTe$_4$ structure was further described in a straightforward way by means of superspace groups (SSG). The (3+1)-dimensional X-ray refinement carried out by van Smaalen et al. [128] resulted in the SSG $W^{P4/mcc}_{1\,\bar{1}11}$, with q=(0.5,0.5,0.691) and a $(a_0 \times c_0)$ BS unit cell and the corresponding dualistic description was given as $^{P4/mcc}_{I4/mmm}$ [134]. All general conclusions of Böhm and Von Schnering [46] regarding different Nb and Te modes as well as the phase difference between neighbouring columns [46,126] were confirmed, except for the nature of the modulation, which was described as IC instead of discommensurate.

In addition, Kusz and Böhm [47] performed a statistical analysis for the RT NbTe$_4$, in which the structural model [128], obtained by the (3+1)-dimensional refinement was compared with a series of models, which were a result of various COM approximations. It was shown that even the simplest COM approximations revealed all important crystallographic character-istics of the discussed MS.

Contrary to some earlier expectations [129], a structure refinement of RT TaTe$_4$ super-structure [130] revealed that the proper SG was either P4/ncc or P4cc, as already predicted by

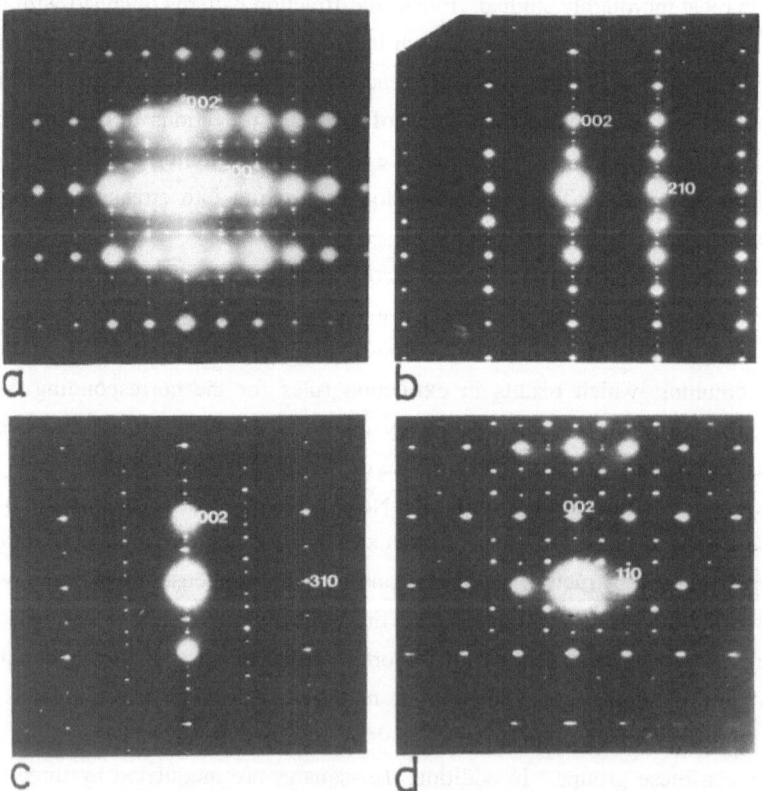

Fig.9 RT electron diffraction patterns from NbTe$_4$, taken with the beam parallel to $[0\,\bar{1}\,0](a), [1\,\bar{2}\,0](b), [1\,\bar{3}\,0](c)$ and $[1\,\bar{1}\,0](d)$ zone axes of the average structure [121].

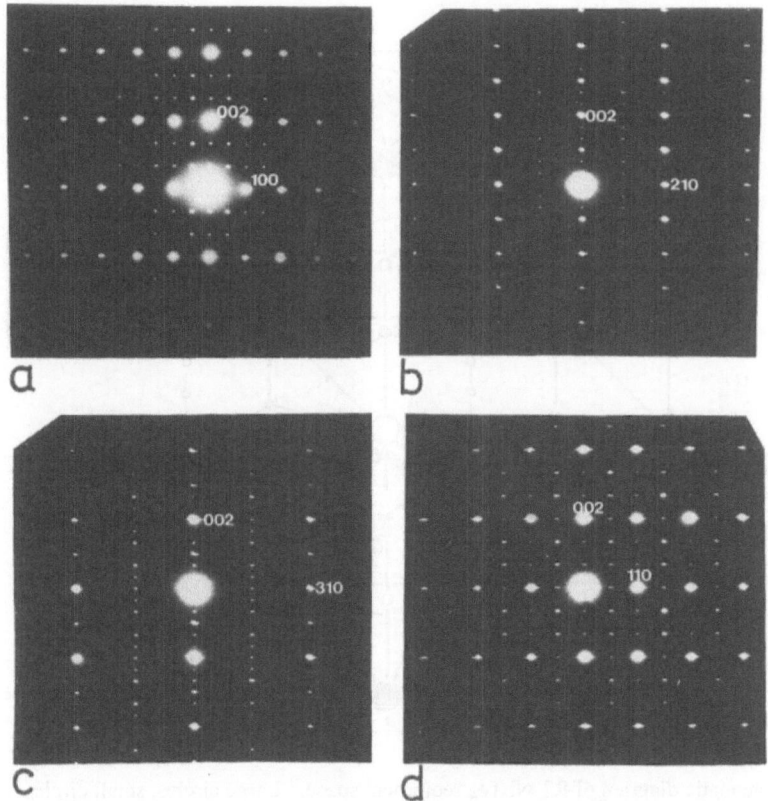

Fig.10 RT electron diffraction patterns from TaTe$_4$, taken with the beam parallel to $[0\bar{1}0](a)$, $[1\bar{2}0](b)$, $[1\bar{3}0](c)$ and $[1\bar{1}0](d)$ zone axes of the average structure [121].

Walker [126]. The first was accepted on basis of structural arguments, although both refinements resulted in very similar atomic positions and R-factors (0.064 and 0.063). In the $(2a_0 \times 3c_0)$ superstructure two columns out of four (I and II) are related by a four-fold axis and the other two (III and IV) by a (pseudo)-center of symmetry. Since Ta-triplets are formed along all columns, one third of Te-squares are contracted and two thirds are expanded. To keep the Ta-Te and Te-Te distances constant, a rotation of the larger squares and (since all atoms are shifted with respect to the ones in the neighbouring columns by $\sim 2\pi/3$) a slight shift in opposite direction of columns III and IV takes place. High resolution electron microscopy [131,132] has confirmed these results. In addition to the conditions placed on the allowed reflections of P4/ncc *(HK0: H+K=2n, 0KL: L=2n, HHL:L=2n)* a further one is obeyed in the diffraction pattern of TaTe$_4$ *(HKL: H=2n, K=2n', L=3n")*. The last was initially ignored, since there was no three-dimensional SG which would include these systematic extinctions. However, as suggested already during the original refinement [130] the structure, in spite of being COM, can conveniently be described by means of a SSG. It was shown [54] that the SSG $P\,{}^{P4/ncc}_{1\,1\bar{1}\bar{1}}$, with $q=(0,0,2/3)$ and a $(2a_0 \times c_0)$ BS cell corresponded to both possible SG's,

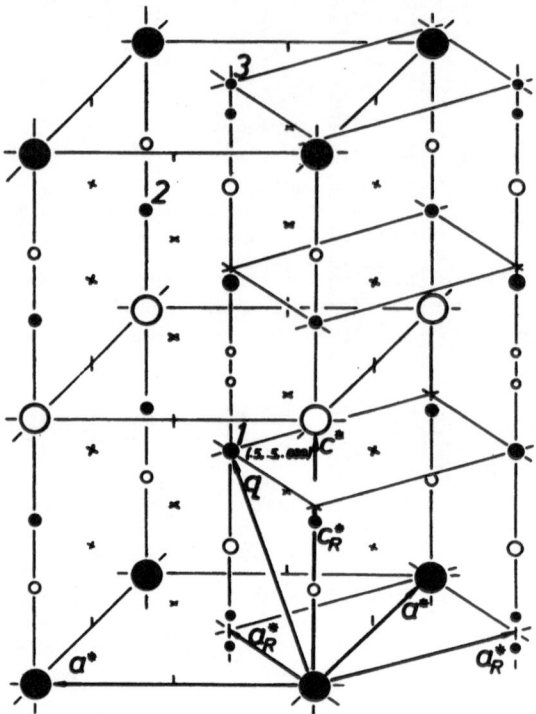

Fig.11 Schematic diagram of RT $NbTe_4$ reciprocal space. Large circles, small circles and crosses show the positions of average structure reflections, RT-modulation and large diffuse satellites, respectively. Empty circles and light crosses indicate forbidden reflections. Average structure (a^*, c^*) and RT-modulation (a_R^*, c_R^*) unit cells as well as the corresponding modulation wave-vector with the first-, second- and third-order satellites are indicated [134].

which appeared as three-dimensional sections through the superspace with different internal parameters $t = x_4 - q \cdot r$ (Figure 13). Contrary to an IC case, a COM supercrystal corresponds to a series of three-dimensional structures with t becoming the phase of the modulation $u^\mu(x_4)$ for a given atom μ. The structural refinement yielded the same R-factor with fewer free parameters and a structural inter-relation of groups of atoms resulted in the additional extinction rule.

3.2(b) *The LT1 and LT2 Transitional Phases of NbTe$_4$*

The LT1 and LT2 phases of $NbTe_4$ appear in a wide temperature range between the RT and LT ones. It was well-known that in addition to the RT IC modulation in $NbTe_4$ a faint, circular and diffuse scattering is always observed at positions (1/2,0,1/3) and equivalent positions of the $(a_0^* \times c_0^*)$ reciprocal unit cell [121]. On cooling below RT these reflections develop gradually into narrow streaks and further into closely spaced satellites, before lock-in occurs at about 50K [133]. The groups of satellites formed can occasionally be divided into two

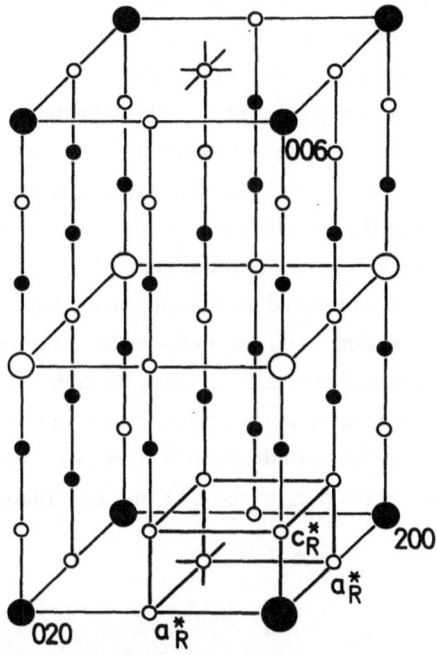

Fig. 12 The reciprocal space of RT TaTe$_4$, indexed with respect to the supercell (a_R^*, c_R^*). Systematically extinct average- (large) and super-structure (small) reflections are indicated by open circles [129].

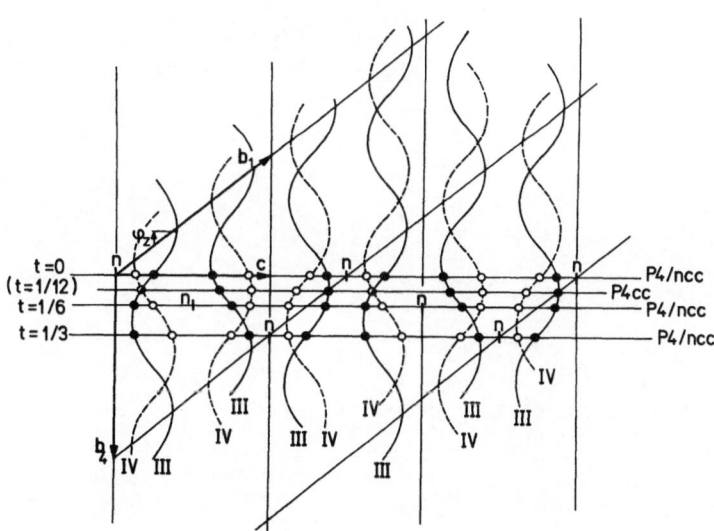

Fig. 13 A two-dimensional section through the supercrystal of TaTe$_4$. The unit cell base is given by $b_4 = e_1$ and $b_1 = a_3 - 2/3\,e_1$. Real space cross sections, perpendicular to e_1 and shown by horizontal lines, correspond to different superstructures. Wavy continuous and doted lines represent Ta atoms of two different column types, III and IV [54].

alternating sets (named LT1 and LT2), Figures 14 and 15. These additional diffraction effects were not included in any of the performed X-ray refinements.

The corresponding phases represent a transitional state between the RT IC and the LT lock-in phases [134], which develops over a wide temperature range. The very sluggish transition into the lock-in $(2a_0 \times 3c_0)$ superstructure indicates that the intermediate transition is accomplished through formation of discommensurations, which have to diffuse out of the sample prior to the transition. The whole process depends on the size of the sample and on the cooling rate, which can even block the transition during LT X-ray experiments (in spite of the lower temperatures achieved compared to transmission electron microscopy experiments) [135]. Since the RT modulation is IC (or long-period commensurate with 11 modulation periods fitting $16c_0 = 32$ trigonal prismatic heights), the transitional state changes on cooling gradually from IC via discommensurate state into the lock-in one. Discommensuration walls, observed by TEM near the transition to the lock-in phase, were reported to be spaced either by $8c_0$ [136] or $16c_0$ [124] corresponding to the LT1 and LT2 phases, discussed below. It was

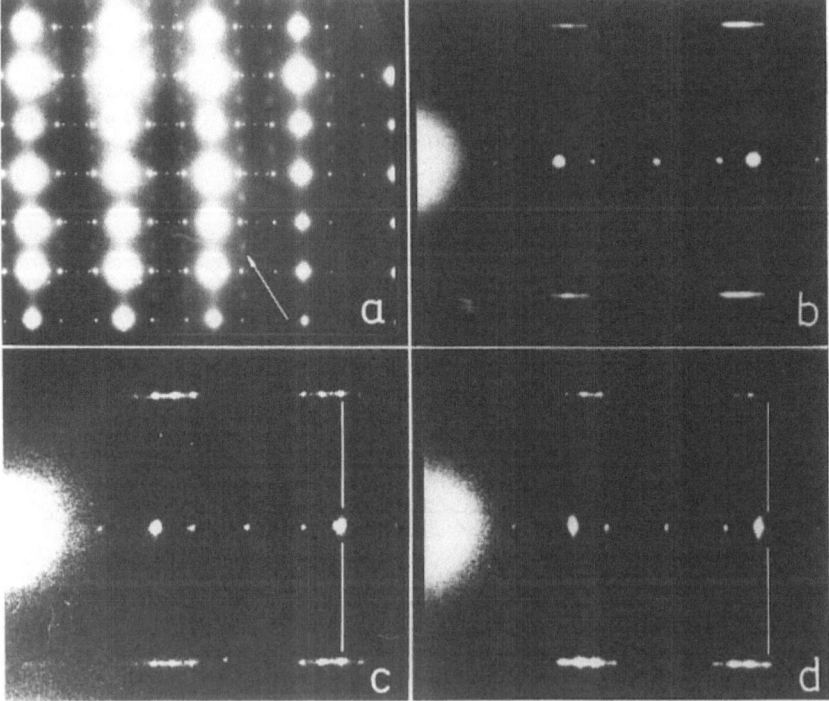

Fig.14 A sequence of diffraction patterns from NbTe$_4$, showing the development of RT diffuse scattering (a) into streaks (b) and further into LT1 (c) and LT2 (d) narrow spaced satellites. Lines in (c) and (d) indicate the position of LT1 and LT2 satellites with respect to those of the RT modulation [133].

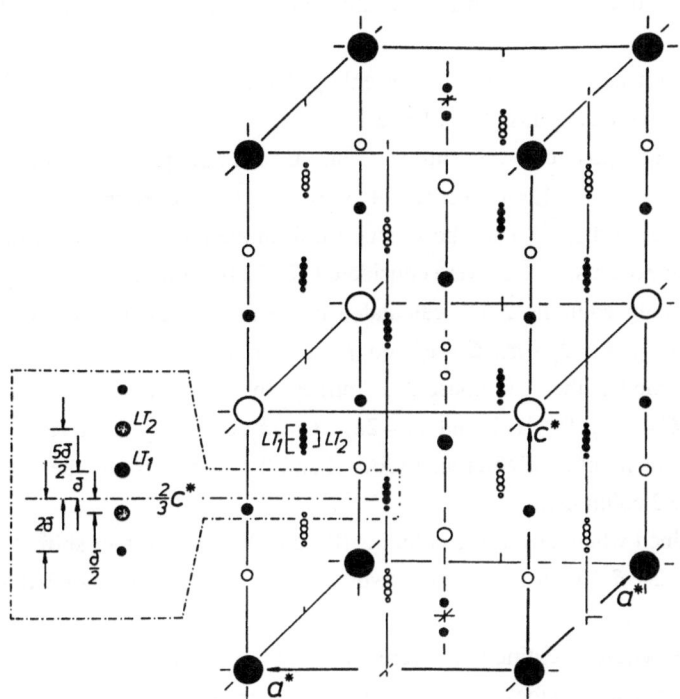

Fig.15 The reconstructed reciprocal space of NbTe$_4$ in the temperature region between 100 K and 50 K. Both LT1 and LT2 satellites are shown superimposed and their shifts with respect to the commensurate 2/3 c* positions are shown [134].

shown [137] that the LT1 set, together with the RT reflections, can be explained on basis of a theory, first developed for shear structures (e.g. in Ti$_n$O$_{2n-1}$) [138], where the periodicity and the R-vector of planar defects were determined from the fractional shifts of the satellites from their COM positions. However, this model did not account for the LT2 satellites.

Three theoretical papers deal with these IC-to-IC phase transitions in NbTe$_4$. On the basis of the Landau free-energy calculation, a model was proposed [71] in which discommensurations separate COM regions. The temperature-dependent phase transition is driven by a competition between nearest-neighbour and next-nearest-neighbour column interactions [72,139]. Both, single-q and double-q states were found to be energetically acceptable, but the former was preferred on the basis of a better fit with the previous experimental results [137]. LT1 was also described in terms of phason and amplitudon symmetry mode distortions of the (3+1)-dimensional RT supercrystal, but only Nb shifts from the average structure positions were taken into account.

Various IC and discommensurate LPS models were examined [140] by comparing computer-simulated electron diffraction patterns with those experimentally observed. It was shown that only IC models give acceptable results. The calculations were carried out in the following way:

- The RT AS unit cell was appropriately enlarged $(A = a_0 + b_0, B = b_0 - a_0, C = 16c_0)$ to include two anti-prismatic columns (types 1 and 2).

- The μ-th atomic position was shifted from its average position according to the modulation function $u_\alpha^\mu(x_4) = \sum_{n=1}(A_{n\alpha}^\mu \cos(nx_4) + B_{n\alpha}^\mu \sin(nx_4))$. Eleven modulation periods fit $16c_0$ and the amplitudes ($A_{n\alpha}^\mu$, $B_{n\alpha}^\mu$) are those obtained from the RT refinement, where only the first two Fourier components ($n=1,2$) were considered [22,128]. With $u_\alpha^\mu(x_4)$ along columns 1 and $u_\alpha^\mu(x_4 + 1/2)$ along columns 2, the calculated intensities correspond to those of IC RT NbTe$_4$ with $A = a_0 + b_0$, $B = b_0 - a_0$, $C = c_0$ and $q_i = (0,0,0.691)$.

- In order to model the LT phases, a further enlargement of the unit cell into $(A = 2a_0, B = 2b_0, C = 16c_0$ for LT1 and $A = 2a_0, B = 2b_0, C = 32c_0$ for LT2) was needed. This also increased the number of columns to four (I and II originate from type 1 columns and III and IV from type 2 columns).

- LT1 was obtained when half of the columns (III and IV) were further shifted in opposite direction $u_\alpha^\mu(x_4 + 1/2 \pm \delta_1)$ with $\delta_1 = 0.00016$ (which approximately corresponded to the shifts in RT TaTe$_4$ [130]).

- For LT2, the required extinctions were obtained for a phase shift of π/2 between neighbouring columns and for clockwise and anti-clockwise rotation of Te-cages at alternating maxima along each column. Additional modulation of the x and y Te-coordinates was included in this way (with an amplitude δ_2, whose absolute value was again comparable to that in RT TaTe$_4$ [130]).

Calculated intensities are shown schematically in Figure 16 and fit very well with the observed electron diffraction patterns. The corresponding SG's for the long-period unit cell were found to be P4/m, P4/n, I4/m and P4$_2$/m for the RT, LT1, LT2 and combined LT1/LT2 modulation structures, respectively.

These structural models allow several conclusions to be drawn [140]. Discommensuration models in which the phase slip between regions is atomically sharp [38] result in too many satellites forming the streaks. Discommensurations must however play an important role in the final stages of the transition between the LT1/LT2 and the lock-in phase, as revealed by HREM. Additional displacements associated with LT2 result in an effective extension of the Te-cages, such that the Te-Te distances change from values found in RT NbTe$_4$ into those of the lock-in LT phase. Further, the LT1 displacements resemble those of RT TaTe$_4$, where such shifts result in constant Te-Te inter-column bonds. We thus conclude that LT1 and LT2 represent precursor states to the formation of the LT NbTe$_4$ COM structure, which is supposed to be isostructural with RT TaTe$_4$.

The structures of the different phases observed in NbTe$_4$ and TaTe$_4$ are a result of several competing mechanisms. The ultimate goal are metal triplets and largest possible phase shifts between adjacent columns. All columns are equivalent and the inter- and intra-column Te-Te distances tend to be kept constant. In case of IC $NbTe_4$ Nb-triplets only cannot form, but phase shifts of π between neighbouring columns can take place. Conversely, in COM $TaTe_4$ only triplets are present, but with neighbouring columns shifted in phase by $2\pi/3$. Since the LT1-like shifts in RT $TaTe_4$ result in $2\pi/3$ phase shifts between columns, and since any discommensurations must involve the same phase shifts, the presence of discommensurations would be expected to support the development of the LT1 phase in $NbTe_4$. Thus a more sophisticated model, incorporating a transitional stage between the IC and discommensurate states, may provide a better fit to the real structure.

3.2(c) *The Mixed $Nb_xTa_{1-x}Te_4\ (0 \leq x \leq 1)$*

As noted above, the LT COM phase of $NbTe_4$ [134] is assumed to be isostructural with RT $TaTe_4$. Lock-in occurs on cooling at about 50K, after a sluggish transition from the RT IC

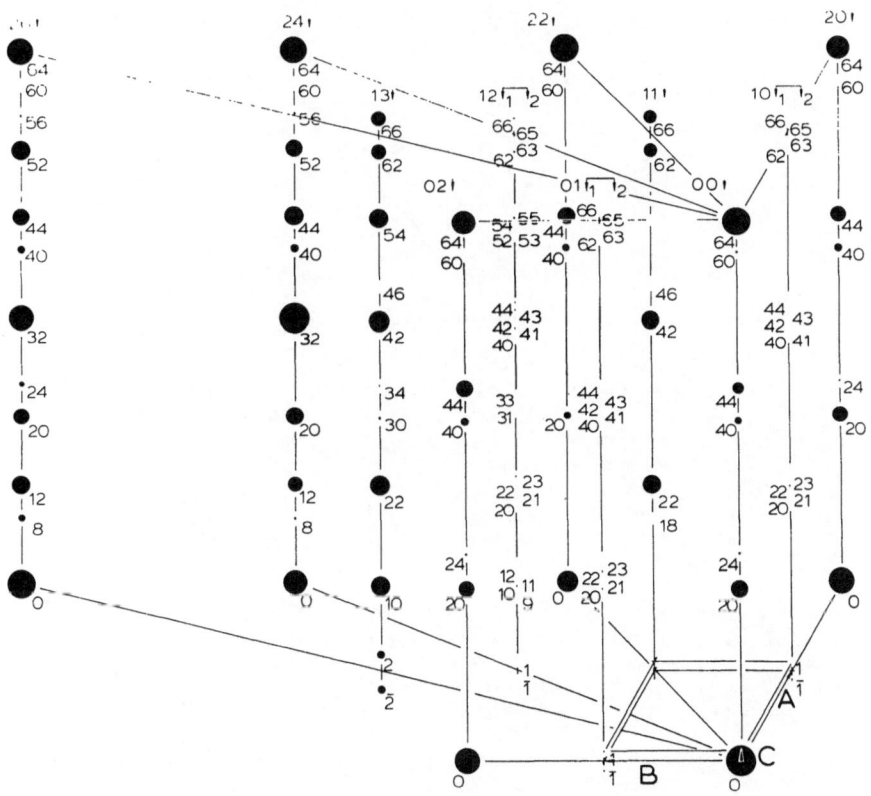

Fig.16 Sections through the calculated reciprocal space of $NbTe_4$ LT1 and LT2 phases. The satellite diameters are approximately proportional to the logarithm of their intensities [140].

phase is accomplished via the two precursor phases LT1 and LT2. A high-temperature (HT) (>790K) modification of $NbTe_4$ was also reported [141]. It is characterized by acentric Nb-positions in the Te-antiprisms and its symmetry is described by the SG P4cc. Disappearance of the RT satellites on heating was also reported in one of the early microscopic studies [122].

$TaTe_4$ also undergoes several phase transitions. At about 450K there is a reversible transition from the RT to a HT superstructure with a corresponding change in unit cell dimensions from $(2a_0 x 3c_0)$ into $(\sqrt{2}a_0 x 3c_0)$ (Figure 17) [129]. However, in addition to this HT COM phase, a further IC HT $TaTe_4$ phase was also reported above 550K (Figure 18) [142]. The transition between these $TaTe_4$ HT COM and IC phases is rapid and without any precursor effects. In spite of this different behaviour as compared to the COM-to-IC transitions in $NbTe_4$, the analogy between both compounds is obvious and it proves that the

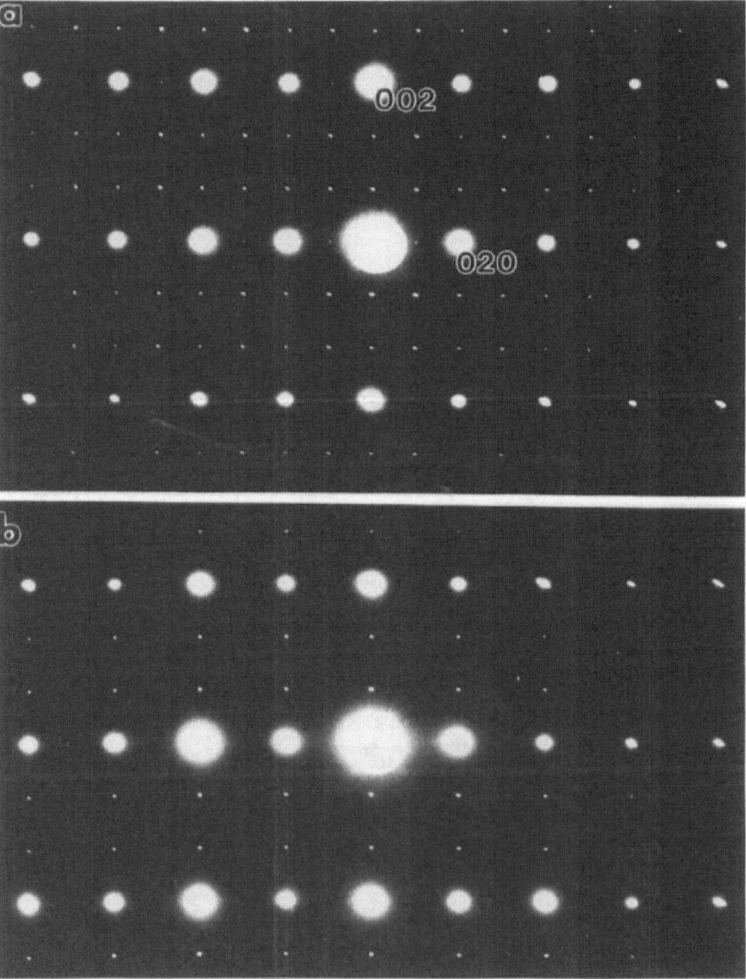

Fig.17 Electron diffraction patterns from $TaTe_4$ taken with the beam along the [100] zone axis of the average structure at RT (a) and at 450 K (b) [129].

competing mechanisms, on which the structures of particular phases depend, are extremely sensitive to temperature. No LT phases were detected in $TaTe_4$ below RT.

In addition to these end-members extensive work was carried out on mixed $Nb_xTa_{1-x}Te_4$ crystals. Preliminary measurements of the wave-vector variation with composition [125] were repeated and it was shown that the q-vector changes with composition in a stepwise fashion rather than continuously [143]. The first step was also confirmed by a (3+1)-dimensional X-ray refinement carried out on a single crystal with composition $Nb_{0.28}Ta_{0.72}Te_4$ [144], which was shown to be commensurately modulated with almost identical crystallographic parameters to those of pure $TaTe_4$. As realized later, the composition of this crystal was close to but still below the first step separating the COM structure of $TaTe_4$ and the IC structure of $Nb_{0.3}Ta_{0.7}Te_4$ (Figure 19).

LPS computer simulations were carried out for a huge unit cell $(2a_0 \times 384c_0)$ to fit all the experimentally observed phases within a single structural model. The calculation procedure was derived from that used for LT1/LT2 $NbTe_4$ [140] using identical parameters δ_1 and δ_2. It was shown, that the LT1/LT2 $NbTe_4$ phases constitute the end members of a family of related structures (see Table I). A common feature of all family members is the appearance of sharp

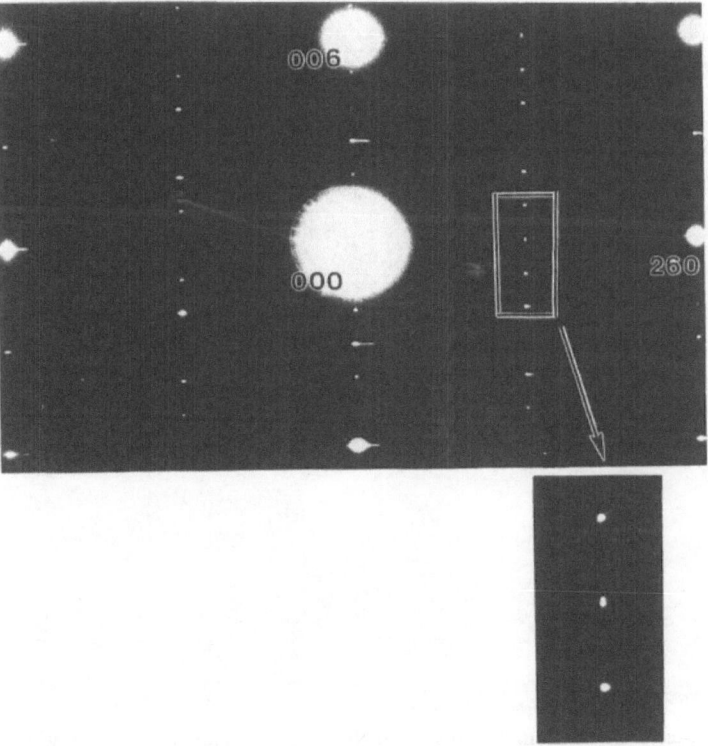

Fig.18 The [130] zone axis electron diffraction pattern from $TaTe_4$ at 550 K. Enlarged is the superlattice doublet revealing incommensurability [142].

satellites in the <100> zones along lines connecting the AS reflections. In addition, satellites occur along lines halfway between the first whose appearance depends on composition. For compositions near $TaTe_4$, these satellites are sharp, becoming increasingly diffuse as the composition of $NbTe_4$ is approached. On cooling these diffuse satellites behave in a similar manner to those of $NbTe_4$, but their transformation into streaks and further into closely spaced satellites is never fully accomplished. For each phase, regardless of composition, all satellites can be described by a single q-vector. The number of LT1 modulation periods fitting C, is 256 (q=0.667), 258 (q=0.672), 260 (q=0.677), 262 (q=0.682) and 264 (q=0.688) for $TaTe_4$, $Nb_{0.3}Ta_{0.7}Te_4$, $Nb_{0.6}Ta_{0.4}Te_4$, $Nb_{0.7}Ta_{0.3}Te_4$ and $NbTe_4$, respectively.

Finally, the variation of the wave-vector is achieved not only by changing the composition of the mixed crystals $Nb_xTa_{1-x}Te_4$, but also by substitution of transition metals Ti, Zr and V for either Nb or Ta. It was shown [145] that these substitutions can both reduce and increase the modulation wave-vectors as compared to pure $NbTe_4$ and $TaTe_4$. These observations were explained in terms of an interplay between bonding and Fermi surface effects.

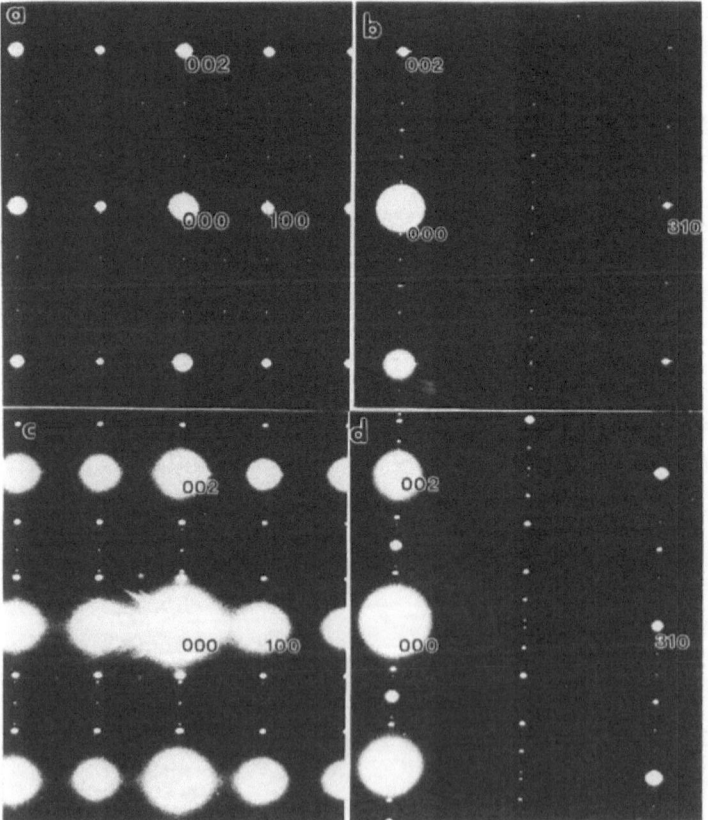

Fig.19 Selected- area diffraction patterns from $Nb_xTa_{1-x}Te_4$, showing the variation of the q-vector with composition. (a) [010] and (b) [130] zone axis patterns for x ~0.3; (c) [010] and (d) [130] zone axis patterns for x ~0.7 [143].

TABLE I

Calculated intensities (I) for Nb$_x$Ta$_{1-x}$Te$_4$. Intensities are rounded up to the first digit δ_1=6.67E-6 for the LT1 type of distortion and δ_2=3.E-2 for the LT2 type of distortion. All calculations were carried out for a (2a x 2a x 384c) unit cell [143].

TaTe$_4$ hkl	I	Nb$_{0.3}$Ta$_{0.7}$Te$_4$ hkl	I	Nb$_{0.6}$Ta$_{0.4}$Te$_4$ hkl	I	Nb$_{0.7}$Ta$_{0.3}$Te$_4$ hkl	I	NbTe$_4$ LT1/LT2 hkl	I
0 0 0	2.E+8	0 0 0	2.E+8	0 0 0	2.E+8	0 0 0	2.E+8	0 0 0	2.E+8
0 0 256	2.E+4	0 0 252	6.E+1	0 0 248	6.E+1	0 0 244	5.E+1	0 0 240	2.E+4
								0 0 288	2.E+0
								0 0 480	7.E+1
0 0 512	1.E+3	0 0 516	2.E+3	0 0 520	2.E+3	0 0 524	2.E+3	0 0 528	9.E+2
								0 0 720	6.E-1
0 0 768	8.E+7	0 0 768	7.E+7	0 0 768	7.E+7	0 0 768	7.E+7	0 0 768	8.E+7
								1 0 12 δ_2	3.E-1
								1 0 240 δ_1	1.E+0
1 0 256 δ_1/δ_2	2.E+4	1 0 255 δ_2	1.E-1	1 0 254 δ_2	1.E-1	1 0 253 δ_2	1.E-1	1 0 252 δ_1	1.E-1
		1 0 258 δ_2	1.E+0	1 0 260 δ_1	1.E+0	1 0 262 δ_1	1.E+0	1 0 264 δ_1	1.E+0
								1 0 276 δ_2	1.E-1
		1 0 510 δ_1	3.E-1	1 0 508 δ_1	4.E-1	1 0 506 δ_1	4.E-1	1 0 504 δ_1	2.E-1
1 0 512 δ_1/δ_2	1.E+3	1 0 513 δ_2	1.E-1	1 0 514 δ_2	1.E-1	1 0 515 δ_2	1.E-1	1 0 516 δ_2	1.E-1
		1 0 516 δ_1	1.E-1	1 0 520 δ_1	1.E-1	1 0 524 δ_1	1.E-1		
								1 0 744 δ_1	1.E+0
								1 0 756 δ_2	3.E-1
		3 0 3 δ_2	3.E+1	3 0 6 δ_2	3.E+1	3 0 9 δ_2	3.E+1	3 0 12 δ_2	3.E+1
3 0 256 δ_1/δ_2	3.E+4	3 0 255 δ_2	3.E+1	3 0 254 δ_2	3.E+1	3 0 253 δ_2	3.E+1	3 0 252 δ_2	3.E+1
		3 0 258 δ_1	2.E+0	3 0 260 δ_1	2.E+0	3 0 262 δ_1	2.E+0	3 0 264 δ_1	2.E+0
		3 0 261 δ_2	1.E-1						
		3 0 510 δ_1	1.E+0	3 0 508 δ_1	1.E+0	3 0 506 δ_1	1.E+0	3 0 504 δ_1	1.E+0
3 0 512 δ_1/δ_2	2.E+3	3 0 513 δ_2	3.E+1	3 0 514 δ_2	3.E+1	3 0 515 δ_2	3.E+1	3 0 516 δ_2	3.E+1
3 0 768 δ_1/δ_2	8.E-1	3 0 765 δ_2	3.E+1	3 0 762 δ_2	3.E+1	3 0 759 δ_2	3.E+1	3 0 756 δ_2	3.E+1
								2 1 12 δ_2	3.E-1
		2 1 126 δ_1	8.E-1	2 1 124 δ_1	8.E-1	2 1 122 δ_1	1.E+0	2 1 120 δ_1	8.E-1
2 1 128 δ_1/δ_2	3.E+3	2 1 129 δ_2	1.E-1	2 1 130 δ_2	1.E-1	2 1 131 δ_2	1.E-1		
								2 1 144 δ_2	1.E-1
								2 1 240 δ_1	1.E-1
2 1 256 δ_1/δ_2	3.E+2							2 1 276 δ_2	7.E-1
								2 1 492 δ_2	7.E-1
2 1 512 δ_1/δ_2	9.E+3	2 1 510 δ_1	7.E-1	2 1 508 δ_1	7.E-1	2 1 506 δ_1	7.E-1	2 1 504 δ_1	4.E-1
		2 1 516 δ_1	1.E-1	2 1 520 δ_1	1.E-1	2 1 524 δ_1	1.E-1	2 1 528 δ_1	4.E-1
								2 1 624 δ_1	1.E+0
2 1 640 δ_1/δ_2	2.E-1	2 1 639 δ_2	1.E-1	2 1 638 δ_2	1.E-1	2 1 637 δ_2	1.E-1		
		2 1 642 δ_1	2.E+0	2 1 644 δ_1	2.E+0	2 1 646 δ_1	2.E+0	2 1 648 δ_1	3.E+0
								2 1 756 δ_2	4.E-1
1 1 256	2.E+4	1 1 258	1.E+4	1 1 260	1.E+4	1 1 262	1.E+4	1 1 264	2.E+4
1 1 512	6.E+3	1 1 510	1.E+4	1 1 508	1.E+4	1 1 506	1.E+4	1 1 504	8.E+4
								1 1 552	1.E+0
1 1 768		1 1 762	4.E+1	1 1 756	4.E+1	1 1 750	4.E+1	1 1 744	4.E+3
2 4 0	2.E+6	2 4 0	2.E+6	2 4 0	2.E+6	2 4 0	2.E+6	2 4 0	2.E+6
2 4 128	7.E+2	2 4 132	2.E+3	2 4 136	2.E+3	2 4 140	2.E+3	2 4 144	2.E+3
2 4 256	8.E+4	2 4 252	2.E+2	2 4 248	2.E+2	2 4 244	2.E+2	2 4 240	1.E+2
2 4 384	7.E+7	2 4 384	7.E+7	2 4 384	7.E+7	2 4 384	7.E+7	2 4 384	7.E+7
2 4 512	1.E+5	2 4 516	3.E+3	2 4 520	3.E+3	2 4 524	3.E+3	2 4 528	4.E+3
2 4 640	2.E+4	2 4 636	2.E+3	2 4 632	2.E+3	2 4 628	2.E+3	2 4 624	2.E+3
2 4 768	1.E+7	2 4 768	1.E+7	2 4 768	1.E+7	2 4 768	1.E+7	2 4 768	1.E+7

4. Concluding Remarks

Incommensurately modulated structures, particularly those based on charge density waves, represent from the crystallographic point of view a real challenge. The classical approach was found to be unsuitable in these cases and new descriptions have had to be developed, of which those used most frequently are described in some detail. These include the superspace group description, the dualistic interpretation, the long-period superstructure approach and normal-mode symmetry analysis. Although different in their origin, we have shown that these approaches are equivalent and largely compatible. Their application was illustrated in two model cases, those of NbS_3 and $Nb_xTa_{1-x}Te_4$, where the temperature and composition dependent phases were described. Finally, it was shown that, although originally developed for the description of incommensurately modulated structures, these approaches are often suitable for the description of commensurate superstructures as well.

Acknowledgments

This work is a result of a continuous collaboration with a number of colleagues from several laboratories, but before all with Prof. F.W. Boswell, Prof. J.M. Corbett and Dr. J.C. Bennett of the Physics Department of the University of Waterloo, Ontario (the latter now at the Physics Department of the Acadia University, Nova Scotia) and with Prof. V. Marinkovic from the J. Stefan Institute and the Department of Metallurgy of the University of Ljubljana. Without their contributions this work could by no means be finished in the present form.

We are indebted to Mrs. Z. Skraba for her assistance and invaluable help in preparing the manuscript.

All figures were reproduced from the previously published papers by the kind permission of the International Union of Crystallography (Figures 1,2,3,11,13,15,16), of the American Physical Society (Figures 4,5,14), of the Institute of Physics (Figures 6-10,18,19 and Table I) and of the Akademie-Verlag (Figures 12,17).

References

1. T. Janssen and A. Janner, *Adv. Phys.* **36**, 519 (1987).
2. S. van Smaalen, *Phys. Rev. Lett.* **70**, 2419 (1993).
3. A. Yoshimori, *J. Phys. Soc. Jap.* **14**, 807 (1959).
4. G.D. Preston, *Proc. Roy. Soc. A* **167**, 526 (1938).
5. E. Brouns, J.W. Visser and P.M. de Wolff, *Acta Cryst.,* **17**, 614 (1964).
6. R.L. Withers and J. A. Wilson, *J. Phys. C* **19**, 4809 (1986).
7. E. Makovicky and B. G. Hyde, *Mat. Sci. Forum* **100&101**, 1 (1992).
8. G.A. Wiegers, A. Meetsma, R.J. Huange, S. van Smaalen, J.L. de Boer, A. Meerschaut, P. Rabu and J. Rouxel, *Acta Cryst.* **B46**, 324 (1990).
9. C. Auriel, R. Roesky, A. Meerschaut and J. Rouxel, *Mat. Res. Bull.* **28**, 247 (1993).
10. J.A. Wilson, J.A. Di Salvo and S. Mahajan, *Adv. Phys.* **24**, 117 (1975).

11. W.C. Koehler, in *Magnetic Ordering of Rare-Earth Metals*, Ed. R.J. Elliott, Plenum Press, New York (1972), p.81. 12.   R.H. Friend and A.D. Joffe, *Adv. Phys.* **36**, 1 (1987).
13. R. Moret and J.P. Pouget, in *Crystal Chemistry and Properties of Materials with Quasi-One-Dimensional Structures*, Ed. J. Rouxel, D. Reidel Publ. Comp., Dordrecht (1986), p.87.
14. F. Hulliger, in *Structural Chemistry of Layered-Type Phases*, Ed. F. Lévy, D. Reidel Publ. Comp., Dordrecht, (1976).
15. A.W. Overhauser, *Adv. Phys.* **27**, 343 (1978).
16. J.A. Steeds, D.M. Bird, D.L. Eaglesham, S. McKernan, R. Vincent and R.L. Withers, *Ultramicroscopy* **18**, 97 (1985).
17. D.M. Bird and R.L. Withers, *J. Phys. C* **19**, 3497 (1986).
18. R.L. Withers and D.M. Bird, *J. Phys. C.* **19**, 3507 (1986).
19. R.V. Coleman, B. Giambattista, P.K. Hansma, A. Johnson, W.W. McNairy and C.G. Clough, *Adv. Phys.* **37**, 559 (1988).
20. S. van Smaalen, J.L. de Boer, A. Meetsma, H. Graafsma, H.-S. Sheu, A. Dorovskikh, P. Coppens and F. Lévy, *Phys. Rev.* **B45**, 3103 (1992).
21. P.M. de Wolff, *Acta Cryst.* **A30**. 777 (1974).
22. A. Yamamoto, *Acta Cryst.* **A38**, 87 and Appendices 1,2,3 (1982).
23. P.M. de Wolff, T. Janssen and A. Janner, *Acta Cryst.* **A37**, 625 (1981).
24. A. Janner, T. Janssen and P.M. de Wolff, *Acta Cryst.* **A39**, 658 (1983).
25. P.M. de Wolff, *Acta Cryst.* **A40**, 34 (1984).
26. T. Janssen and A. Janner, *Physica* **126A**, 163 (1984).
27. J.M. Pérez-Mato, G. Maradiaga and M.J. Tello, *J. Phys. C.* **19**, 2613 (1986).
28. A.A. Maradudin and S.H. Vosko, *Rev. Mod. Phys.* **40**, 1 (1968).
29. A. Janner and T. Janssen, *Phys. Rev.* **B15**, 643 (1977).
30. A. Yamamoto, T. Janssen, A. Janner and P.M. de Wolff, *Acta Cryst.* **A41**, 528 (1985).
31. S. van Smaalen, *Acta Cryst.* **A43**, 202 (1987).
32. *International Tables of Crystallography, Vol.I*, The Kynoch Press, Birmingham (1952), or Vol.A, D. Reidel Publ. Comp., Dordrecht, (1983).
33. J.M. Pérez-Mato, G. Madariaga, F.J. Zuñinga and A. Garcia Arribas, *Acta Cryst.* **A43**, 216 (1987).
34. P.M. de Wolff, in *Modulated Structure Materials*, Ed. T. Tsakalatos, NATO ASI Series E: Appl. Sci. No.83, M.Nijhoff Publ. Comp., Dordrecht (1984), p.133.
35. K. Gesi and M. Iizumi, *J. Phys. Soc. Japan* **45**, 1777 (1978).
36. M. Iizumi and K. Gesi, *J. Phys. Soc. Japan* **52**, 2526 (1983).
37. P. Bak, *Phys. Today* **39**, No. 12, 38 (1986).
38. P. Bak, *Rep. Progr. Phys.* **45**, 587 (1982).
39. A.C.R. Hogervorst and P.M. de Wolff, *Sol. St. Comm.* **43**, 179 (1982).
40. A. Loiseau, G. van Tendeloo, R. Portier and F. Ducastelle, *J. Phys. (Paris)* **46**, 595 (1985).
41. D. Gibbs, D.E. Moncton, K.L. D'Amico, J. Bohr and B.H. Grier, *Phys. Rev. Lett.* **55**, 234 (1985).
42. M.J. Winokur and R. Clarke, *Phys. Rev. Lett.* **56**, 2072 (1986).
43. Y. Komura and Y. Kitano, *Acta Cryst.* **B33**, 2496 (1977).
44. L. Bernard, R. Currat, P. Delanoye, C.M.E. Zeyen, S. Hubert and R. de Kouchkovsky, *J. Phys. C* **16**, 433 (1983).
45. H. Cailleau, F. Moussa, C.M.E. Zeyen and J. Bouillot, *Solid St. Comm.* **33**, 407 (1980).
46. H. Böhm and H.-G. von Schnering, *Z. Kristallogr.* **162**, 26 (1983); **171**, 41 (1985).
47. J. Kusz and H. Böhm,. *Z. Kristallographie* **201**, 9 (1992).
48. A. Yamamoto, *Acta Cryst.* **B38**, 1446 (1982).
49. P. Bak and R. Bruinsma, *Phys. Rev. Lett.* **49**, 249 (1982).
50. R. Bruinsma and P. Bak, *Phys. Rev.* **B27**, 5824 (1983).
51. J. von Boehm and P. Bak, *Phys. Rev.Lett.* **42**, 122 (1979).
52. P. Bak and J. von Boehm, *Phys. Rev.* **B21**, 5297 (1980).
53. A. Yamamoto, *Acta Cryst.* **B39**, 17 (1983).
54. A. Budkowski, A. Prodan, V. Marinkovic, D. Kucharczyk, I. Uszynski and F.W. Boswell, *Acta Cryst.* **B45**, 529 (1989).
55. B. Dam and A. Janner, *Acta Cryst.* **B42**, 69 (1986).
56. A. Yamamoto and H. Nakazawa, *Acta Cryst.* **A38**, 79 (1982).
57. L.D. Landau and E.M. Lifshitz, *Statistical Physics*, Pergamon Press, Oxford (1980).
58. J.M. Pérez-Mato, G. Madariaga and M.J. Tello, *Phys. Rev.* **B30**, 1534 (1984).
59. M. Iizumi, J.D. Axe, G. Shirane and K. Shimaoka, *Phys. Rev.* **B15**, 4392 (1977).

60. D.E. Moncton, J.D. Axe and F.J. DiSalvo, *Phys. Rev.* **B16**, 801 (1977).
61. J.M. Pérez-Mato, *J. Phys.: Condens. Matt.* **3**, 391 (1991).
62. J. Warren, *Rev. Mod. Phys.* **40**, 38 (1968).
63. J.M. Pérez-Mato, F. Gaztelua, G. Madariaga and M.J. Tello, *J. Phys. C* **19**, 1923 (1986).
64. R.L. Withers, B.G. Hyde, A. Prodan and F.W. Boswell, *J. Phys.: Condens. Matt.* **2**, 4051 (1990).
65. T.R. Welberry, R.L. Withers and J.C. Osborn, *Acta Cryst.* **B46**, 267 (1990).
66. R.L. Withers, B.G. Hyde and J.G. Thompson, *J. Phys. C* **20**, 1653 (1987).
67. W.L. McMillan, *Phys. Rev.* **B14**, 1496 (1976).
68. P. Bak and V.J. Emery, *Phys. Rev. Lett.* **36**, 978 (1976).
69. P. Bak, *Phys. Rev. Lett.* **37**, 1071 (1976).
70. P. Bak and T. Janssen, *Phys. Rev.* **B17**, 436 (1978).
71. M.B. Walker and R. Morelli, *Phys. Rev.* **B38**, 4836 (1988).
72. R. Morelli and M.B. Walker, *Phys. Rev. Lett.* **62**, 1520 (1989).
73. A. Janner and T. Janssen, *Acta Cryst.* **A36**, 399 (1980).
74. A. Prodan, A. Budkowski, F.W. Boswell, V. Marinkovic, J.C. Bennett and J.M. Corbett, *J. Phys.: Condens. Matt.* **21**, 4171 (1988).
75. A. Budkowski, V. Marinkovic, A. Prodan and F.W. Boswell, *Phys. Stat. Sol.(a)* **117**, 351 (1990).
76. R.L. Withers, Y.C. Feng and G.H. Lu, *J. Phys. Cond. Matt.* **2**, 3187 (1990).
77. R.L. Withers, J.G. Thomson and B.G. Hyde, *Acta Cryst.* **B47**, 166 (1991).
78. A. Janner, T. Janssen, P.M. de Wolff, *Acta Cryst.* **A39**, 671 (1983).
79. V. Heine and E.H. Simmons, *Acta Cryst.* **A43**, 289 (1987).
80. O.V. Kovalev, in *Irreducible Representations of the Space Groups*, Gordon and Breach Sci. Publ., N.Y. (1965).
81. D.A. Rabson, N.D. Mermin, D.S. Rokhsar and D.C. Wright, *Rev. Mod. Phys.* **63**, 699 (1991).
82. N.D. Mermin, *Rev. Mod. Phys.* **64**, 3 and 1163 (1992).
83. N.D. Mermin, *Phys. Rev. Lett.* **68**, 1172 (1992).
84. N.D. Mermin and R. Lifshitz, *Acta Cryst.* **A48**, 515 (1992).
85. A. Bienenstock and P.P. Ewald, *Acta Cryst.* **15**, 1253 (1962).
86. H. Brown, R. Bülow, J. Neubüser, H. Wondratschek and H. Zassenhaus, in *Crystallographic Groups in Four-Dimensional Space*, John Wiley and Sons Inc., New York, 1978, p.80.
87. J. Neubüser, H. Wondratschek and R. Bülow, *Acta Cryst.* **A27**, 517 (1971).
88. R. Bülow, J. Neubüser and H. Wondratschek, *Acta Cryst.* **A27**, 520 (1971).
89. D. Weigel, Th. Phan and R. Veysseyre, *C.R. Acad. Sci.* **298**, 825 (1984).
90. R. Veysseyre, Th. Phan and D. Weigel, *C.R. Acad. Sci.* **300**, 51 (1985).
91. D. Weigel, Th. Phan and R. Veysseyre, *Acta Cryst.* **A43**, 294 (1987).
92. D. Grebille, D. Weigel, R. Veysseyre and Th. Phan, *Acta Cryst.* **A46**, 234 (1990).
93. R. Veysseyre and D. Weigel, *Acta Cryst.* **A45**, 187 (1989).
94. Th. Phan, R. Veysseyre, D. Weigel and D. Grebille, *Acta Cryst.* **A45**, 547 (1989).
95. A. Janner, T. Janssen and P.M. de Wolf, *Acta Cryst.* **A39**, 667 (1983).
96. D. Weigel, R. Veysseyre and Th. Phan, *Acta Cryst.* **A46**, 463 (1990).
97. R. Veysseyre, Th. Phan and D. Weigel, *Acta Cryst.* **A47**, 233 (1991).
98. Th. Phan, R. Veysseyre and D. Weigel, *Acta Cryst.* **A47**, 549 (1991).
99. J.D.C. McConnel and V. Heine, *Acta Cryst.* **A40**, 473 (1984).
100. E.H. Simmons and V. Heine, *Acta Cryst.* **A43**, 626 (1987).
101. B.L. Litvin, *Phys. Rev.* **B21**, 3184 (1980).
102. V.A. Koptsik, *Ferroelectrics* **21**, 499 (1978).
103. P. Monceau, in *Electronic Properties of Inorganic Quasi-One-Dimensional Materials, II*, Ed. P. Monceau, D. Riedel Publ. Comp., Dordrecht, (1985), p.139.
104. W. Krönert and K. Plieth, *Z. Anorg. Allg. Chem.* **336**, 207 (1965).
105. M.H. Whangbo, in *Crystal Chemistry and Properties of Materials with Quasi-One-Dimensional Structures*, Ed. J .Rouxel, D. Riedel Publ. Comp., Dordrecht, (1986), p.27.
106. K. Endo, H. Ihara, K. Watanabe and S. Gonda, *J. Sol. St. Chem.* **39**, 215 (1981).
107. A. Meerschaut, *J. Phys. (Paris)* **44-C3**, 1615 (1983).
108. A. Meerschaut and J. Rouxel, in *Crystal Chemistry and Properties of Materials with Quasi-One-Dimensional Structures,* Ed. J. Rouxel, D. Riedel Publ. Comp., Dordrecht, (1986), p.205.
109. F.W. Boswell and A. Prodan, *Physica* **99B**, 361 (1980).
110. T. Cornelissens, G. van Tendeloo, J. van Landuyt and S. Amelinckx, *Phys. Stat. Sol.(a)* **48**, K5 (1978).
111. T. Iwazumi, M. Izumi, K. Uchinokura, R. Yoshizaki and E. Matsuura, *Physica* **143B**, 255 (1986).

112. J. Rijnsdorp and F. Jellinek, *J. Sol. St. Chem.* **25**, 325 (1978).
113. J.A. Wilson, *Phys. Rev.* **B19**, 6456 (1979).
114. J.A. Wilson, *J. Phys. F* **12**, 2469 (1982).
115. D.W. Bullet, *J. Sol. St. Chem.* **33**, 13 (1980).
116. S. van Smaalen, *Phys. Rev.* **B38**, 9594 (1988).
117. S. van Smaalen and J.L. de Boer, private communication.
118. C.J. Roucau, *J. Phys. (Paris)* **44-C3**, 1725 (1983).
119. K. Selte and A. Kjekshus, *Acta Chem. Scand.*, **18**, 690 (1964).
120. E. Bjerkelund and A. Kjekshus, *J. Less-Common Met.*, 7, 231 (1964).
121. F.W. Boswell, A. Prodan and J.K. Brandon, Coll. Int. CNRS, E1, Paris, (1981); *J. Phys. C*, **16**, 1067 (1983).
122. J. Mahy, J. van Landuyt and S. Amelinckx, *Phys. Stat. Sol.(a)*, **77**, K1 (1983).
123. J. Mahy, G.A. Wiegers, J. van Landuyt and S. Amelinckx, 8.ECM, Liège,(1983); *Mat. Res. Soc. Symp. Proc.*, **21**, 181 (1984).
124. D.J. Eaglesham, D. Bird, R.L. Withers and J.W. Steeds, *J. Phys. C* **18**, 1 (1985).
125. F.W. Boswell and A. Prodan, *Mat. Res. Bull.*, **19**, 93 (1984).
126. M.B. Walker, *Can. J. Phys.*, **63**, 46 (1985)
127. R. Morelli, D. Sahu and M.W. Walker, *Phys. Rev.*, **B33**, 4843 (1986).
128. S. van Smaalen, D.K. Bronsema and J. Mahy, *Acta Cryst.*, **B42**, 43 (1986).
129. F.W. Boswell, A. Prodan, J.C. Bennett, J.M. Corbett and L.G. Hiltz, *Phys. Stat. Sol.(a)*, **102**, 207 (1987).
130. K.D. Bronsema, S. van Smaalen, J.L. de Boer, G.A. Wiegers and F. Jellinek, *Acta Cryst.*, **B43**, 305 (1987).
131. J. Mahy, J. van Landuyt and S. Amelinckx, *Phys. Stat. Sol.(a)*, **102**, 609 (1987).
132. J.M. Corbett, L.G. Hiltz, F.W. Boswell, J.C. Bennett and A. Prodan, *Ultramicroscopy*, **26**, 43 (1987).
133. F.W. Boswell and A. Prodan, *Phys. Rev.*, **B34**, 2979 (1986).
134. A. Prodan and F.W. Boswell, *Acta Cryst.*, **B43**, 165 (1987).
135. J. Kusz and H. Böhm, *Z. Kristallogr.*, **208**, 187 (1993).
136. J. Mahy, J. van Landuyt, S. Amelinckx, Y. Uchida, K.D. Bronsema and S. van Smaalen, *Phys. Rev. Lett.*, **55**, 1188 (1985).
137. J. Mahy, J. van Landuyt, S. Amelinckx, K.D. Bronsema and S. van Smaalen, *J. Phys. C* **19**, 5049 (1986).
138. J. van Landuyt, R. de Ridder, R. Gevers and S. Amelinckx, *Mat. Res. Bull.*, **5**, 353 (1970).
139. C.Y. Chen and M.B. Walker, *Phys. Rev.*, **B40**, 8983 (1990).
140. A. Prodan, F.W. Boswell, J.C. Bennett, J.M. Corbett, T. Vidmar, V. Marinkovic and A. Budkowski, *Acta Cryst.*, **B46**, 587 (1990).
141. H. Böhm, *Z. Kristallogr.*, **180**, 113 (1987).
142. J.C. Bennett, F.W. Boswell, A. Prodan, J.M. Corbett and S. Ritchie, *J. Phys. C.* **3**, 6959 (1991).
143. J.C. Bennett, S. Ritchie, A. Prodan, F.W. Boswell and J.M. Corbett, *J. Phys. C* **4**, 2155 (1992).
144. D. Kucharczyk, A. Budkowski, F.W. Boswell, A. Prodan and V. Marinkovic, *Acta Cryst.*, **B46**, 153 (1990).
145. J.C. Bennett, F.W. Boswell, A. Prodan, J.M. Corbett and S. Ritchie, *Austr. J. Chem.*, **45**, 1363 (1992).

X-RAY CRYSTALLOGRAPHIC ANALYSIS OF THE CHARGE-DENSITY WAVE MODULATED PHASES IN THE NbTe$_4$-TaTe$_4$ SYSTEM

HORST BÖHM

Institut f. Geowissenschaften, Universität Mainz
D-55099 Mainz, Germany

1. Introduction

NbTe$_4$ and TaTe$_4$ belong to the important structural family CuAl$_2$ (tl20) characterized by a tetragonal antiprismatic coordination of the minority component. The basic structures were first described by Selte and Kjekshus [1] and Selte and Bjerkelund [2]. In these structures the Te atoms form 1-D infinite columns of stacked tetragonal antiprisms along [001] which are centered by the metal atoms (Figure 1). The arrangement of the antiprismatic columns in the tetragonal $\mathbf{a_1a_2}$-plane yields short Te-Te contacts, forming Te$_2$ pairs with covalent bonds of about 2.92 Å. Thus, following the usual chemical arguments MTe$_4$ (M=Nb,Ta) is a ditelluride of M(IV), i.e. M^{4+}(Te$_2^{2-}$)$_2$. The structure is strongly related to that of the mineral patronite V(S$_2$)$_2$ [3,4] where, in addition, covalently bonded V$_2$ pairs are formed along the columns of the *quasi-tetragonal* antiprisms. The formation of M$_2$ pairs is very common in the chemistry of M^{4+} compounds with a d^1 electron configuration. Selte and Kjekshus [1] also observed weak satellite reflections which indicated a twofold superstructure along [110] and a threefold superstructure along [001]. However, they did not account for these additional reflections in their structure determination. Thus, the reported structure is really an average structure (Table I).

The crystals can be grown from the elements in evacuated quartz tubes, using iodine or bromine as a transport agent [1]. The tubes are heated at 800–900 °C for several days and slowly cooled to room temperature over a period of one week. The crystals generally grow as bundles of parallel needles with the crystallographic **c**-axis being the needle axis and are metallic in appearance with flat faces parallel to the needle axis. Crystals up to 1 mm length have good quality for single crystal studies.

The basic structures of NbTe$_4$ and TaTe$_4$ are isostructural. The initial study of the system $(Ta_{1-x}Nb_x)Te_4$ with $0 \leq x \leq 1$ revealed the existence of a series of solid solutions [5]. The main interest in these structures arose when it was found that the room temperature superstructures were incommensurate for NbTe$_4$ [6,7] and commensurate for TaTe$_4$ [5]. Since

41

F. W. Boswell and J.C. Bennett (eds.),
Advances in the Crystallographic and Microstructural Analysis of Charge Density Wave Modulated Crystals, 41–67.
© 1999 *Kluwer Academic Publishers.*

a

b

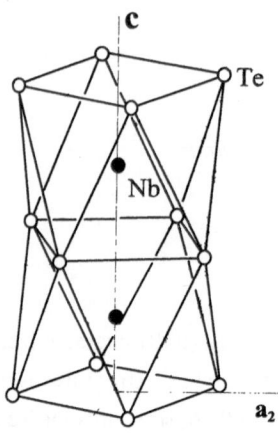

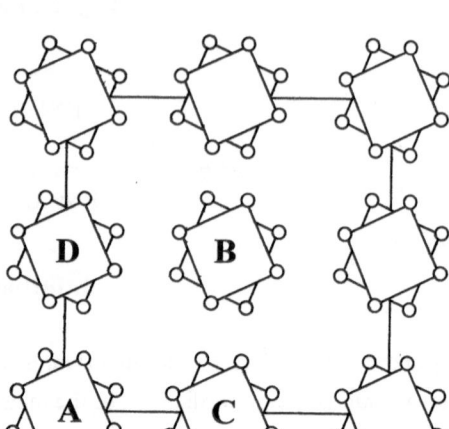

Fig.1 The basic structure of NbTe$_4$ after [1]; (a) the content of the unit cell; (b) a projection of the structure along the tetragonal c-axis on the basal plane of (2ao × 2ao). Only the Te-squares are shown.

TABLE I

Crystal data and atomic parameters for the basic structure of the system

$\left(Ta_{1-x}Nb_x\right)Te_4$ with $0 \leq x \leq 1$

NbTe$_4$ [1]

        SG P4/mcc (no. 124)

        a_0=6.499 Å, c_0=6.837 Å, Z=2 (Guinier film, CuKa$_1$ radiation)

        2 Nb in 2a: $\pm(00¼)$

        8 Te in 8m: $\pm\left(xy\,0;\ \bar{y}x\,0;\ \bar{x}y\,½;\ yx\,½\right)$

(Ta$_{0.72}$Nb$_{0.28}$)Te$_4$ [26]

        SG P4/mcc (no.124)

        a_0=6.512(3) Å, c_0=6.818 Å (Philips Diffractometer PW 1100)

atomic parameters are the same as above

TaTe$_4$ [2]

        SG P4cc (no.103), (P4/mcc after [28])

        a_0= 6.514 Å, c_0=6.809 Å, Z=2

atomic parameters are the same as above

---

the structures exhibit isolated one-dimensional chains of metal atoms, the modulation of the structures is correlated with the occurrence of charge-density waves [6].

2. Modulations

The modulation of the structure is associated with the occurrence of satellite reflections in the diffraction pattern. The satellite pattern is characterized by the following q-vectors (Figure 2):

$$\mathbf{q_1} = (1/2, 1/2, q_z)$$
$$\mathbf{q_2} = (1/2, 0, q_z), \ \mathbf{q_3} = (0, 1/2, q_z)$$

The q-vector notation is based on the cell of the basic structure. The indexing of reflections in the reciprocal lattice, however, is for convenience based on a 3-fold superstructure in the incommensurate case as well.

At room temperature $NbTe_4$ is incommensurate ($\mathbf{q_1}$ –satellites with a value of $q_z = 0.688$) whereas $TaTe_4$ is commensurate (with $q_z = 1/3$). In Figure 2, the positions of first, second and fourth order $\mathbf{q_1}$ –satellites are shown in a schematic diagram. The first order satellites only occur adjacent to positions of systematically absent main reflections, whereas, second order satellites are found around observed main reflections (Figure 2b). Close to the second order satellites, those of fourth order can also be observed in electron or X-ray diffraction patterns (Figure 2b); for a commensurate modulation both orders must coincide. At room temperature, the satellites related to $\mathbf{q_2}$ and $\mathbf{q_3}$ are diffuse for $NbTe_4$ but are sharp and commensurate for $TaTe_4$ ($q_z = 2/3$). These satellites appear in the diffraction pattern at roughly the same value of $\mathbf{c^*}$ and at the midpoint between two $\mathbf{q_1}$-satellites (Figure 2b). For $NbTe_4$, the peak intensity of the $(\mathbf{q_2}, \mathbf{q_3})$–satellites is also at the commensurate value in a X-ray experiment. In the mixed crystal series $(Ta_{1-x}Nb_x)Te_4$ with $0 \le x \le 1$, electron diffraction studies [8] revealed that for the room temperature structure the periodicity along $\mathbf{c}$ remains commensurate as x increases until a threshold concentration ($x \approx 0.3$) is exceeded and then jumps discontinuously to an incommensurate value.

3. Phase Transitions

For many modulated structures there is a sequence of phase transitions from a high-symmetry undistorted (high temperature) phase to an incommensurate phase and further to a commensurate (low temperature) phase. The first transition is of second order while the second is a first order phase transition ("lock-in transition"). Such a sequence of phase transitions is also observed in the $NbTe_4/TaTe_4$ system; it depends on temperature and composition which structure is stable.

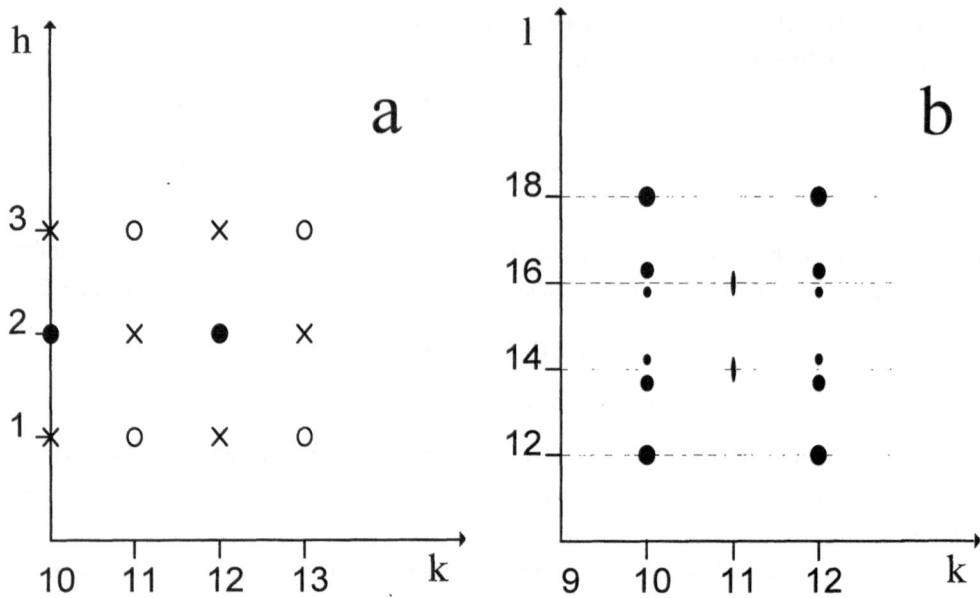

Fig. 2. Schematic diagram of the diffraction pattern: (a) The projection on the $(\mathbf{a_1}^*$-$\mathbf{a_2}^*)$-plane; main reflections (solid circles), $\mathbf{q_1}$-satellites (open circles) and $(\mathbf{q_2}, \mathbf{q_3})$-satellites (crosses). (b) The $(\mathbf{a_2}^*$-$\mathbf{c}^*)$-plane; $\mathbf{q_1}$-satellites of 2nd order and 4th order are drawn separately, the $(\mathbf{q_2}, \mathbf{q_3})$-satellites are elongated [14].

3.1 NbTe$_4$: INCOMMENSURATE (RT) TO NORMAL (HT) PHASE TRANSITION

The undistorted phase (basic structure) exhibits no satellite reflections. An intermediate phase is characterized by satellite reflections $\mathbf{q_1}$; in most phases of the system they are observed at incommensurate positions. If no satellites are present, the unit cell $a_0 \times a_0 \times c_0$ only consists of identical columns of antiprisms (A = B = C = D in Figure 1b). In a phase with $\mathbf{q_1}$-satellites, the metric of the unit cell is given by $\sqrt{2}a_0 \times \sqrt{2}a_0 \times \mu c_0$ (A = B and C = D in Figure 1b); $\mu = 3$ for a commensurate structure and $\mu \neq 3$ for an incommensurate structure. If, in addition to the $\mathbf{q_1}$-satellites, $(\mathbf{q_2}, \mathbf{q_3})$ – satellites are also present, the metric of the cell is $2a_0 \times 2a_0 \times \mu c_0$ and the unit cell is characterized by four independent columns A, B, C, D. However, in this case the tetragonal symmetry (i.e. the fourfold axis in column A) requires that C = D (A $\neq$ B). Other symmetry elements may impose further restrictions.

If the crystal is heated, NbTe$_4$ exhibits the undistorted high-temperature (HT) modification with unit cell $a_0 \times a_0 \times c_0$. A reversible phase transition (2nd order) is observed from the incommensurately modulated (IC) room temperature phase to the HT phase [9]. In Figure 3 the variation of the intensity of the $\mathbf{q_1}$-satellite (3 3 2) is shown for heating and cooling; there is no detectable hysteresis between the cooling and the heating cycle. Note, however, that in order to prevent decomposition the crystals must be sealed in quartz capillaries filled with Ar gas before heating. A fit with the equation:

$$I = \alpha (T_c - T)^{2\beta} \qquad (1)$$

yields the parameters: $T_c = 793.0 \pm 0.6$ K and $\beta = 0.35$ (dashed curve in Figure 3). This corresponds to the critical exponent of a 3-d Ising model for a second order phase transition.

3.2 NbTe$_4$: INCOMMENSURATE (RT) TO INCOMMENSURATE (LT) TRANSITION

On cooling below room temperature, a further phase transition has been observed for NbTe$_4$ in electron [6,10,11] and X-ray diffraction experiments [14]; it is associated with the occurrence of the additional $(\mathbf{q}_2, \mathbf{q}_3)$ – satellites (corresponding to an expansion of the basal plane of the unit cell to $2a_0 x 2a_0$). These $(\mathbf{q}_2, \mathbf{q}_3)$ – satellites can be anticipated even at room temperature as diffuse spots in the electron or X-ray diffraction pattern. On cooling the diffuse $(\mathbf{q}_2, \mathbf{q}_3)$ – satellites gradually sharpen. At about 140–120 K, they are sharp within the $\mathbf{a}_1^* \mathbf{a}_2^*$ –plane but remain diffuse streaks elongated along $\mathbf{c}^*$ (Figure 2b). Electron diffraction experiments [12] reveal that the diffuse streaks develop into closely spaced sharp satellites on further cooling; the separation of these satellites is beyond the resolution of conventional X-ray diffraction methods. The low temperature satellites can be separated into two sets LT$_1$ and LT$_2$ [13,14]. According to Prodan and Boswell [13], these sets are correlated to corresponding structures LT$_1$ and LT$_2$ (see section 4.1).

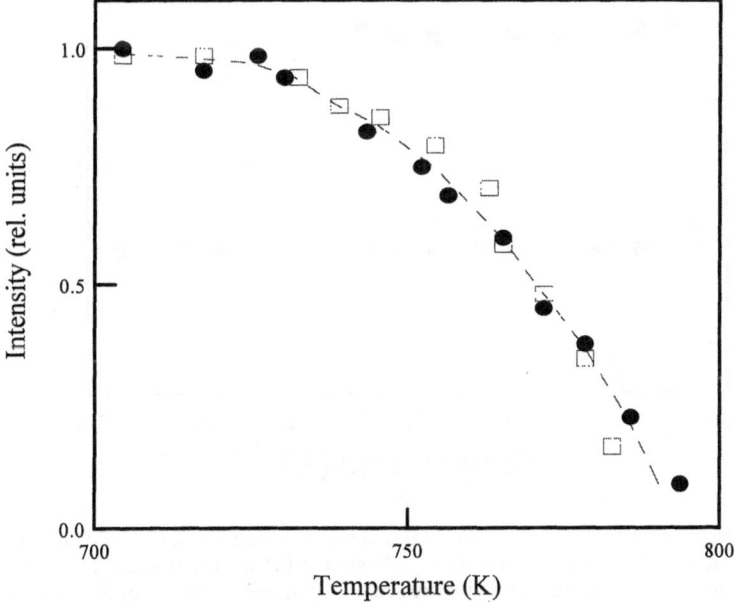

Fig. 3. The intensity variation of the $\mathbf{q}_1$-satellite (3 3 2) of NbTe$_4$ on heating (●) and cooling (□) [9].

On further cooling the $(\mathbf{q_2},\mathbf{q_3})$ – satellites transform to sharp superstructure reflections at ~50 K [6]. At this temperature the structure undergoes a first order (lock-in) phase transition to a commensurate (C) low temperature structure; both $\mathbf{q_1}$- and $(\mathbf{q_2},\mathbf{q_3})$ – satellites are commensurate ($q_z = 2/3$).

The behaviour is different for an X-ray experiment [15]. The first step of the phase transition, i.e. the sharpening of the $(\mathbf{q_2},\mathbf{q_3})$ – satellites in two dimensions, is the same as described for the electron diffraction experiment. The intensity profile of the diffuse streaks, which are elongated along $\mathbf{c^*}$, has a maxima at the commensurate value of $q_z = 2/3$, whereas that of the $\mathbf{q_1}$-satellites remains incommensurate. However, in the X-ray experiment, on further cooling below 50 K the picture remains unchanged. Even when the crystal was held at 15 K for 20 days it did not transform to the commensurate structure. Apparently the transformation is inhibited for kinetic reasons in the macroscopic crystal of the X-ray experiment. In Figure 4, the temperature variation of the full width at half maximum (FWHM) normalized to the

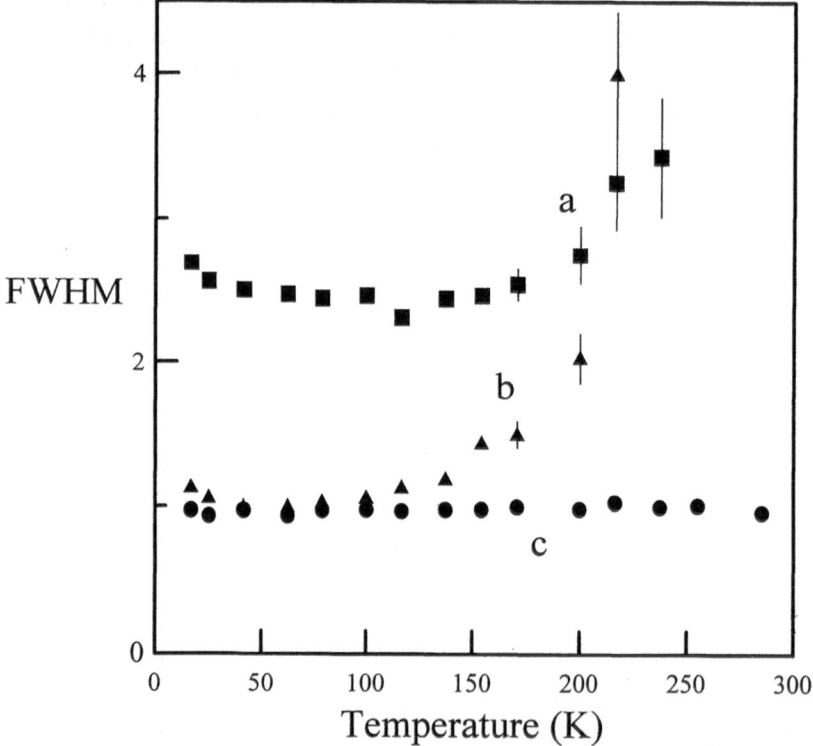

Fig. 4. The variation with temperature of the full width at half maximum (FWHM) normalized to the FWHM of the main reflection (0 10 12) of NbTe$_4$: (a) for the $\mathbf{q_3}$-satellite (0 11 14) scanned parallel to $\mathbf{c^*}$, (b) for the same satellite scanned parallel to $\mathbf{a^*}$ and (c) for the (2nd order) $\mathbf{q_1}$-satellite (0 10 14) [14].

FWHM of the main reflection (0 10 12) is shown for a $\mathbf{q_3}$ – satellite scanned parallel to $\mathbf{c}^*$, for the same satellite scanned parallel to $\mathbf{a}^*$ and for a $\mathbf{q_1}$-satellite. At about 140–120 K, the FWHM along $\mathbf{a}^*$ of the $\mathbf{q_3}$ – satellite is the same as that of the sharp $\mathbf{q_1}$-satellite. The first step in the phase transition, i.e. correlations in planes perpendicular to $\mathbf{c}$ is completed.

In contrast, in the TEM experiments the intermediate LT phase is characterized by incommensurate $\mathbf{q_1}$-satellites and two sets of $(\mathbf{q_2},\mathbf{q_3})$ – satellites, LT_1 and LT_2. On further cooling the phase transition proceeds until the $(\mathbf{q_2},\mathbf{q_3})$ – satellites become superstructure reflections at the lock-in transition (~50 K) [6]. However, the low temperature phase only develops short range order for the macroscopic crystal in the X-ray experiment.

3.3 HIGH TEMPERATURE PHASE TRANSITIONS IN $TaTe_4$

For the other end member of the system, $TaTe_4$, an unmodulated HT phase has not been observed prior to decomposition of the crystal [16]. At high temperatures the structure is incommensurate and characterized by incommensurate $\mathbf{q_1}$-satellites [12,16].

On cooling, a transition to a commensurate (C*) structure (unit cell $\sqrt{2}a_0 \times \sqrt{2}a_0 \times 3c_0$) with the $\mathbf{q_1}$-satellites lying at commensurate positions ($q_z = 2/3$) is observed. In the TEM, the transition occurs on heating at about 550 K [16]. In the X-ray experiment the transition temperature is 650 K [17]. However, the conditions prevailing in these two experiments were quite different: for the X-ray diffractometer, a $TaTe_4$ needle (of approximate dimensions 0.2 × 0.2 × 0.6 mm) was sealed in a capillary filled with argon. Note that in the IC phase the crystal was not heated up until it decomposed; thus it was not verified that the HT phase occurs under these conditions. On cooling, the C* structure undergoes another reversible phase transition (C* to C transition) which is characterized by the occurrence of $(\mathbf{q_2},\mathbf{q_3})$ – satellites (unit cell $2a_0 \times 2a_0 \times 3c_0$); however, these satellites are not diffuse as in $NbTe_4$. Again, the observed transition temperatures are different: 450 K in the TEM experiment [16] and ~580 K for the X-ray experiment [17]. The intensity variation of the $\mathbf{q_3}$-reflection (0 11 2) in an X-ray experiment is shown in Figure 5. Since no hysteresis for the C* to C transition is observed [17] the transition may be of second order, however, TEM results [18] indicate a first order phase transition since both phases can coexist. If the intensity variation in Figure 5 is fitted to equation (1) the parameters are: $T_c = 579 \pm 4$ K, $\beta = 0.176$. This indicates that the order parameter does not vary with a critical exponent β of a second order phase transition.

Overall, the sequence of phases in both end members $NbTe_4$ and $TaTe_4$ is very similar. The transition from the high temperature IC phase to the low temperature C phase occurs through intermediate phases. However, the transition temperatures for $TaTe_4$ are shifted to much higher temperatures compared to those for $NbTe_4$; the lock-in transition temperatures into the C phase being 580 K and 50 K, respectively. For $TaTe_4$, the intermediate phase is commensurate (C*). For $NbTe_4$, the situation is more complicated: for the intermediate phase

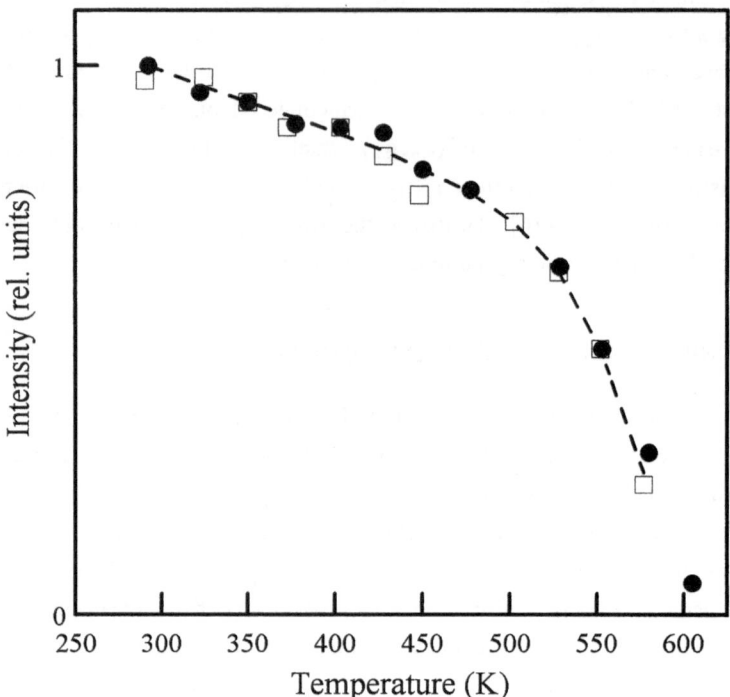

Fig. 5: The intensity variation of the q$_3$-satellite (0 11 2) of TaTe$_4$ in an X-ray experiment on heating(
●) and cooling (□) [17].

(between ~120 K–50 K) Prodan et al. [13, 14] conclude that two deformation systems LT$_1$ and LT$_2$ must be postulated. In the LT structure either both deformation systems occur simultaneously or they exist as domains of two intergrown phases LT$_1$ and LT$_2$.

3.4 PHASE TRANSITIONS IN $\left(Ta_{1-x}Nb_x\right)Te_4$

The mixed crystals $\left(Ta_{1-x}Nb_x\right)Te_4$, $0 \leq x \leq 1$, have been studied extensively in TEM experiments by Bennett et al. [8] and, for two compositions, also by X-ray diffraction [17]. Below the critical Nb concentration ($x < 0.3$), the diffraction patterns observed at room temperature are identical to those of RT TaTe$_4$. As x increases, a commensurate to incommensurate (C to IC) phase transition (at about $x \sim 0.3$) is marked by the abrupt change of q$_z$ for the q$_1$-satellites to an incommensurate value while the (q_2, q_3)–satellites remain centered at commensurate positions; the diffraction pattern is comparable with the low temperature (LT) phase of NbTe$_4$. Further changes occur in the positions and appearance of the satellites in the electron diffraction patterns (EDP) with increasing Nb content x. The (q_2, q_3)–satellites become progressively fainter and more diffuse; this may be interpreted as a second order phase transition as x is increased. The $\mathbf{q_1}$-satellites in all cases remain sharp. There is an indication in the q_z versus x curve that an increase of x in the incommensurate region leads to a discontinuous variation of q_z which proceeds through a sequence of discrete jumps [8] (Figure 6). These plateaus imply the existence of a series of long-period commensurate phases [8]. In the composition range near that of $NbTe_4$ lock-in was not observed in TEM experiments for cooling below RT.

To date, X-ray investigations of the mixed crystals are limited to the two compounds $x = 0.25$ and $x = 0.5$ and to temperatures above room temperature [17]. These compositions are nominal; a range of compositions about the nominal value as a mean is found by TEM microanalysis for crystallites prepared from cleaving a single macrocrystal.

Phase transitions from the IC phase into a LT phase, indicated by the occurrence of $(\mathbf{q_2}, \mathbf{q_3})$–satellites, are observed at ~480 K ($x = 0.25$) and at ~390 K ($x = 0.5$). The $\mathbf{q_1}$-satellites are at incommensurate positions in both the LT and IC phase. The intensity variation of the $\mathbf{q_3}$-reflection (0 11 4) with temperature is shown in Figure 7 for $x = 0.25$. The fit according to equation (1) yields a value of $T_c = 480.0 \pm 4$ K of $\beta = 0.21$. As for $TaTe_4$ β indicates a first order phase transition. The same value of β and $T_c = 390.0 \pm 4$ K is obtained for the $x = 0.5$ composition.

The temperature variation of the component q_z of the $\mathbf{q_1}$-satellite is depicted in Figure 8. For the three compounds containing tantalum, the temperature increase was intentionally

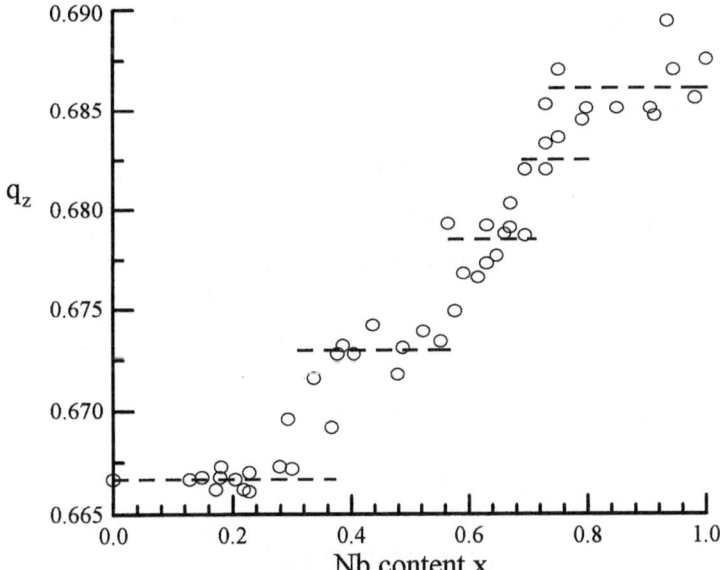

Fig. 6: Variation of the q_z-component of the $\mathbf{q_1}$-satellite with Nb content in $(Ta_{1-x}Nb_x)Te_4$ [8].

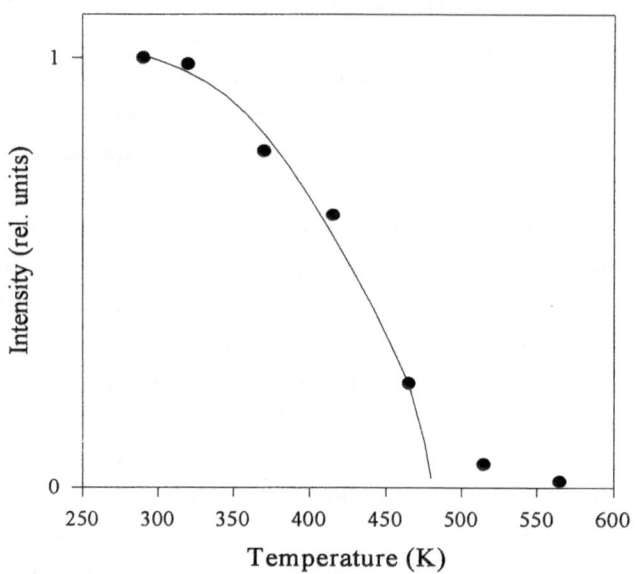

Fig. 7: The intensity variation of the q_3-satellite (0 11 4) on heating for (Nb$_{0.25}$Ta$_{0.75}$)Te$_4$ [17].

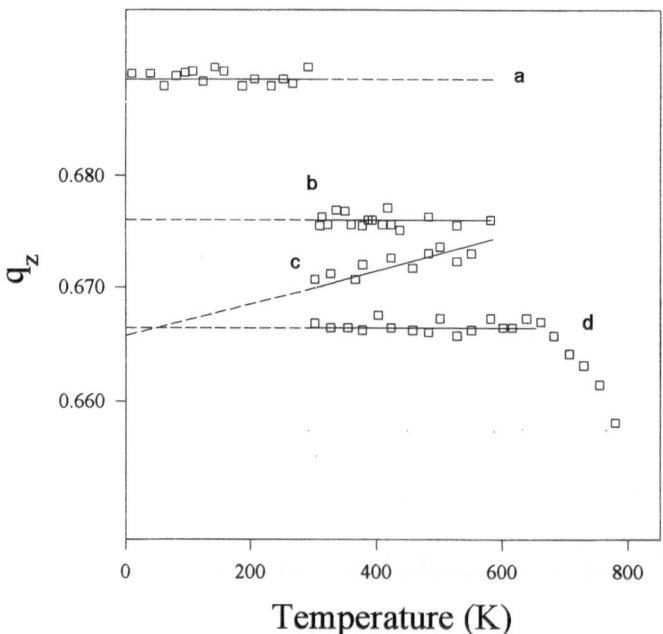

Fig. 8: Variation of the q_z-component of the q_1-satellite with temperature for four compositions (nominal values): (a) x=1, (b) x=0.5, (c) x=0.25 and (d) x=0 [17].

stopped to avoid decomposition, i.e. for none of these compounds was an attempt made to verify a HT phase. For $TaTe_4$ (curve d), the q-vector has the commensurate value $q_z = 2/3$ and is temperature independent up to the C* to IC phase transition at 650 K. Above that temperature q_z is incommensurate, temperature dependent and has a value of less than 2/3. This is in contrast to the $NbTe_4$ case (curve a), where q_z is greater than 2/3. The value of q_z for the specimen with $x = 0.5$ (curve b) exhibits the same temperature behaviour as $NbTe_4$, whereas, the crystal with $x = 0.25$ (curve c) shows a significant temperature dependence. From an extrapolation of curve c to lower temperature, one would expect a lock-in transition below 100 K which might not be inhibited in a macroscopic crystal as in $NbTe_4$. Note for the composition $x = 0.25$, the structure is still incommensurate (LT phase) at RT while according to Bennett et al. [8] the C to LT transition occurs for $x \geq 0.3$. However, the compositions quoted here are nominal values and the actual composition may be slightly different; the observed behaviour is very sensitive to the actual composition.

From the known facts about the system $(Ta_{1-x}Nb_x)Te_4$ with $0 \leq x \leq 1$, one is tempted to propose a tentative phase diagram. Such a diagram is depicted in Figure 9. The following phases can be distinguished:

HT phase: Only verified for $x = 1$; basic unit cell $a_0 \times a_0 \times c_0$; no satellites.

IC phase: Verified for all compositions; unit cell $\sqrt{2}a_0 \times \sqrt{2}a_0 \times \mu c_0$, incommensurate q_1-satellites.

LT phase: Verified for several compositions $x > 0$; unit cell $2a_0 \times 2a_0 \times \mu c_0$, incommensurate q_1-satellites., (q_2, q_3) – satellites .

C phase*:   Verified for $x = 0$, unit cell $\sqrt{2}a_0 \times \sqrt{2}a_0 \times 3c_0$, commensurate q_1-satellites.

C phase: Verified for $x = 0$ and $x = 1$, unit cell $2a_0 \times 2a_0 \times 3c_0$, commensurate q_1-satellites, commensurate (q_2, q_3) – satellites .

3.5 SUMMARY

In the system $(Ta_{1-x}Nb_x)Te_4$, $0 \leq x \leq 1$ there is a sequence of phase transitions from a high-symmetry undistorted (HT) phase to an incommensurate phase and further to a commensurate (low temperature) phase as is observed for many other modulated structures. It depends on temperature and composition which structure is stable. In general, the undistorted basic structure exhibits no satellite reflections. It transforms to a phase which is characterized by satellite reflections at incommensurate positions (IC-structure); in our system these are the q_1-satellites: $q_1 = (0, 1/2, 2/3 + \delta)$. This phase transition is different for each of the two end members of the system: $\delta \leq 0$ for $TaTe_4$ and $\delta \geq 0$ for $NbTe_4$. When the composition is varied there is one range of compositions which behaves like $NbTe_4$ and another range which behaves ike $TaTe_4$. We note that if an appropriate temperature is chosen where both end members are

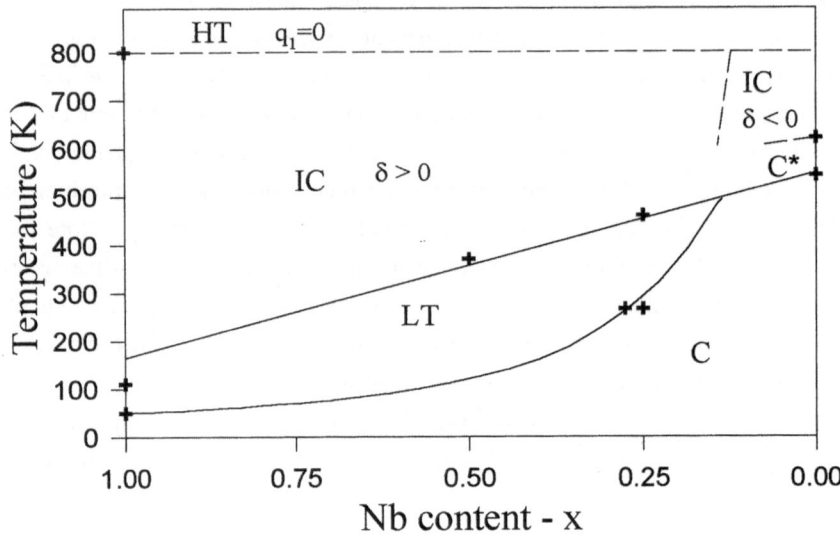

Fig. 9: A tentative phase diagram of the system $(Ta_{1-x}Nb_x)Te_4$ with $0 \leq x \leq 1$; $q_1 = (0, 1/2, 2/3 + \delta)$; crosses are verified compositions (nominal values) [17].

incommensurate, a continuous variation of the composition spanning the range from $\delta \leq 0$ to $\delta \geq 0$ will cross a line where δ happens to be zero (Figure 9). This, however, has not yet been verified by experiment. On cooling, the IC-phase undergoes a reversible phase transition. Again, it is different for the two end members: for TaTe$_4$ the q_1-satellites become commensurate ($\delta = 0$, IC to C* transition) while for NbTe$_4$ an intermediate low temperature phase ($\delta > 0$, LT phase) is formed before the q_1-satellites become commensurate. This intermediate phase is characterized by the occurrence of (q_2, q_3) – satellites. These satellites are diffuse at RT and become sharp below 120 K in the $a_1^* a_2^*$ –plane but remain diffuse streaks elongated along c^* in an X-ray experiment. However, the higher resolution of the electron microscope reveals that the diffuse streaks consist of closely spaced sharp satellites with incommensurate q-vectors. On further cooling the phase transforms to a commensurate structure (lock-in transition into the C phase). For NbTe$_4$ this transformation can only be verified in the electron microscope where the specimens have very small dimensions; in the X-ray experiment the LT to C transformation is inhibited. The IC phase transforms to the commensurate C phase only after forming the intermediate incommensurate LT phase. On the other hand, the IC to C* transition in TaTe$_4$ does not require an intermediate incommensurate phase. The q_1-satellites are commensurate (C* phase), when the (q_2, q_3) – satellites occur

(C* to C transition); therefore they are also commensurate. The C phase exists in the entire range of compositions. However, the Nb-rich range of the system and the Ta-rich range transform through different mechanisms from the IC to the C phase. Therefore, a multicritical point must be postulated in the phase diagram (Figure 9) which separates the two ranges. Additional experiments are necessary to confirm the proposed phase diagram and to verify the existence of the undistorted HT phase for compositions other than $NbTe_4$.

4. Structures

4.1 STRUCTURES OF $NbTe_4$

4.1(a) *Room Temperature Structure*

It is common to solve incommensurately modulated structures using the method developed by de Wolff [19] for higher dimensional space and applying superspace groups [20]. Often a faster way of solving a modulated structure is by a simple superstructure approximation. In the case of $NbTe_4$, the first model of the incommensurate structure was given using this approach as an approximation of a 3-fold superstructure [7], it must be described in a $\sqrt{2}a_0 \times \sqrt{2}a_0 \times 3c_0$ cell. The true modulation period ($\mu = 3.20$) was thus approximated by $\mu = 3$ i.e. by shifting the satellites to a commensurate position in reciprocal space. Applying standard refinement methods the essential features of the structure could then be derived: "The modulating function causes a longitudinal-optical wave of the Nb atoms along the direction of c which is coupled with a transverse shifts of the Te atoms (breathing mode). The longitudinal wave produces sequences of isolated Nb atoms, Nb_2 pairs and Nb_3 chains. Superimposed on the breathing mode of the Te squares is an additional libration which rotates the stacked Te squares against each other" [7]. This result was achieved in two steps: (i) by solving the average structure with the main reflections alone. The average structure, which is the projection of the modulated structure into the cell of the basic structure should, in general, reproduce the original basic structure of Selte and Kjekshus [1]. A detailed analysis, however, indicates how the individual atoms are affected by the modulation: e.g. the atomic shifts of all subcells of the modulated structure, which are superimposed in the cell of the average structure appear as the elongation of the "thermal" ellipsoids; a Fourier synthesis calculated with $\left\| F_c \right| - \left| F_o \right\|$ reveals the difference between the true average structure and an idealized basic structure (F_o = observed and F_c = calculated structure factors of the idealized model). (ii) The second step is to elaborate the complementary structure with the satellite reflections alone. The latter is a difference structure between the true modulated structure and the average structure; it consists of positive and negative density. Thus, a Fourier map calculated with the satellites alone will show how the true structure deviates from the average structure and provides a good starting model for the refinement calculations. Since 35 out of 323 main reflections violate the systematic absence rule for c-glide planes, the space group was taken to be P4 rather than P4/mcc. Double reflections as well as the λ/2-effect could be excluded as the origin of the violations. In any case, glide planes parallel to [001] are not compatible with the incommensurateness of the structure. Therefore, the space group P4 was postulated [21].

The result of the structure determination as a 3-fold superstructure is shown in Figure 10; the refinement parameters are listed in Table II. The characteristic features of the structure are obvious: there is one longitudinal distortion mode along [001] which gives rise to linear Nb$_3$ groups. Three characteristic Nb-Nb bond lengths may be distinguished: $d_1 \cong 3.07$ Å, $d_2 \cong 3.17$ Å and $d_3 \cong 3.25$ Å. One Nb$_3$ unit has d_1 and d_2 separations, the other d_2 and d_3 separations. This may be an indication of the presence of two different electronic systems, e.g. Nb$_3$ (3e) and Nb$_3$ (4e). The Nb$_3$ units are separated by non-bonding distances of about 3.9 Å. In the two columns at 00z and ½½z, the Nb$_3$ groups are arranged in such a way that a non-bonding distance in one column is adjacent to a short bond length in the neighbouring column. Thus the modulations are $2\pi/3$ out of phase on nearest-neigbour columns.

The Te$_4$ squares of the basic structure are modulated by three modes: (i) a breathing mode in the a_1a_2-plane, (ii) a libration mode about [001] which is coupled to (i) and (iii) a small longitudinal shift of the Te$_4$ squares. Very important is the strong coupling of the longitudinal shifts of the Nb atoms with the transverse modulation of the Te$_4$ squares. The formation of covalently bonded Te$_2$ pairs with constant bond length of about 2.94 Å to 2.91 Å between adjacent columns seems to be a boundary condition for the structure. The coupling of the libration and breathing modes is necessary to maintain these constant Te-Te bond lengths. The weighted R-value for the modulated structure with 1227 unique main and satellite reflections in P4 was $R_w = 0.068$ for anisotropic temperature factors, which were linked to be the same in the refinement calculation within each type of atoms.

It should be emphasized, however, that this is a commensurate approximation which cannot show the origin of the incommensurate character of the structure. The advantage of this solution is that the model exhibits the basic distortion modes and that it is easily adapted to a crystal chemical interpretation where bond lengths, spheres of coordination, etc. can be perceived. The disadvantage is that the solution of Figure 10 is not unique [26]. If the starting conditions for the refinement are changed, a second solution of equal quality is obtained with Nb$_3$-groups in one chain and Nb$_2$-groups together with isolated Nb-atoms in the second chain. This solution is shown in Figure 11. The R-value is very similar ($R_w = 0.064$) for this alternative solution; obviously there are two minima in the refinement of the residuum. Since both models are approximations of the exact solution, each model apparently reflects a different aspect of the exact structure.

One may suspect that in the real incommensurate structure the features of both solutions are present along each chain; i.e. a sequence of Nb$_3$ groups (approx. 75%) together with Nb$_2$

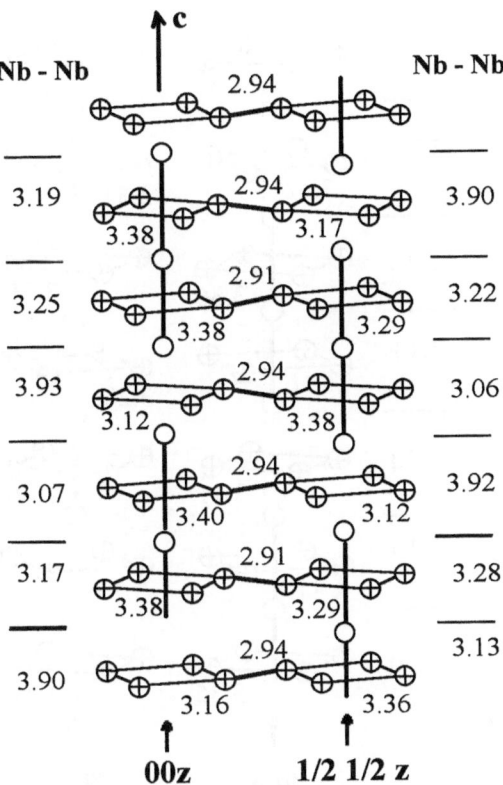

Fig. 10: The superstructure approximation of the modulated RT phase of NbTe$_4$ with interatomic distances Nb-Nb and Te-Te [21].

TABLE II

Atomic parameters of incommensurate RT structure (3-fold superstructure approximation) of NbTe$_4$ [21]. SG = P4, R= 0.043, unique satellite reflections N.Ref = 904, $U_i \times 10^4$ [Å^2], $0.0001 \leq \sigma_z(Te) \leq 0.0002$, $0.0003 \leq \sigma_{xy}(Te) \leq 0.0005$

	x	y	z		x-1/2	y-1/2	z
Te11	0.2222	0.0975	-0.0016	Te12	0.2442	0.0849	-0.0016
Te21	0.0906	0.2433	0.1683	Te22	0.0944	0.2346	0.1627
Te31	0.2471	0.0855	0.3293	Te32	0.2199	0.0964	0.3307
Te41	0.0979	0.2194	1/2	Te42	0.0840	0.2464	0.4999
Te51	0.2433	0.0913	0.6680	Te52	0.2339	0.0940	0.6623
Te61	0.0864	0.2450	0.8330	Te62	0.0955	0.2240	0.8326
Nb11	0	0	0.0969	Nb21	0	0	0.0774
Nb21	0	0	0.2517	Nb22	0	0	0.2376
Nb31	0	0	0.4014	Nb32	0	0	0.4286
Nb41	0	0	0.5929	Nb42	0	0	0.5778
Nb51	0	0	0.7512	Nb52	0	0	0.7350
Nb61	0	0	0.9067	Nb62	0	0	0.9250

U_i(Te)=127(30) U_i(Nb)=126(30)

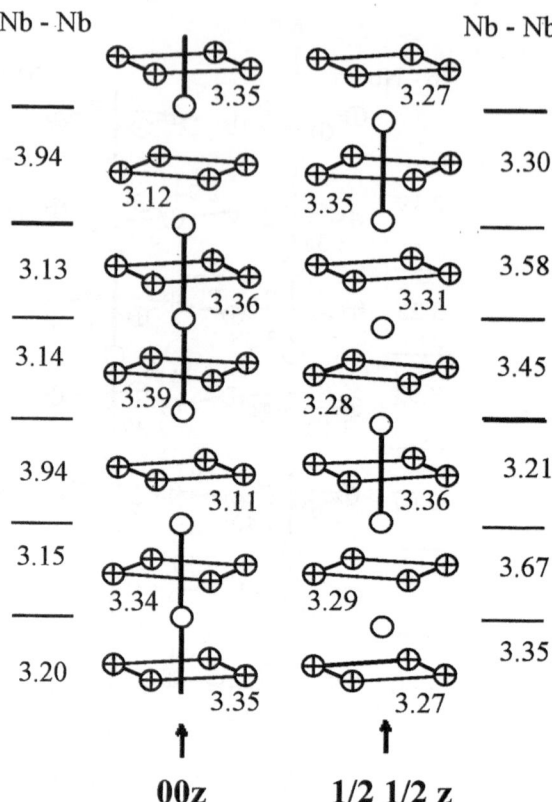

Fig. 11: Alternative solution: the superstructure approximation of the modulated RT phase of NbTe$_4$
with interatomic distances Nb-Nb and Te-Te [26].

groups and isolated Nb atoms (approx. 25%). This sequence should have a period of 16c$_o$ of
the basic structure (i.e. $\mu = 3.2$). This would imply that the incommensurate structure is a
string of 8 Nb$_3$ triplets bordered by a sequence of Nb$_2$-Nb-Nb$_2$-Nb-Nb$_2$. As q$_1$-satellites of
odd order are observed around the position of extinct main reflections and those of even order
are found around the observed main reflections, a string of Nb$_3$-Nb$_2$-Nb groups in 00z should
have a phase shift of π with respect to the one in ½½z.

The refinement of the incommensurate structure was published by van Smaalen et al. [22].
This solution accounts for the true non-periodic character along $\mathbf{c}$ of the incommensurate
structure in the 3-dimensional space. The refinement was carried out in the (3+1)-dimensional
space in the superspace group W$^{P4/ncc}_{1\bar{1}11}$. The modulation function of the individual atom μ is
described by a Fourier series expansion:

$$u_i^{\mu}(t) = \sum_{n>0} \left\{ u_{i,n}^{\mu,c} \cos\left[2\pi n\left(t_0^{\mu} + t\right)\right] + u_{i,n}^{\mu,s} \sin\left[2\pi n\left(t_0^{\mu} + t\right)\right] \right\} \qquad (2)$$

where $u_{i,n}^{\mu,c}$ and $u_{i,n}^{\mu,s}$ are the amplitudes of the n-th order Fourier component of the atom μ; and

$t_0^\mu = q \cdot x_i^\mu$. In this case, the determination of the modulation function was limited to first and second order harmonics; in the data set only q_1-satellites up to the second order were included. The atomic parameters are shown in Table III. The solution derived is exact in a sense that it reflects the correct incommensurate character of the modulation, it may have deficiencies, however, in a case where the true modulation function is strongly anharmonic and the experimental determination of the Fourier coefficients is limited to one or two harmonics.

The characteristic features of the superstructure model (Figure 10) are confirmed by the incommensurate solution. The deformation modes of Nb and Te are the same as in Figure 10: the Nb string 'condenses' into triplets and the Te-squares are affected by breathing and libration. However, Nb_3 groups alone would only lead to a 3-fold superstructure. Figure 11 indicates that after a sequence of Nb_3 groups, Nb_2 groups or Nb atoms should also be present within the period $16c_0$. Boswell et al. [6] speculated that, in fact, such discommensurations might be causing the incommensurate character following the argument of Wilson [23] for $NbSe_3$. Based on the analysis of the X-ray diffraction data, Böhm et al.[20] also postulated Nb_2 groups as discommensurations which should follow periodically after 10 Nb_3 groups. The same model was favoured by Mahy et al. [10,24] and Eaglesham et al. [11] from their interpretation of TEM results. The discommensurate character can be inferred from satellite dark field images. The fact that the q_1-satellites can be observed up to the fifth order in TEM experiments also supports the model of a rather anharmonic modulation function.

Since the incommensurate solution is non-periodic along c, no simple model can be drawn showing the true sequence of Nb groups. Therefore, the exact incommensurate solution was analyzed by Kusz et al. [25] with respect to the Nb-sequences along c. This statistical analysis was carried out in order to determine the most probable inter-atomic Nb-Nb distances within a

TABLE III
Atomic parameters of the incommensurate RT structure of $NbTe_4$ [21]
$SSG = W^{P4/mcc}_{1\bar{1}11}$, N.Ref=3894; R-value=0.106

Nb	x	0.0	$u^s_{z,1}$	0.03899(22)		
	y	0.0	$u^s_{z,2}$	0.01640(24)		
	z	0.25				
	β_{33}	0.00463(13)				
	$\beta_{11}=\beta_{22}$	0.00188(4)				
Te	x	0.23589(4)	$u^c_{x,1}$	0.01389(5)	$u^c_{x,2}$	-0.00386(6)
	y	0.09195(4)	$u^c_{y,1}$	-0.00669(5)	$u^c_{y,2}$	-0.00103(6)
	z	0.0	$u^s_{z,1}$	-0.00637(8)	$u^s_{z,2}$	-0.00236(7
	β_{11}	0.00192(3)				
	β_{22}	0.00187(3)				
	β_{33}	0.00408(4)				
	β_{12}	-0.00008(2)				

string of, say, 10^4 subcells in the incommensurate solution. From this analysis, the occupation probability of Nb was plotted in a superstructure of $16c_0$ (Figure 12). If just the *most probable* positions are taken, the result is a string of Nb$_3$ groups which is interrupted by one Nb$_2$ group as a discommensuration (bottom of Figure 12); with less probability other Nb-Nb$_2$ sequences are possible discommensurations. Model calculations also support the proposal that a chain composed of 10 Nb$_3$ groups and one Nb$_2$ group is the most likely one: the Fourier transform of this chain exhibits maximal amplitudes at 11/32 and 21/32 in reciprocal space. This corresponds approximately to the observed positions of the satellite reflections. Other sequences of Nb$_3$-Nb$_2$ groups would yield somewhat different values.

4.1(b) *High Temperature Structure*

From the study of many modulated structures, it is known that, in general, a low temperature modulated phase transforms to a high temperature phase of greater symmetry via a second order phase transition; at the transformation point the modulation disappears. For NbTe$_4$ such a transformation can indeed be observed if the crystal is sealed in a capillary filled with argon to prevent decomposition [9].

The transformation temperature is ~793 K. The small thermal hysteresis and the critical exponent β (see section 3.1) indicate that the transformation is probably of second order (Figure 3). The structure should correspond to the basic structure of Selte and Kjekshus [1]. The result of the refinement calculations indicates that the structure is likely to be slightly acentric (SG P4cc) rather than centric (Figure 13). The structural parameters are given in Table IV.

4.1(c) *The Incommensurate LT Phase*

The structure must be described in a $2a_0 \times 2a_0 \times \mu c_0$ cell. Below 120 K in an X-ray experiment the diffuse $(\mathbf{q_2},\mathbf{q_3})$ – satellites become sharp in two dimensions (i.e. in the $\mathbf{a_1}^*\mathbf{a_2}^*$-plane) and remain diffuse along $\mathbf{c}^*$ with maximum intensity at $2/3c^*$ [15]; the $\mathbf{q_1}$-satellites remain incommensurate. An electron diffraction experiment reveals that the $(\mathbf{q_2},\mathbf{q_3})$ – satellites develop into some closely spaced satellites [15]; the separation is beyond the resolution of conventional X-ray diffraction methods. Therefore, a structure determination from X-ray data does not exist for the LT phase with sharp $(\mathbf{q_2},\mathbf{q_3})$ – satellites. Prodan et al. [13] have shown that the closely spaced $(\mathbf{q_2},\mathbf{q_3})$ – satellites can be correlated with two different types of deformations which they labeled LT$_1$ and LT$_2$. The LT$_1$ deformation is generated with the same period as the incommensurate RT phase, i.e. ~$16c_0$. In LT$_2$ the modulation length is doubled, i.e. ~$32c_0$. As yet the actual structure remains unknown but the model calculations of these authors indicate that the LT$_1$ deformation corresponds to a modulation similar to that of the RT phase but with slight opposite longitudinal shifts of the columns A and B in Figure 2. The LT$_2$ deformation is characterized by a clockwise and anticlockwise rotation of the Te-squares

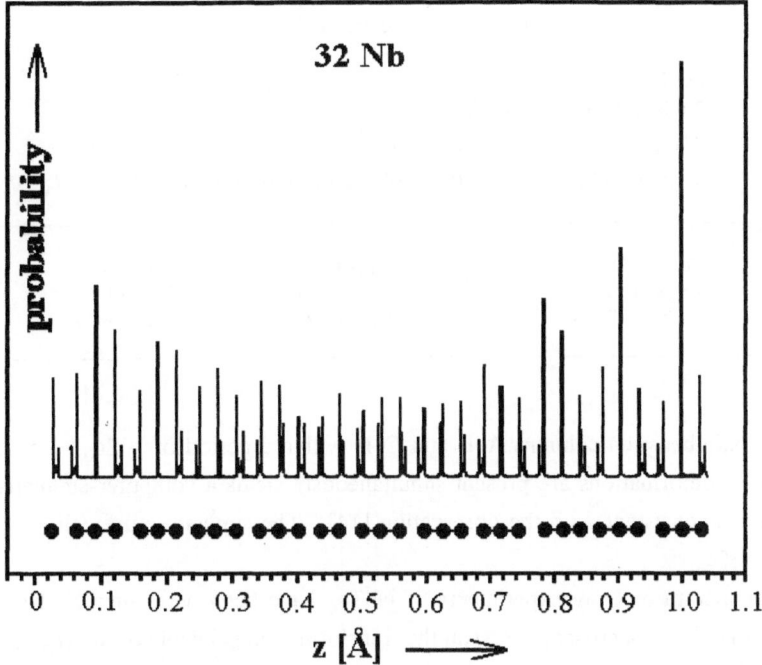

Fig. 12: The occupation probability of niobium in the chain along z of the incommensurate structure within a cell of $16c_O$. The positions with the highest probability are indicated by circles (bottom) [25].

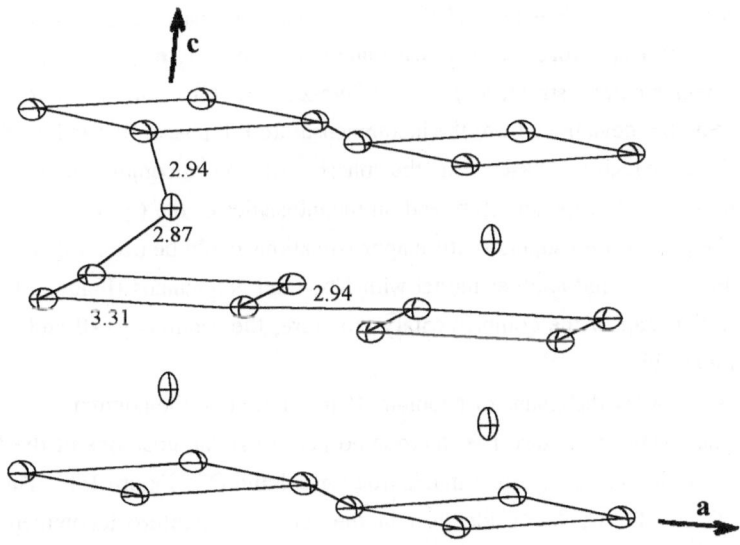

Fig. 13: The high temperature structure of $NbTe_4$ in P4cc with the columns in (00z) and (01z) [9].

TABLE IV

High temperature structure of NbTe$_4$ after [9].

SG P4cc;

2 Nb in 2a: $(00z);(001/2+z)$.

8Te in 8d: $(xyz;\overline{xy}z;\overline{y}xz;y\overline{x}z;\overline{x}y\,1/2+z;x\overline{y}\,1/2+z;yx\,1/2+z;\overline{yx}\,1/2+z)$.

origin in z(Te) =0; U$_{ij}$ ($\times10^3$) [Å^2]; unique reflections N.Ref=100; R$_w$=0.061.

Nb z=0.25	U$_{11}$=15(2)	U$_{33}$=48(3)
Te x=0.1445(6)	U$_{11}$=26(2)	U$_{12}$=-1(1)
y=0.3279(6)	U$_{22}$=25(2)	U$_{33}$=18(1)

which alternate between columns A–B and C–D within a period of ~32c$_0$. The model where LT$_1$ and LT$_2$ deformations are present simultaneously yields a computer-simulated diffraction pattern that is consistent with the observation [14]. The superstructure cell for this model is $2a_0 \times 2a_0 \times 32c_0$.

Low temperature X-ray experiments of NbTe$_4$ have been performed by Kusz et al. [15] who show that for a macroscopic crystal the transformation is inhibited; the (q_2,q_3) – satellites remain elongated in direction of c^* even at 10 K. Since neither the positions nor the intensities have changed in the range 10–120 K, refinement calculations were carried out with a data set taken at 100 K [26], including the diffuse (q_2,q_3) – satellites. If all satellites are put at the commensurate position of 2/3c*, the structure is calculated as a 3-fold superstructure approximation. In order to determine the deformations of the low temperature (LT) phase one may ask in what respect does it deviate from the incommensurate phase at room temperature. The answer is given by a structure which is calculated with the (q_2,q_3) – satellites alone. This structure ("complementary structure") is a difference structure with positive and negative density exhibiting the deviations from the incommensurate RT phase. It can be calculated by a Fourier synthesis which is based on the phases of the incommensurate structure as approximated by a 3-fold superstructure and on the intensities of the (q_2,q_3) – satellites at 100 K [26]. For the phases, two superstructure approximations could be used: one model with two Nb$_3$-chains (Figure 10) and another model with Nb$_3$- and Nb$_2$-chains (Figure 11). Figure 14 shows the Fourier map of the complementary structure; the sections y = 0 and y = 0.5 in the (a–c) plane are exhibited.

These sections show the chains of Niobium atoms along c in the columns A–C and D–B of Figure 1b, respectively. In Figure 14a, there is no density at the positions of the Nb-atoms in column C (crosses in Figure 14); the same is true for column D in Figure 14b. This means that the columns C and D which are identical in the room temperature incommensurate phase remain unchanged. The columns A and B show positive and negative densities about the positions of Niobium (crosses in Figure 14). They refer to the shifts of Nb-atoms from the positions of the incommensurate phase, which is here described as a 3-fold superstructure approximation. These shifts are antiphase in A and B; when the density is positive in the complementary structure of A it is negative in that of B. Therefore, the approximation to the LT phase is characterized by a structure in which A and B are different while C and D are identical. In the same way the sections $z =$ constant parallel to the $\mathbf{a_1 a_2}$-plane can be analyzed showing the correlated Te-shifts. The Te-squares perform clockwise-anticlockwise rotations which are coupled either between columns A and B or between C and D. If there is a strong rotation in A and B in the complementary structure, there is little rotation in C and D and vice versa. These rotations alternate between A-B and C-D for the six Te-squares along $\mathbf{c}$. The resulting structure shown in Figure 14 is the same, no matter from which model (Figure 10 or Figure 11) the phases are taken.

It is interesting to note that Prodan et al. [14] earlier proposed these atomic shifts for the LT phase on the basis of model calculations in order to reproduce the electron diffraction pattern with sharp $(\mathbf{q_2, q_3})$ – satellites (LT_1 and LT_2). These authors conjectured that their LT_1/LT_2 model is a precursor phase of the commensurate C phase. In fact, their model differs from the present superstructure approximation for the LT phase by the postulated existence of discommensurations in the LT phase in order to account for the incommensurate (long period commensurate) superstructure of $\sim 32 c_0$. The sequence of Nb_3 groups between the discommensurations is the same as in the present commensurate model for the LT phase which is derived from the X-ray data.

Based on the model derived from the complementary structure, the LT structure has been refined in the SG P4 with all 4604 reflections (main, $\mathbf{q_1}$-satellites and $(\mathbf{q_2, q_3})$ – satellites) [26]; the R-value converged to $R_W = 0.077$ (the refinement parameters are shown in Table V). In the SG P4 there are twofold axes in columns C and D and fourfold axes in A and B. Therefore, the tellurium atoms in A and B must form squares by symmetry. On the other hand, the tellurium atoms in C and D may form rectangles; in fact, slight distortions occurred. The anisotropic temperature coefficients were taken to be the same by constraints within each species of atoms. The Nb-shifts in columns A and B are, in fact, small in order to change a bonding to a non-bonding distance and vice versa. With the precondition C = D the result yields new sequences of Nb_3 groups with a phase shift of $2\pi/3$ between adjacent columns. If a reference point of A is on, say, $z = 0$ it is on $z = 1/3$ in C, on $z = 2/3$ in B and again on $z = 1/3$ in D. The breathing and libration of the Te-squares in all columns is correlated with the Nb-shifts in order to maintain a constant intercolumn Te-distance (~ 2.92 Å). Different starting configurations for the refinement yielded the same columns of Nb_3-groups with a phase shift of $2\pi/3$ with respect to each other.

It must be emphasized, however, that the structure described above has long range order only in two dimensions. It is a short range model for the LT structure along $\mathbf{c}$; since the $(\mathbf{q_2, q_3})$ – satellites of the X-ray experiment still remain elongated along $\mathbf{c^*}$.

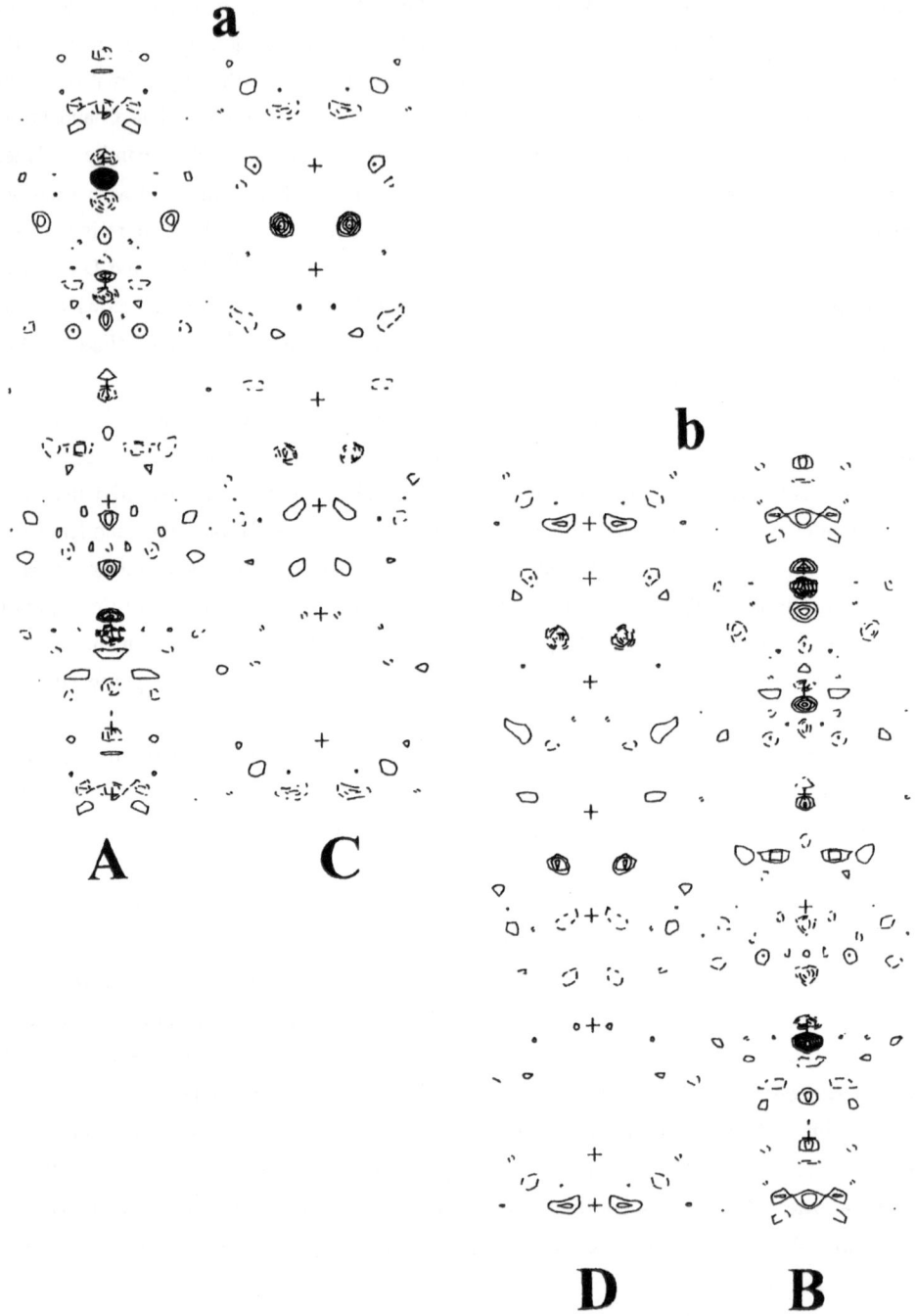

Fig. 14: Fourier synthesis in the (a-c)-plane of the 3-fold superstructure (complementary structure) with (q_2, q_3)-satellites at 100K; (a) the section y=0 and (b) the section y=0.5 [26].

TABLE V

Atomic parameters of the LT phase (commensurate approximation) of $NbTe_4$ [25]

SG=P4, Uij[$Å^2$]; unique reflections N.Ref = 4604; R_w = 0.060 .

	x	y	z		x	y	z
NbA1	0.0	0.0	0.0915(5)	TeA1	0.0631(3)	0.1560(3)	0.0000(2)
NbA2	0.0	0.0	0.2409(5)	TeA2	0.1693(3)	0.0734(3)	0.1691(2)
NbA3	0.0	0.0	0.3978(5)	TeA3	0.0802(3)	0.1635(3)	0.3290(2)
NbA4	0.0	0.0	0.5925(5)	TeA4	0.1557(3)	0.0626(3)	0.4996(2)
NbA5	0.0	0.0	0.7513(5)	TeA5	0.0749(3)	0.1691(3)	0.6683(2)
NbA6	0.0	0.0	0.9022(5)	TeA6	0.1633(3)	0.0762(3)	0.8294(2)
NbB1	0.5	0.5	0.0693(5)	TeB1	0.5684(3)	0.6670(3)	-0.0033(3)
NbB2	0.5	0.5	0.2458(5)	TeB2	0.6615(3)	0.5705(3)	0.1643(3)
NbB3	0.5	0.5	0.4091(5)	TeB3	0.5791(3)	0.6670(4)	0.3319(3)
NbB4	0.5	0.5	0.5700(6)	TeB4	0.6663(3)	0.5683(3)	0.4974(3)
NbB5	0.5	0.5	0.7561(5)	TeB5	0.5716(3)	0.6619(3)	0.6647(3)
NbB6	0.5	0.5	0.9110(6)	TeB6	0.6666(3)	0.5818(3)	0.8321(3)
NbC1	0.5	0.0	0.0758(3)	TeC11	0.5798(3)	0.1668(4)	-0.0005(2)
NbC2	0.5	0.0	0.2364(2)	TeC12	0.3349(4)	0.0751(4)	-0.0002(2)
NbC3	0.5	0.0	0.4288(3)	TeC21	0.6668(3)	0.0754(3)	0.1622(2)
NbC4	0.5	0.0	0.5844(4)	TeC22	0.4271(3)	0.1668(4)	0.1629(2)
NbC5	0.5	0.0	0.7339(3)	TeC31	0.5621(3)	0.1605(3)	0.3316(2)
NbC6	0.5	0.0	0.9241(3)	TeC32	0.3414(3)	0.0629(3)	0.3305(2)
				TeC41	0.6665(4)	0.0765(4)	0.5
Nb	$U_{11}=U_{22}=0.0021(1)$			TeC42	0.4203(3)	0.1653(4)	0.4999(2)
	$U_{33}=0.0052(2)$			TeC51	0.5730(3)	0.1682(3)	0.6626(2)
TeA1	$U_{11}=0.0059(1)$	$U_{22}=0.0024(1)$		TeC52	0.3346(3)	0.0747(3)	0.6621(3)
	$U_{33}=0.0048(3)$	$U_{12}=0.0016(1)$		TeC61	0.6589(3)	0.0610(3)	0.8300(2)
	$U_{13}=0.0000(2)$	$U_{23}=-0.0000(2)$		TeC62	0.4358(3)	0.1603(4)	0.8317(2)

4.1(d) *The Low Temperature Commensurate (C) Structure*

Below ~50 K the structure becomes a 3-fold commensurate superstructure $2a_0 \times 2a_0 \times 3c_0$. The lock-in transition has been observed in electron diffraction experiments [6]. Since $NbTe_4$ is isostructural with $TaTe_4$, many authors conclude that the C phase of $TaTe_4$ at room temperature has the same structure as the C phase of $NbTe_4$ at low temperatures. As low temperature X-ray experiments could not be carried out for the C phase, no structure determination is yet available. Kusz and Böhm [15] conjecture that discommensurations (e.g. Nb_2-groups and single Nb-atoms) remain metastable as defects if a macroscopic crystal of $NbTe_4$ is cooled below 50 K, thus pinning the incommensurate LT structure. If the discommensurations were to diffuse to the surface the proposed commensurate C structure would be formed. However, the observed sequence of Nb_3 groups and the phase shift of $2\pi/3$ between neighbouring columns in the LT phase reveal a strong similarity to the room temperature phase of $TaTe_4$ (see next section). Therefore, the LT phase is indeed a precursor phase to the C phase which, on the other hand, apparently is isostructural with the C phase of $TaTe_4$.

4.2 TaTe$_4$ AND MIXED CRYSTALS NbTe$_4$/TaTe$_4$

4.2(a) The RT Structure of TaTe$_4$ (C structure)

Initial electron diffraction studies reported by Boswell et al. [6] and Eaglesham et al. [11] revealed that the room temperature structure is a commensurate 3-fold superstructure generating a $2a_0 \times 2a_0 \times 3c_0$ unit cell. Subsequent X-ray studies [28] have shown that the space group of the basic structure (unit cell $a_0 \times a_0 \times c_0$) is P4/mcc, while that of the superstructure is P4/ncc.

The structure is modulated in such a way that Ta$_3$ clusters are formed. The Te atoms respond to the shifts of the Ta atoms by forming two types of Te "squares": one third of these have expanded while the remaining two thirds have contracted relative to the squares of the basic structure. Since in columns A and B there are only twofold axes, the Te-squares of the basic structure slightly distort to form rectangles while keeping the bonds of the Te dimers, which link the columns, equal in length. The parameters are listed in Table VI. There is a phase shift of $2\pi/3$ between triplets in neighbouring columns. The structure is very similar to the superstructure approximation of the LT phase.

4.2(b) The RT Structure of (Ta$_{0.72}$Nb$_{0.28}$)Te$_4$ (C structure)

The mixed crystals $\left(Ta_{1-x}Nb_x\right)Te_4$, $0 \le x \le 1$, have been studied extensively by TEM experiments [8]. The experiments show that the diffraction patterns at room temperature correspond to those of TaTe$_4$ below the critical Nb concentration ($x < 0.3$). Therefore the structure can be described as a threefold commensurate superstructure in a $2a_0 \times 2a_0 \times 3c_0$ cell. The SG is P4/ncc as for TaTe$_4$. Only two of the four columns A, B, C and D are independent since C and D are related by the center of symmetry and A and B by the 4-axis. Kucharczyk et al. [27] have determined the RT structure by X-ray diffraction for $x = 0.28$. The two independent columns A and C are depicted in Figure 15.

TABLE VI

Atomic parameters of the superstructure (room temperature phase) of TaTe$_4$ after [22].
SG=P4/ncc, Uij[Å^2] secondary extinction = 2.25 10^{-4}; N.Ref = 6422; R= 0.064.

	x	y	z	U_{11}	U_{22}	U_{33}	U12	U_{13}	U_{23}
Ta1	0.0	0.0	0.09652(4)	0.0064	0.0065(2)	0.0068(2)	0.0002(2)	0	0
Ta2	0.0	0.0	0.25	0.0063		0.0068(3)	0	0	0
Ta3	0.0	0.5	0.41827(5)	0.0065		0.0056(3)	0	0	0
Ta4	0.0	0.5	0.26351(6)	0.0063		0.0070(3)	0	0	0
Ta5	0.0	0.5	0.07030(5)	0.0064		0.0072(3)	0	0	0
Te1	0.0611(1)	0.1579(1)	0.0690(5)	0.0090(4)	0.0095(4)	0.0085(3)	-0.0006(3)	-0.0004(4)	-0.0005(3)
Te2	0.5802(1)	0.1655(1)	0.2540(5)	0.0097(4)	0.0077(3)	0.0096(4)	-0.0007(3)	-0.0001(4)	-0.0000(3)
Te3	0.1656(1)	0.0789(1)	0.4066(5)	0.0076(3)	0.0096(4)	0.0092(4)	-0.0005(3)	0.0005(3)	-0.0003(4)
Te4	0.6665(1)	0.5747(1)	0.5696(5)	0.0076(3)	0.0086(3)	0.0095(4)	0.0001(2)	0.0000(3)	0.0000(3)
Te5	0.6687(1)	0.0700(1)	0.7552(5)	0.0083(4)	0.0098(3)	0.0085(4)	-0.0005(2)	0.0003(3)	0.0001(3)
Te6	0.1561(1)	0.5648(1)	0.9118(6)	0.0102(4)	0.0085(3)	0.0088(3)	-0.0008(3)	0.0008(3)	0.0006(3)

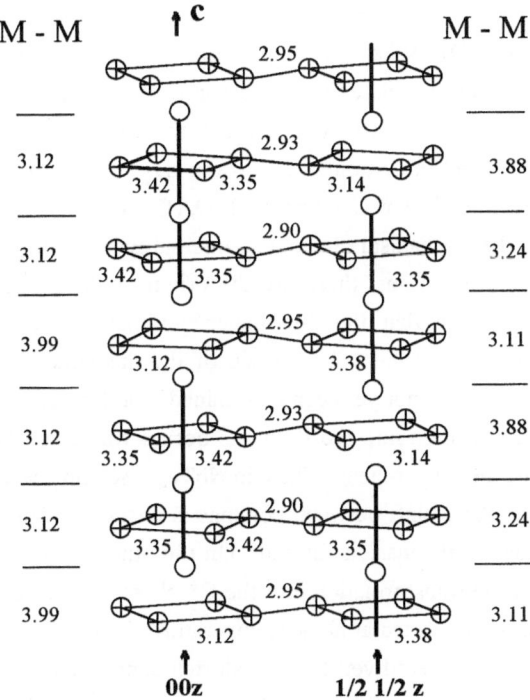

Fig. 15: The superstructure of $(Ta_{0.72}Nb_{0.28})Te_4$ (RT phase) with inter-atomic distances M-M and Te-Te after [27].

4.3 DISCUSSION

The basic structure of equally spaced Nb-atoms in Te-antiprisms is only realized at very high temperatures (for $NbTe_4$, above 792 K). The characteristic feature at lower temperatures is the sequence of Nb-triplets along the c-axis, which are separated by non-bonding distances of about 3.9 Å. This longitudinal modulation of Nb-atoms is coupled with transverse shifts of Tellurium by forming two types of Te 'squares': two thirds have expanded, one third have contracted relative to the squares of the basic structure. Another boundary condition for the modulation seems to be the formation of covalently bonded Te-dimers of constant bond length (2.90–2.94 Å) between neighbouring columns of antiprisms. With these preconditions the structures could be formed as 3-fold superstructures along the c-axis. If we also presume that the non-bonding Nb-Nb gap of adjacent columns (A, B, C, D in Figure 1) cannot be at the same level of z, a phase shift of $2\pi/3$ will require a unit cell of basal dimensions $2a_0 \times 2a_0$. In fact, this structure is realized at low temperatures (C phase) for all compositions. The question remains, why Nb-triplets are more favoured than doublets or quartets?

   If the temperature is raised from the range of stability of the C phase, the structures must relax internal stress. The mechanism, however, is different for the two groups of the phase diagram. TaTe$_4$ can relax the internal stress by changing the phase shifts between adjacent columns (phase C*, A = B, C = D). When the temperature is raised for NbTe$_4$ the commensurate C phase transforms into an intermediate, incommensurate (LT) phase. In the electron microscope the specimens have a domain structure in which the discommensurations have different periods. Prodan et al. [14] call these domains LT$_1$ and LT$_2$. Thus, the LT phase still has the structure of the C phase with different columns A, B, C, (C = D) but, in addition, also with discommensurations. Thus the mechanism of avoiding internal stress at the Nb-rich end is through the formation of discommensurations; these are the origin of the incommensurateness. For NbTe$_4$ there is some evidence that the discommensurations are sequences of Nb-triplets followed by doublets or single atoms. The origin of the incommensurateness, however, in TaTe$_4$ at higher temperatures has not yet been determined. On further heating of the LT phase of NbTe$_4$ the next phase transition into the IC phase occurs where A = B, C = D and the discommensurations all have the same repeat. Thus, in NbTe$_4$ discommensurations are formed (in the LT phase) before the phase shift between columns is changed. The phase shift occurs at higher temperatures when the IC phase is formed. On the other hand, in TaTe$_4$ change of the phase shift occurs first at the transformation into the C* phase (A = B, C = D). It is unknown yet, if discommensurations are formed at higher temperatures in the IC phase.

One question cannot be answered yet: for the Nb-rich compounds why is the repeat of the discommensurations (i.e. the period of the modulation) temperature independent (curves a and b in Figure 8), while for a Nb-poor compound it varies with temperature (curve c in Figure 8)? Is this an indication that for compounds $x \geq 0.5$ the discommensurations are inhibited to move? In this case, a phase transition on cooling with LT$_1$ /LT$_2$ domains and a transition into the C phase would not be observed for macroscopic crystals in an X-ray experiment. This , however, has only been verified for the compound $x = 1$.

References

1. K. Selte and A. Kjekshus, *Acta Chem Scand* **18**, 690 (1964).
2. E. Bjerkelund and A. Kjekshus, *J. Less-Common Met.* 7, 231 (1964).
3. R. Allmann, I. Baumann, A. Kutoglu, H. Rösch. and E. Hellner, *Naturwissenschaften* **51**, 263 (1964).
4. W. Klemm. and H. G. von Schnering,. *Naturwissenschaften* **52**, 12 (1965).
5. F.W. Boswell, and A. Prodan, *Mater. Res. Bull.* **19**, 93 (1984).
6. F.W. Boswell, A. Prodan and J.K. Brandon, *J. Phys. C: Solid State Physics*, **16**, 1067 (1983).
7. H. Böhm and H.G. von Schnering, *Z. Kristallogr.* **162**, 26 (1983).
8. J.C. Bennett, S. Ritchie, A. Prodan, F.W. Boswell, and J.M. Corbett, *J. Phys.: Condens. Matter* **4**, 2155 (1992).
9. H. Böhm, *Z. Kristallogr.* **180**,113 (1987).
10. J. Mahy, J. van Landuyt, S. Amelinckx, Y. Uchida, K.D. Bronsema and S. van Smaalen, *Phys. Rev. Lett.* **55**, 1188 (1985).
11. D.J. Eaglesham, D. Bird, R.L. Withers. and J.W. Steeds, *J. Phys. C: Solid State Phys.* **18**, 1 (1985).
12. F.W. Boswell and A. Prodan, *Phys. Rev.* **B34**, 2979 (1986).
13. A. Prodan and F.W. Boswell, *Acta Cryst.* **B43**, 165 (1987).

14. A. Prodan, F.W. Boswell, J.C. Bennett, J.M. Corbett, T. Vidmar, V. Marinkovic and A. Budkowski, *Acta Cryst.* **B46**, 587-591 (1990).
15. J. Kusz and H. Böhm, *Z. Kristallogr.* **208**, 187 (1993).
16. J.C. Bennett, F.W. Boswell, A. Prodan, J.M. Corbett and S. Ritchie, *J. Phys.: Condens. Matter* **3**, 6959 (1991).
17. J. Kusz, H. Böhm and J.C. Bennett, *J. Phys.: Condens. Matter* **7**, 2775-2782 (1995).
18. F.W. Boswell, A. Prodan, J.C. Bennett, J.M. Corbett and L.G. Hiltz, *Phys. Stat. Sol. A*, **102**, 207 (1987)
19. P.M. de Wolff, *Acta Cryst.* **A30** 777 (1974).
20. A. Janner, T. Janssen and P.M. de Wolff, *Acta Cryst.* **A39**, 671 (1983).
21. H. Böhm and H.G. von Schnering, *Z. Kristallogr.* **171**, 41 (1985).
22. S. van Smaalen, K.D. Bronsema and J. Mahy, *Acta Cryst.* **B42**, 43 (1986).
23. J.A. Wilson, *Phys. Rev.* **B19**, 6456 (1979).
24. J. Mahy, J. van Landuyt, S. Amelinckx, K.D. Bronsema, and S. van Smaalen, *J. Phys. C: Solid State Phys.* **19**, 5049 (1986).
25. J. Kusz and H. Böhm, *Z. Kristallogr.* **201**, 9 (1992).
26. J. Kusz and H. Böhm, *Acta Cryst.* **B50**, 649, (1994).
27. D. Kucharczyk, A. Budkowski, F.W. Boswell, A. Prodan and V. Marinkovic, *Acta Cryst.* **B46**, 153 (1990).
28. K.D. Bronsema, S. van Smaalen, J.L. de Boer, G.A. Wiegers, F. Jellinek and J. Mahy, *Acta Cryst.* **B43**, 305 (1987).

CHARGE DENSITY WAVE PHASE TRANSITIONS AND MICROSTRUCTURES IN THE TaTe$_4$ - NbTe$_4$ SYSTEM

J. C. Bennett

Department of Physics, Acadia University, Wolfville, Nova Scotia, Canada B0P 1X0

F.W. Boswell

Department of Physics, University of Waterloo, Waterloo, Ontario, Canada N2L 3G1

1. Introduction

Charge density waves (CDW) and their accompanying periodic lattice distortions are a unique and fascinating feature of low-dimensional systems. Although first predicted by Peierls [1] several decades earlier, it was only in the 1970's that quasi-one- and quasi-two-dimensional metals were first synthesized and CDW-related phenomena observed. Since that time, CDW have been identified in an extremely diverse collection of materials including organic charge transfer salts such as TTF-TCNQ [2], numerous oxides such as the blue bronzes K$_{0.3}$MoO$_3$ and Rb$_{0.3}$MoO$_3$ and the Magneli phase η-Mo$_4$O$_{11}$ [3], metallic elements like α-U [4] and Cr [5], and martensitic precursor phases occurring in certain intermetallic alloys [6,7]. However, many of the CDW systems that have attracted the most interest to date are found among the transition metal chalcogenides. These compounds possess both a high degree of structural anisotropy and the partially filled energy bands necessary for CDW formation. Over the past two decades, CDW have been reported in various transition metal dichalcogenides (MX$_2$) [8,9], trichalcogenides (MX$_3$) [10-12], tetrachalcogenides (MX$_4$) [13], triniobium tetra-chalcogenides Nb$_3$X$_4$ [14-16] and numerous related compounds (here M refers to the transition metal elements V, Ti, Zr, Nb, Ta or Hf while X refers to the chalcogenides S, Se and Te). In many of these systems, temperature or compositional variations lead to transformations between distinct CDW phases accompanied by the formation and interaction of novel structural defects such as discommensurations and CDW domain boundaries. Using trans-mission electron microscopy (TEM), an extremely diverse range of microstructural phenomena has been observed in the transition metal chalcogenides and a complete survey of the field is beyond the scope of this article. A comprehensive review of the application of TEM techniques to the study of CDW is presented elsewhere in this volume. Here, we will attempt to highlight some of the more unusual aspects through an examination of the

F. W. Boswell and J.C. Bennett (eds.),
Advances in the Crystallographic and Microstructural Analysis of Charge Density Wave Modulated Crystals, 69–120.
© 1999 *Kluwer Academic Publishers.*

microstructural phenomena accompanying CDW-driven phase transitions occurring in the transition metal chalcogenides $TaTe_4$ and $NbTe_4$ and derived compounds. In the $NbTe_4$-$TaTe_4$ system, certain characteristic behaviours may be distinguished which are general features of the microstructural development of CDW states. The results reviewed represent the most complete structural studies of a CDW compound by TEM. Since many of the outstanding questions regarding CDW including the phenomena of "sliding CDW" and "phase slip" are intimately linked to microstructural features and defect behaviour, these TEM studies constitute an important step forward in the development of a complete microscopic theory of the CDW state.

Extensive experimental and theoretical investigations of the tetrachalcogenides $TaTe_4$ and $NbTe_4$ have demonstrated that these compounds constitute a nearly prototypical system for the study of CDW. The compounds possess a quasi-one-dimensional crystal structure in which chains of metal atoms are centered within extended cages of Te atoms in square anti-prismatic coordination [17]. The observed lattice modulations, involving mainly longitudinal motions of the metal atoms along a chain correspond to that of the classic CDW model. In addition, the compounds exhibit the full spectrum of possible CDW-driven phase transitions: commensurate to incommensurate (C↔IC), commensurate to commensurate (C↔C) and incommensurate to incommensurate (IC↔IC) [13,18-21]. In contrast to the dichalcogenides and trichalcogenides, the relatively simple subcell structure of the tetrachalcogenides tends to limit the number of lattice defects such as stacking faults and renders these crystals ideal for structural studies using x-ray diffraction and transmission electron microscopy.

The transmission electron microscope offers several advantages for the characterization of CDW modulated phases and the study of the phase transitions occurring in these materials. Electron diffraction and microscopy provide a direct method of studying the modulated structures on a length scale consistent of that of the CDW. The strong interaction of the incident electrons with the specimen allows the subtle diffraction effects associated with the structural modulations to be more readily detected than in conventional x-ray or neutron diffraction experiments. In many cases, due to the difficulty in obtaining crystals of suitable size and perfection, electron microscopy provides the only feasible method for structural studies. In addition, structural defects intrinsic to the CDW state may be directly imaged using satellite dark-field techniques and readily interpreted in light of well-established diffraction contrast mechanisms. Finally, using variable temperature stages in the TEM, the formation of and transitions between CDW modulated phases may be continuously observed *in situ*.

Before dealing with the microstructural behaviour of the tetrachalcogenides, the theoretical description of CDW-driven structural modulations in low-dimensional compounds will be reviewed in the next section. This is followed by a brief description of the TEM techniques used to study CDW modulated structures. A detailed survey of the microstructural phenomena accompanying CDW-driven phase transformations in the $NbTe_4$-$TaTe_4$ family of compounds is then presented. It will be shown that defects intrinsic to the CDW modulations, including discommensurations and antiphase boundaries, play a central role in mediating the phase transitions. Finally, the factors responsible for determining the relative stability of CDW phases are summarized.

1.1 CHARGE DENSITY WAVES IN LOW-DIMENSIONAL METALS

Peierls first predicted [1] that a one-dimensional metal is inherently unstable to the formation of a coupled, normally static, distortion of the conduction electron density and the ion lattice now referred to as a charge-density wave. The coupling is mediated by electron-phonon interactions and, as a result, the mathematical description of the CDW state is similar to the BCS theory of superconductivity [9,22]. However, most of the essential aspects of the theory can be illustrated using a simple model and physical arguments. In a metal, the electron states are described by Bloch waves whose energies ε_k near the bottom of the energy bands varies according to a dispersion relation $\varepsilon_k = h^2k^2/2m$. For an ideal one-dimensional chain of metal atoms, the Fermi surface consists of parallel planes at $\pm k_F$. A structural modulation with wavevector $q = 2 k_F$ will couple electron states lying on the two planes of the Fermi surface, creating a gap 2Δ in the energy spectrum at ε_F. A reduction in the electronic energy results from a lowering of the occupied band state energies below the new energy gap while the states that are correspondingly raised in energy remain unoccupied (Figure 1). For a one-dimensional metal, the reduction in electronic energy will always exceed the elastic energy cost of the lattice modulation, stabilizing the CDW. Note for temperatures above 0 K, thermal excitations of electrons lead to an eventual closing of the CDW energy gap at a transition temperature T_p. The relation between the gap energy and T_p is BCS-like [23] and has the form:

$$\Delta = 2H\exp\left\{\frac{-A}{N}\right\} = 1.76e^{-1}k_BT_p \qquad \text{... (1)}$$

where N is the density of states, A is the electron-lattice coupling constant and H is an electronic parameter on the order of ε_F. Since ε_F is much larger than the corresponding quantity in the BCS theory, onset temperatures for CDW are generally an order of magnitude

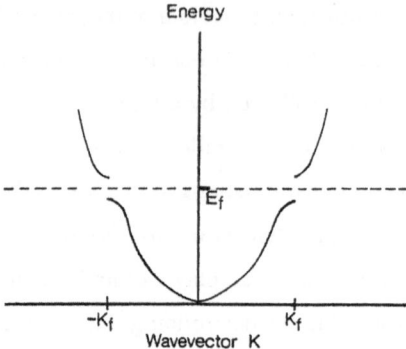

Fig.1: Energy spectrum of a one-dimensional metal. The presence of a periodic potential with
wavevector $2k_F$ results in an opening of a gap at the Fermi level.

higher, in the range of 100 K to room temperature.

The preceding discussion of an ideal one-dimensional metal is somewhat misleading in that, due to thermal effects, the lattice must possess some three-dimensional character for a CDW to occur at a finite temperature. For a crystal composed of completely independent linear chains, thermal flucuations delocalize the electron states and, as a result, decouple the planar Fermi surfaces. As a result, above 0 K, there can be only short-range order in a one-dimensional system and a CDW transition is not possible. However, if the lattice has some limited three-dimensional character such that only a small degree of curvature is introduced into the Fermi surfaces, a single **q**-vector may still couple sufficient states to enable a CDW to form at finite temperature. In most CDW compounds, interchain bonding interactions provide the required higher degree of dimensionality and are an important factor in achieving electron-phonon coupling. Tunneling measurements [24] often give an energy gap an order of magnitude higher than the mean field value of eqn. (1), illustrating the degree of deviation from an ideally one-dimensional situation in these systems.

The periodicity of the CDW is, according to the model described above, determined by the level of filling of the conduction band and may therefore be either commensurate or more generally incommensurate with the unperturbed lattice. In the latter case, the CDW has no preferred phase relationship with the lattice and the resultant crystal structure lacks true three-translational symmetry along the direction of the modulation. The concept of translational symmetry is recovered in the so-called "superspace" description of incommensurately modulated structures in which the observed structure is derived from a projection of a four-dimensional lattice [25-27]. It is possible for a commensurate modulation to adjust its phase to minimize the potential energy and this drives a process wherein an incommensurate modulation "locks-in" to a nearby commensurate periodicity [28]. Since the lock-in energy is a function of the CDW amplitude and this increases at lower temperatures, a sequence of reversible phase transitions is anticipated on cooling: a normal to incommensurate (N↔IC) transition followed by an incommensurate to commensurate (IC↔C) lock-in transition. On the basis of Landau theory, McMillan [29] has developed a detailed model for the lock-in behaviour of these materials.

In actual practice, the observed sequence of phase transitions may be considerably more complicated than the independent chain model would indicate due to the influence of interchain interactions. These interactions are primarily responsible for determining the relative phasing of the CDW modulations along neighbouring chains. Since the strength of the interactions is also subject to variations with temperature, transitions between modulated phases, associated with rearrangement of the relative phasings of the CDW along the chains, may occur on cooling.

The theoretical work of McMillan [29,30] also showed how strong electron-phonon coupling may lead to a *discommensurate* rather than uniformly incommensurate state. In this discommensurate state, the CDW arranges itself to take advantage of the lock-in energy by forming essentially commensurate domains separated by regions across which the phasing slips by a lattice translation. The discommensurations formed are one-dimensional defects appearing periodically along an individual chain although their exact nature at the microscopic level is of course left unspecified by the Landau theory. Interchain interactions may result in an alignment of the discommensurations to form arrays of planar defects, referred to as discommensuration walls [31,32]. A discommensuration wall is a translational interface closely related to an antiphase boundary (APB) however, while the phase slip occurs for an APB occurs abruptly, the width of the discommensuration walls depends on the relative strength of the lock-in and gradient terms in the Landau free energy. The theory would thus appear to demand sharply defined discommensuration walls only near lock-in transition.

In summary, the periodicity ultimately adopted by a CDW modulation is determined by a number of competing processes. The Peierls instability favours a wavevector $q = 2k_F$, which is generally incommensurate with the lattice, while the lock-in energy favours the adoption of a nearby commensurate periodicity. Coulomb and interchain coupling interactions dictate the relative phasing of the CDW on adjacent chains. However, elastic strain energy associated with the lattice distortion opposes all of the above processes and, as a result, the observed

CDW modulation is the product of a subtle balancing of competing interactions.

2. Transmission Electron Microscopy of CDW Phase Transitions

It is abundantly clear that the details of the crystal structure, and the lattice defects which are inevitably present, play a central role in CDW phenomena. Direct observation of the microscopic details of these processes thus assumes a critical importance. Transmission electron microscopy offers a unique combination of capabilities for these studies. The strong interaction of electrons with matter coupled with the capacity to obtain diffraction patterns corresponding to various cross-sections of reciprocal space enables the subtle diffraction effects associated with the CDW modulation to be more readily detected and interpreted than is generally the case using conventional x-ray techniques. In addition, selected area electron diffraction provides spatially localized structural information on a nanometer scale which may be directly correlated with the microstructure of the crystal as observed by diffraction contrast and high-resolution imaging. Of particular importance for the study of CDW-driven phase transitions, the structural changes can be observed *in situ* with the use of variable temperature specimen holders. Finally, convergent beam techniques for the symmetry determination of modulated structures have also been developed by Tanaka and co-workers [33,34]. However since the application of TEM techniques to the study of CDW compounds is reviewed in detail elsewhere in this volume, only the essential elements will be presented here.

The CDW modulation periodicity may be directly determined from electron diffraction patterns through measurements of the positions of the associated satellite reflections. This is easily demonstrated for the case of a displacive modulation. In the simplest kinematical approximation, for a sinusoidal modulation of amplitude $\mathbf{A}$ and wavevector $\mathbf{q}$ occurring in a primitive Bravais lattice, the amplitude of the diffracted beam is given by:

$$A(\mathbf{K}) = \sum_{n=1}^{N} f_k \exp\{i\mathbf{K} \cdot \mathbf{r}^n\} = \sum_{n=1}^{N} f_k \exp\{i\mathbf{K} \cdot [\mathbf{r}_0^n + A\sin(\mathbf{q} \cdot \mathbf{r}_0^n)]\} \qquad \dots (2)$$

where $\mathbf{r}$ represents the atomic positions in the modulated structure, f_k are the atomic form factors, N is the number of unit cells and $\mathbf{K}$ is the scattering vector. The diffracted amplitude may be re-expressed as [35]:

$$A(\mathbf{K}) = \sum_{m} J_m(\mathbf{K} \cdot \mathbf{A}) \sum_{n} f_k^n \exp\{i(\mathbf{K} + m\mathbf{q}) \cdot \mathbf{r}_0^n\} = N \sum_{m} J_m(\mathbf{K} \cdot \mathbf{A}) \sum_{G} \delta(\mathbf{K} + m\mathbf{q} - \mathbf{G}) \qquad \dots (3)$$

where J_m is the Bessel function of the first kind for integer m, and $\mathbf{G}$ is a reciprocal lattice vector for the subcell. The diffraction pattern then consists of Bragg spots at $\mathbf{K} = \mathbf{G}$ along with an infinite series of satellites at $\mathbf{K} = \mathbf{G}\text{-mq}$ for a sinusoidal modulation. However, the diffracted amplitude is proportional to the Bessel function J_m, which for small displacements decreases approximately as [36]:

$$J_m(\mathbf{K} \cdot \mathbf{A}) = \frac{(\mathbf{K} \cdot \mathbf{A})^m}{2m!} \qquad \text{... (4)}$$

As a result, only first order satellites at $\pm\mathbf{q}$ are typically observed for a sinusoidal modulation. If the modulation is anharmonic, higher order satellites will also appear in electron diffraction patterns although care is usually necessary to distinguish genuine higher order reflections from those associated with double diffraction.

Similar diffraction effects are produced by interface modulated structures although the origin of the structural distortion is of course quite different. In the simplest approximation, an interface modulated structure may be regarded as being composed of blocks of perfect crystal having width D separated by translational interfaces such that each block is displaced relative to its neighbours over a distance $\mathbf{R}$. As shown by Van Landuyt et al. [37], the diffraction pattern from a crystal containing a regular array of planar interfaces contains satellites at positions given by:

$$\mathbf{G} = \mathbf{K} + \frac{1}{D}(m - \mathbf{K} \cdot \mathbf{R})\mathbf{e}_D \qquad \text{... (5)}$$

where $\mathbf{e}_D$ is the unit normal to the plane of the interfaces. The diffraction pattern thus consists of lines of equally spaced satellites (spacing 1/D) running parallel to $\mathbf{e}_D$ around each of the subcell reflections but shifted with respect to these positions by a fraction $\mathbf{K} \cdot \mathbf{R}$ of the subcell spot spacing. Note this is in contrast to a displacive modulation where the satellites are centered with respect to the generating Bragg reflection. The satellite intensity may again be shown to fall off rapidly for higher orders [38]. In the diffraction pattern, the fractional shift $\mathbf{K} \cdot \mathbf{R}$ may be measured directly and hence $\mathbf{R}$ determined.

Conventional TEM images, in which only one beam is allowed to pass through the electron optical system, exhibit diffraction contrast in either bright-field (BF) mode (if the transmitted beam is used to form the image) or dark-field (DF) mode (using a diffracted beam). If sufficient intensity is attainable, a satellite reflection may be similarly employed to form a satellite dark-field (SDF) image. For CDW phases, these SDF images reveal defects in the modulation structure and are generally dramatically different from the corresponding BF or matrix DF images. For example, CDW domain structures may be readily distinguished in SDF images since only domains so orientated as to diffract strongly into the chosen satellite reflection will appear with strong contrast. Also the observed defects, such as antiphase boundaries and discommensuration walls, exhibit characteristic diffraction contrast effects that may often be interpreted intuitively.

High-resolution electron microscopy (HREM) is another very useful tool for the study of CDW modulated structures [20,39-41]. Typically, a large number of both subcell and satellite reflections are used to form the image and atomic positions are inferred from the resulting interference pattern. Since the extinction distances of the satellites are normally much longer than those of the subcell reflections, HREM images from thicker crystals tend to show the effects of the modulation more strongly. Interpretation of these images requires reference to the dynamical theory and is accomplished via a comparison with computer simulations [42]. However, HREM images of CDW modulation defects and antiphase boundaries may often still be interpreted intuitively, even for relatively thick specimens.

3. CDW Phase Transitions in the Transition Metal Tetrachalcogenides $TaTe_4$ - $NbTe_4$

The structural anisotropy and partly filled conduction bands necessary for CDW formation are often present among compounds of the transition metal chalcogenides. However, in many of these compounds, such as quasi-one-dimensional compound $NbSe_3$ and the quasi-two-dimensional $NbSe_2$, numerous subcell stacking faults and other defects are typically observed. On the other hand, the transition metal tetrachalcogenides $TaTe_4$ and $NbTe_4$ are generally free of lattice defects. For this reason, along with the relative simplicity of the crystal structure and low sensitivity to electron irradiation, the tetrachalcogenides constitute an ideal system for the study of CDW modulations by transmission electron microscopy.

The tetrachalcogenides are normally prepared by vapour transport as described elsewhere [13] and typically grow in the form of small bundles of needles. For most studies, TEM specimens have been obtained by cleaving these needles between glass slides to produce numerous fragments with electron transparent edges parallel to the chain axis. Ion beam etching has also been used but the formation of the CDW modulation is strongly suppressed, presumably as a consequence of irradiation damage [18].

Kjekshus and coworkers first reported the subcell structures of $TaTe_4$ [43] and $NbTe_4$ [17] in the early 1960's. However, relatively little attention was paid to these compounds until Boswell et al. [13] identified the presence at room temperature of a commensurate superstructure for $TaTe_4$ and an incommensurate structure for $NbTe_4$. Several TEM [18-20,44] and x-ray diffraction [45-49] studies subsequently confirmed these results. It was also shown that a complete range of solid solutions exist for $(Ta_{1-x}Nb_x)Te_4$ with $0 \leq x \leq 1$ and that various new incommensurate phases are stabilized at room temperature [50,51]. In addition, a variety of thermodynamically stable CDW phases have been reported in the $NbTe_4$ - $TaTe_4$ system upon heating or cooling. These studies together represent the most detailed characterization by transmission electron microscopy of the microstructural phen-omena accompanying CDW transitions. The surprisingly rich spectrum of behaviour observed in this relatively simple system has also inspired several theoretical studies using Landau theory, notably by Walker and coworkers [32]. The results of these investigations have established the transition metal tetrachalcogenides as a model system of critical importance, leading to the development of a better understanding of the factors influencing the stability of CDW modulated phases and the mechanisms by which phase transitions occur.

In the following sections, the results of experimental and theoretical studies of the $TaTe_4$ - $NbTe_4$ system are reviewed with emphasis on CDW phase transitions occurring in these compounds. First, a description of the subcell crystal structure is given followed by an analysis of the CDW modulated phases and microstructural features found in $NbTe_4$ at, and below, room temperature. This is followed in turn by a discussion of the modulated phases and CDW-driven structural transformations observed in $TaTe_4$. In the final sections, the effects of isoelectronic and nonisoelectronic substitutions on the stability of the CDW phases is reviewed with regard to the solid solutions $(Ta_{1-x}Nb_x)Te_4$ and $(M_{1-x}A_x)Te_4$ with M = Ta, Nb and A = Ti, Zr or V. The direct observations of the CDW modulations and their thermal evolution reveal a wide variety of novel features. However, overall the observed CDW behaviour in these compounds provides a truly remarkable confirmation of the Landau theoretical models originally proposed by McMillan [29].

3.1 THE SUBCELL CRYSTAL STRUCTURE

$TaTe_4$ and $NbTe_4$ are structurally the simplest inorganic compounds for which the formation of CDWs has been reported. The subcell structure consists of linear chains of equidistant metal atoms centered within extended columnar cages composed of square antiprisms of Te atoms (Figure 2). These columns are arranged on a tetragonal $(a \times a \times c)$ lattice with space group P4/mcc where the chains run parallel to the c-axis. Intrachain metal-metal bond lengths are ~3.4 Å which is about 20% greater than the spacing in elemental Ta (~2.5 Å) and much less than the average interchain spacing of ~6.5 Å [47]. As a result, metallic bonding occurs

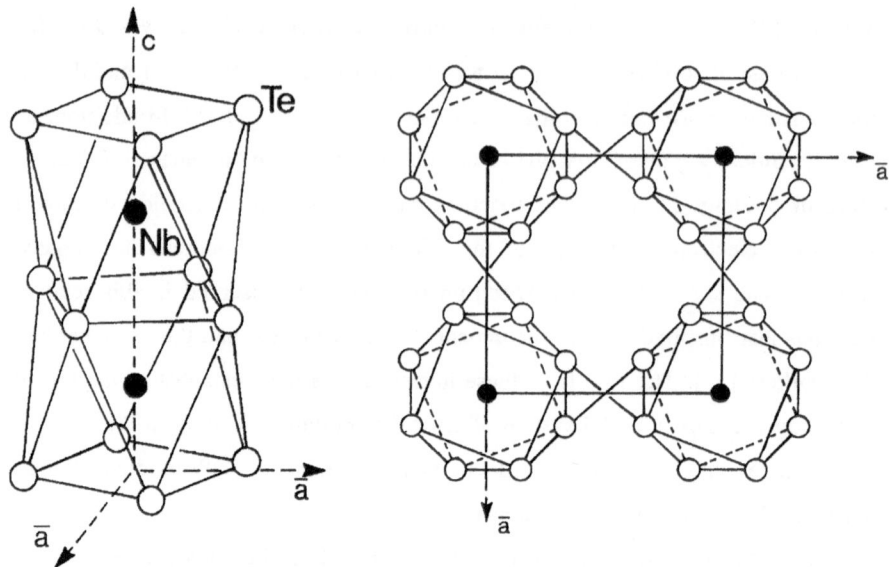

Fig. 2: The subcell structure of the tetrachalcogenides MX_4 (M=Ta or Nb). The structure consists of
linear chains of metal atoms surrounded by Te atoms in square antiprismatic coordination.

only along the chains and the structure may be considered quasi-one-dimensional. Band structure calculations indicate the Fermi surface is nearly planar [52] and therefore conducive to
the formation of a CDW.

The shortest Te-Te separation (~2.9 Å) occurs between neighbouring columns, only slightly exceeding the interatomic separation in tellurium crystals, and strongly suggests the presence of covalent bonding (i.e. the formation of Te_2 dimers). As a result, the subcell may be
visualized as composed of chains of metal atoms separated by isolated Te_2 molecules.
Although the strong interchain coupling reduces the inherent one-dimensionality of the structure, electrical measurements reveal the conductivity along the chains exceeds that perpendicular to the chains by an order of magnitude [53,54] and indicates the electronic environment
remains anisotropic.

3.2 NIOBIUM TETRATELLURIDE $NbTe_4$

3.2.1 Room Temperature Modulated Structure

Since the initial studies of Boswell et al., the incommensurate crystal structure of $NbTe_4$ at
room temperature has been the subject of several x-ray diffraction [45-47] and TEM studies
[18,20]. Electron diffraction patterns (EDPs) reveal strong satellites due to the CDW

ulation; Figure 3 shows a representative survey of the reciprocal space obtained from the four principle zones containing the c-axis. Due to the preferred cleavage of NbTe$_4$ on the {100} planes, EDPs with the beam parallel to the c-axis are very difficult to obtain. The complete reciprocal lattice is sketched in Figure 4. At room temperature, all the satellite reflections may be indexed with a single incommensurate modulation wavevector $\mathbf{q} = [1/2\mathbf{a}^*, 1/2\mathbf{b}^*, 0.688\mathbf{c}^*]$ defined relative to the subcell with satellites up to fifth order being easily distinguishable. In addition to the sharp satellites lying along the subcell rows, regions of very faint, circular,

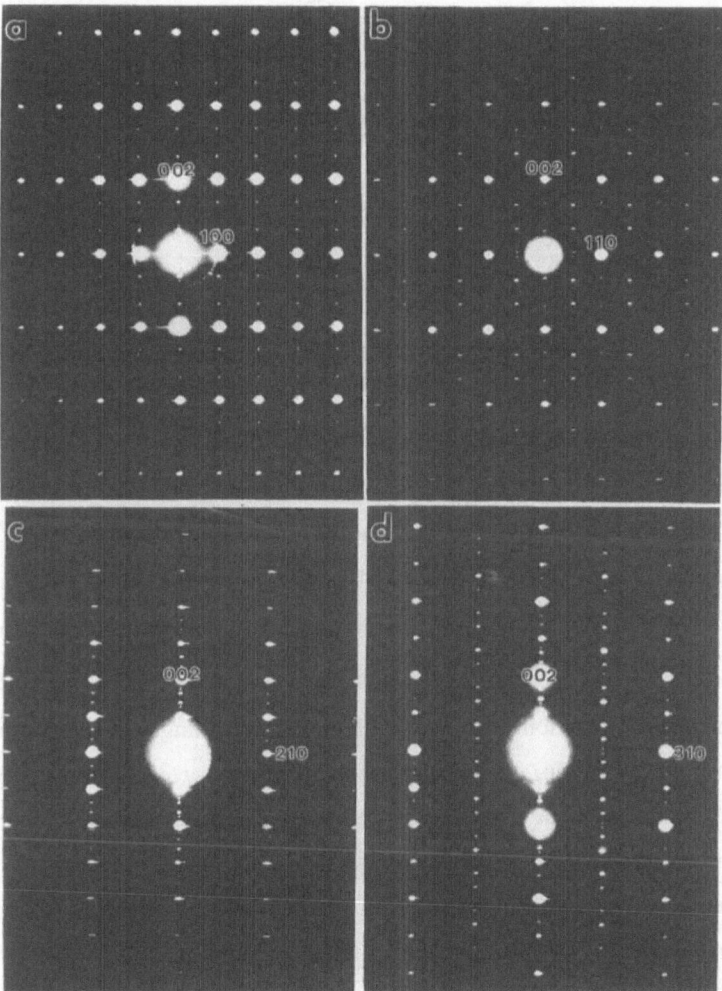

Fig. 3: Electron diffraction patterns from NbTe$_4$ at room temperature. The zone axes are labelled with respect to the tetragonal (a × a × c) subcell. (a) [010] zone axis, (b) [110] z.a., (c) [120] z.a. and (d) [130] z.a..

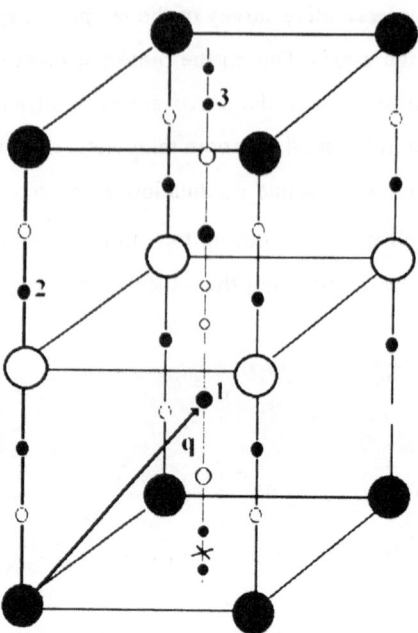

Fig. 4: Reciprocal lattice of room temperature NbTe$_4$. The larger filled circles, smaller filled circles and open circles represent subcell reflections, superlattice reflections and systematically extinct reflections, respectively. Up to third order satellites are shown.

diffuse intensity is observed in the room temperature EDPs at positions midway between the subcell rows. Neglecting these diffuse reflections, the 4-dimensional Bravais class of the room temperature modulation may be found from the symmetry of the EDPs to be $W_{1\bar{1}11}^{P4/mcc}$, in agreement with superspace group analysis of van Smaalen et al. [47]. The superstructure consists of two energetically equivalent columns bearing antiphased incommensurate CDW which gives rise to a $(\sqrt{2}\,a \times \sqrt{2}\,a \times \sim 16c)$ supercell. A schematic representation of the room temperature modulation function is shown in Figure 5. According to the symmetry requirements of the superspace group, the Nb displacements are restricted to occur only along the chain direction, producing both elongated and contracted intrachain Nb distances relative to the subcell. Taking into account satellites up to second order, the x-ray structural refinement of van Smaalen et al. has determined the form of the displacement function of the room temperature modulation [47]. It was shown that the modulation mainly produces "triplets" of Nb atoms with contracted interatomic spacings along each chain. However, due to the presence of the higher harmonics in the incommensurate modulation function, other types of multiplets, i.e. "doublets" and "singlets", also appear periodically along the chains. As sub-

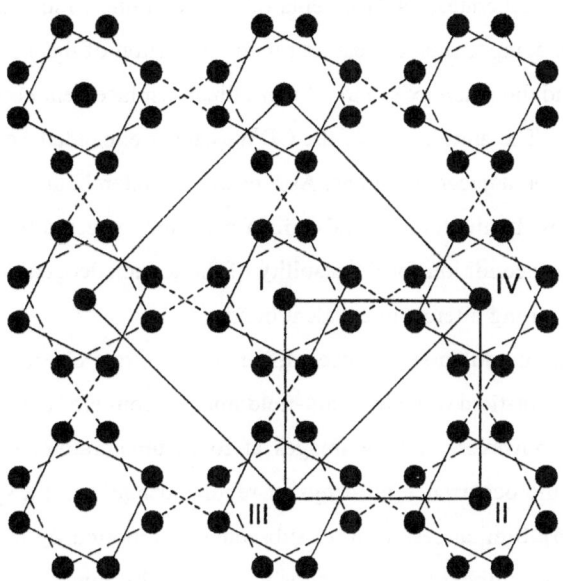

Fig. 5:Schematic representation of the room temperature modulation structure of $NbTe_4$ projected along the c-axis. The column phasings are $\phi_I = \phi_{II} = \pi$ and $\phi_{III} = \phi_{IV} = 0$, resulting in basal plane unit cell dimensions $\left(\sqrt{2}a \times \sqrt{2}a\right)$.

sequently shown by Bronsema et al. [48], this contrasts with the commensurate room temperature phase of $TaTe_4$ where the modulation generates triplet sequences exclusively. One may therefore identify "discommensurations" in the room temperature modulation of $NbTe_4$ as regions where the Nb atoms form doublet-singlet combinations instead of triplet groupings. The Landau theoretical treatment of discommensurations places no formal restrictions on their spatial extent and here the phase slip associated with the discommensurations is clearly gradual rather than abrupt, occurring over several tens of Angströms. Mahy et al. [20] have explicitly shown that the observed EDPs may be interpreted either from the point of view of a deformation modulation of the subcell structure, as described above, or a periodic interface modulated structure derived from the corresponding commensurate superstructure (assumed isostructural with room temperature $TaTe_4$). This latter approach is based on the method of "fractional shifts" [37] and assumes the two-dimensional discommensuration walls generated through an alignment of the one-dimensional discommensurations on adjacent columns may be treated as "stacking faults" in the modulation. Therefore, the classification of the room temperature modulation in $NbTe_4$ as either "discommensurate" or "purely incommensurate" is somewhat arbitrary since both approaches lead to equivalent structures.

The modulation also generates displacements of the Te atoms which can be decomposed into the following three components: a breathing mode distortion of the Te squares, a rotation of the Te squares around the chain axis, and a longitudinal displacement coupled to that of the neighbouring Nb atoms. The antiphasing of the CDWs causes expanded Te squares to line up with contracted squares on adjacent columns. As a result, the interchain Te-Te separations are essentially unchanged by the modulation and remain near that expected for a Te_2 dimer. This appears to be a necessary condition for the stability of the tetrachalcogenides and presumably reflects the presence of strong interchain covalent bonds.

The incommensurate modulation produces diffraction contrast effects which are directly observable by TEM. Bright-field or matrix dark-field images from $NbTe_4$ show no evidence of the CDW modulation, however, in SDF images at room temperature, Eaglesham and co-workers [18] reported the occurrence of arrays of regular fringes with a spacing of approximately 52 Å (Figure 6). Similar results were subsequently reported by Mahy et al. [20,44]. Although these fringes resembled those previously observed by Fung and coworkers [55] in incommensurately modulated $2H\text{-}TaSe_2$ and attributed to the presence of discommensuration walls, an unambiguous interpretation is not possible in the case of $NbTe_4$ since invariably two diffraction spots (i.e.. first and fifth order or second and fourth order satellites) are present within the objective aperture.

HREM images of the room temperature phase of $NbTe_4$ have been obtained by Eaglesham et al. [18] and Mahy et al. [20,44]. Generally speaking, the effects of the incommensurate CDW are quite subtle and difficult to distinguish in these images. A complicated variation in the fringe intensity is evident in Figure 7 where the pattern of strong and weak fringes shows a repeat periodicity of 16c, consistent with the structural model described above. Here, the CDW is clearly not strongly discommensurate since there are no regions which consistently exhibit the commensurate structure (compare with Figure 16 for commensurate $TaTe_4$). In Figure 8, Mahy et al. have identified a reversal in fringe contrast occurring across rather diffuse lines perpendicular to the c-axis with an average spacing of ~52 Å and attributed this to the presence of broad discommensuration walls.

3.2.2 The Low Temperature Incommensurate to Incommensurate Phase Transition

Upon cooling, the room temperature modulation gradually evolves through one or more intermediate phases until lock-in to a commensurate state occurs at ~50 K. The development of the intermediate incommensurate phases was first observed in electron diffraction experiments by Boswell et al. [13] and subsequently in diffraction contrast images [18], lattice

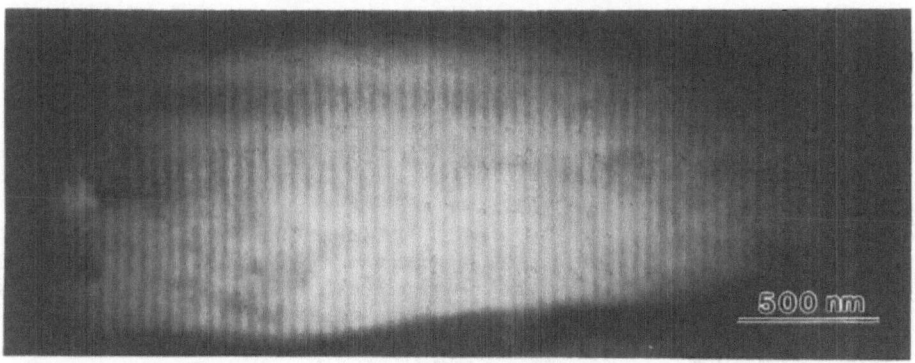

Fig. 6: SDF image of NbTe$_4$ at room temperature showing an array of regular fringes 52 Å apart.

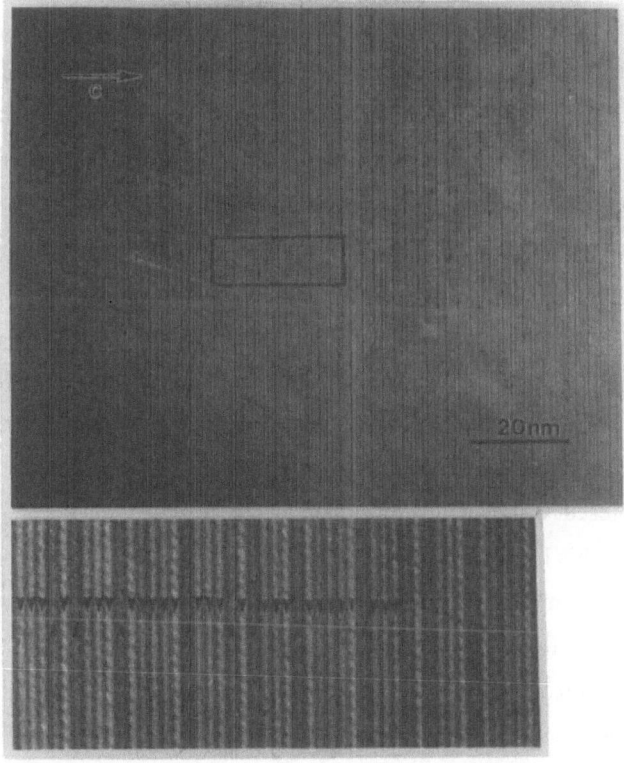

Fig. 7: [130] z.a. HREM image of NbTe$_4$ at room temperature. A periodic variation in the intensity of the interference fringes is evident with a wavelength of 16c. The marked area is shown magnified in the lower part of the figure. Strong and weak fringes are indicated by inverted and upright arrows, respectively.

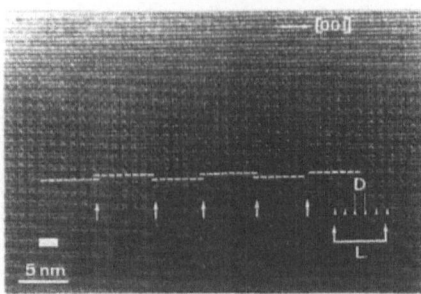

Fig. 8: A [110] z.a. HREM image of NbTe₄ at room temperature showing the L = 52 Å pattern of the discommensuration array. The antiphase relation between adjacent columns is visible as a contrast change perpendicular to the c-axis (indicated by dashed lines) [20].

images [20] and x-ray diffraction experiments [56] by several workers. The electron diffraction experiments revealed that, during cooling below ~200 K, the weak, diffuse regions of intensity observed at room temperature develop into strong, relatively narrow streaks which are extended along the c-axis direction (Figure 9). At ~100 K, regularly spaced spots developed along the streaks becoming stronger as the temperature decreases [13,20]. Boswell and Prodan [57] report these satellites occur in two equally spaced sets, denoted LT_1 and LT_2; in several other studies [20,44] only the LT_1 satellites were observed. Neither the existing satellites nor subcell reflections change in sharpness or intensity when the LT_1/LT_2 satellites appear, suggesting an independent mechanism is involved.

The presence of the additional LT_1/LT_2 satellites may be accounted for by the presence of a new incommensurate phase (or phases) with basal plane dimensions (2a x 2a) existing in the temperature range 50 to 200 K. Based on a comparison of the observed electron diffraction patterns with those calculated from various models, Prodan et al. [58] have concluded that the LT_1 phase mainly involves additional longitudinal shifts of the Nb atoms while the LT_2 phase is related to a further breathing mode distortion of the Te squares. The LT_1/LT_2 distortions thus appear to represent a precursor to the lock-in phase transition. Recent x-ray diffraction studies [56] support this hypothesis. Since both distortions manifest elements of the commensurate modulation present in TaTe₄, it is reasonable to expect that the lock-in phase of NbTe₄ is isostructural with the room temperature phase of TaTe₄, although this has yet to be verified by an x-ray diffraction experiment. Thus the following picture of the structural changes occurring in NbTe₄ has emerged: a gradual transition from an purely incommensurate $\left(\sqrt{2}\, a \times \sqrt{2}\, a \times \sim 16\, c\right)$ structure at room temperature through an intermediate precursor stage where the unit cell is enlarged to $\left(2\, a \times 2\, a \times 32\, c\right)$, i.e. the LT_1/LT_2 phase, which is followed by lock-in to a commensurate $\left(2\, a \times 2\, a \times 3\, c\right)$ structure. The precursor stage represents a period of

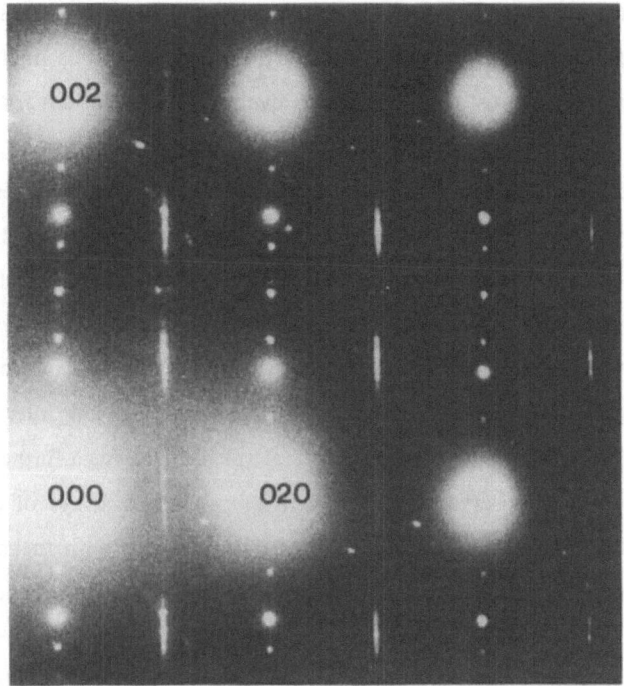

Fig. 9: [100] z.a. EDP of NbTe$_4$ at 30 K. Additional streaks develop between the subcell rows. Along the streaks, closely spaced satellites for the LT$_1$ and LT$_2$ phases are observed.

adjustment of the CDW phasing on the various columns with attendant changes in the Te inter- and intra-column bond lengths.

The LT$_1$/LT$_2$ models proposed by Prodan et al. are derived from the x-ray structural refinement of van Smaalen et al. [47]. As discussed above, in this model the discommensuration walls are sufficiently broad that the room temperature modulation is essentially indistinguishable from the purely incommensurate case. From Landau theory, a strongly discommensurate state is predicted prior to lock-in. Thus, it is not unreasonable to expect that discommensurations play a role in the LT$_1$/LT$_2$ phases of NbTe$_4$. However, Prodan et al. have shown that, for model structures which assume sharp discommensuration walls, some calculated satellite positions do not agree with those observed for either the LT$_1$ or LT$_2$ phases. It therefore appears that for both the room temperature and LT modulations, the phase slip associated with discommensurations is continuous rather than abrupt. Although the lock-in transition must involve the annihilation of discommensurations, these would appear to be sharp only very near the transition temperature.

SDF images of NbTe$_4$ obtained on cooling below room temperature have revealed a wealth of unusual microstructural phenomena which occur during the approach to the lock-in.

In images formed with the streaked LT satellites, Eaglesham et al. [18] observed lock-in during a sluggish first-order transition involving the heterogeneous nucleation of domains, resembling stripes, normal to the c-axis which eventually grew to cover the entire crystal at lock-in with the exception of a few isolated faults (Figure 10). On warming from the commensurate state, these isolated faults evolved into multiple lines of variable orientation and spacing which ultimately expand throughout the crystal to form an array of fine fringes (Figure 11). Similar results were reported by Mahy et al. [20] who further observed that the commensurate to LT transition proceeds through the growth of islands of well-defined parallel discommensuration walls within the commensurate phase which, by a temperature of ~100 K, develop into irregular moiré-like fringes (Figure 12). The apparently second-order transition from the LT phase to the room temperature incommensurate phase involves a further broadening of the moiré-like fringes. The observation of defects, interpreted as arrays of discommensuration walls, in the SDF images may appear somewhat at variance with the model proposed by Prodan et al. for the LT phases. It should be noted however that the diffraction contrast expected for discommensuration walls in these compounds is unknown and thus the SDF observations do not exclude models in which the phase slip occurs gradually.

Based on their TEM observations, Mahy and coworkers [20] have also outlined two different approaches to the description of the LT_1 phase (LT_2 satellites were apparently not observed by these authors) as either an interface modulation, generated by a periodic array of discommensurations, or, alternatively, as a displacive modulation. For the case of spatially extended discommensurations or a displacive modulation containing higher harmonics of the primary deformation wave, these two approaches are equivalent. However, the underlying physical assumptions of the two approaches are obviously quite different and Prodan et al. [58] have demonstrated that, given the precursor nature of the LT_1/LT_2 distortions, a model derived from a displacive modulation is the more physically meaningful,

It is interesting to consider how the model presented here compares with the results of the Landau theoretical analysis of the NbTe$_4$ system by Walker and coworkers [59-64]. Briefly, the approach rests on the assumption that only weak coupling exists between the CDW on different columns and it is therefore sufficient to model the CDW independently while treating the inter-column interactions using perturbation theory. In principle the order parameter of the incommensurate phase contains an infinite series of harmonics, however near the normal to incommensurate transition, the first harmonic is dominant and the modulation is approximately sinusoidal. In this case, minimizing the free energy associated with the inter-

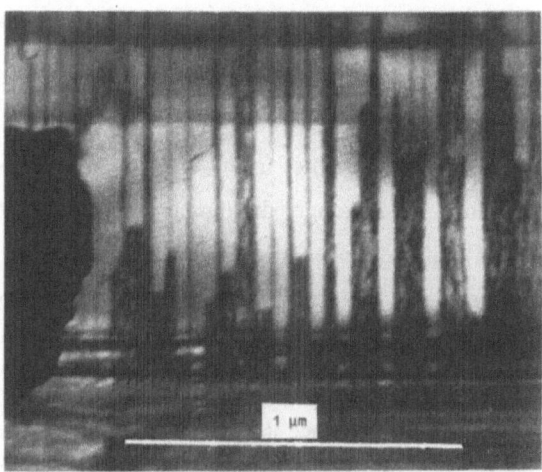

Fig. 10: SDF image of NbTe$_4$ using the LT satellites during lock-in showing bright stripes of commensurate phase driving across the crystal [18].

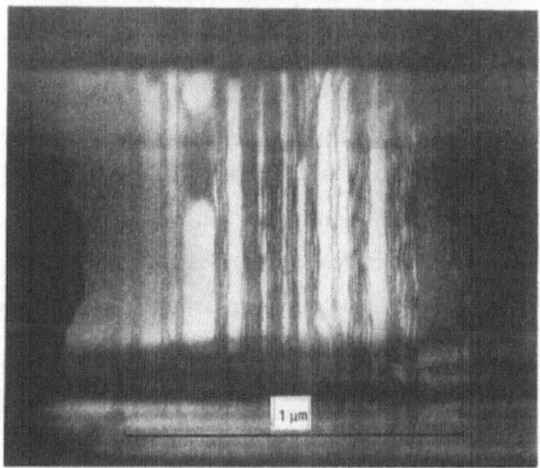

Fig. 11: SDF image of NbTe4 using the LT satellites showing the evolution of the defect micro-structure during warming out of the commensurate phase [18].

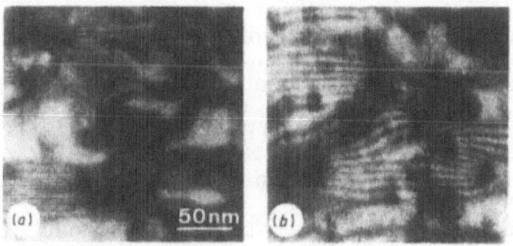

Fig. 12: SDF images of NbTe$_4$ using the LT satellites showing the 3c commensurate superlattice fringes and (a) at 50 K, islands of well defined parallel discommensuration walls; (b) at 100K, moire-like arrays of discommensuration walls [20].

column interactions gives an optimum phasing of π for CDW on nearest neighbour columns and predicts a $\left(\sqrt{2}a \times \sqrt{2}a\right)$ base for the unit cell, in agreement with the available experimental data.

At lower temperatures, higher harmonics in the order parameter are assumed to become more important resulting in a more strongly discommensurate modulation. To examine this situation, Walker et al. [61] defined a new order parameter in terms of the normal mode distortions of the discommensuration wall array found in the room temperature phase. A comparison of the model of the LT_1 structure derived from the analysis of the electron diffraction patterns by Prodan et al. and the "double-q" state described by Walker et al. [63] shows that these are identical. The LT_2 distortion, since it involves shifts of the Te atoms, is outside the scope of the free energy analysis where only displacements of the Nb atoms were considered. In order to provide a physical basis for their phenomenological model, Morelli and Walker [62] proposed a mechanism in which competition between temperature dependent nearest neighbour and next-nearest neighbour column interactions accounts for the LT phase transition. Note that the free energy analysis applies whether the discommensurations are considered to be broad or sharp; however, the TEM observations indicate these defects only become sharp as the lock-in temperature is closely approached.

In light of the above discussion, it is now clear that the different phases observed in $NbTe_4$ are the result of several competing mechanisms. On one hand, there is a general tendency for triplets of metal atoms to form with the largest possible phase shift between neighbouring columns. At the same time, due to their strong covalent character, the structure attempts to maintain the inter-column Te-Te bonds as nearly constant as possible. These factors, related to the Coulomb and lattice strain energies, act in opposition to the Fermi surface in determining the CDW modulation structure. In the room temperature phase of $NbTe_4$, the formation of a structure containing metal atom triplets only is frustrated, presumably reflecting the dominance of the electronic factors and producing an incommensurate modulation. In consequence, the largest possible phase difference between columns is adopted, i.e. π, with the antiphasing serving to minimize the Coulomb contribution to the interaction energy between CDW on adjacent columns. As a result, only two column types are possible leading to a structure with a $\sqrt{2}a$ base. The structure does however seek to retain the energy advantages of commensurability, forming predominantly triplets or near-triplets of Nb atoms wherever possible, as is shown by the x-ray refinements [46,47]. For the commensurate lock-in phase of $NbTe_4$, it is possible for triplets only to be formed. In this case, there are three energetically equivalent but physically distinct phasings of the CDW along a column, leading to phase shifts of $2\pi/3$ between columns and generating a structure with a $(2a \times 2a)$ base. The CDW phasing on the columns is determined mainly by the strengths of the various inter-column interactions, the dominant interaction being mediated by the short Te -Te bonds. This in turn determines the unit cell basal dimensions. Of course the actual situation is more complicated than the above discussion indicates since all of these factors are interdependent; for example, the inter-column Te bonding directly affects the Fermi surface [65].

3.3 TANTALUM TETRATELLURIDE TaTe$_4$

TaTe$_4$ is isostructural to NbTe$_4$ however the CDW occurring in these compounds exhibit strikingly different behaviour. Both subcell structures are tetragonal and consist of chains of metal atoms centered within Te antiprismatic columns. Despite this similarity, TaTe$_4$ is commensurately modulated at room temperature while NbTe$_4$ is incommensurately modulated. TaTe$_4$ also undergoes a series of CDW-driven phase transitions but these differ fundamentally in character from those observed for NbTe$_4$. As a result, comparison of the two tetrachalcogenides provides an opportunity to gain new insight into the factors influencing the stability of CDW phases.

3.3.1 Room Temperature Modulated Structure

Selected-area electron diffraction patterns (EDPs) of the four principal zones containing the c-axis are shown in Figure 13 and the corresponding reciprocal lattice in Figure 14. At room temperature, the superlattice spots appear in the commensurate positions and may be indexed using three modulation wavevectors: $\mathbf{q}_1 = \left[\frac{1}{2}\mathbf{a}^*, \frac{1}{2}\mathbf{b}^*, \frac{2}{3}\mathbf{c}^*\right]$, $\mathbf{q}_2 = \left[\frac{1}{2}\mathbf{a}^*, 0, \frac{2}{3}\mathbf{c}^*\right]$ $\mathbf{q}_3 = \left[0, \frac{1}{2}\mathbf{b}^*, \frac{2}{3}\mathbf{c}^*\right]$ defined relative to the subcell. The electron diffraction observations indicate TaTe$_4$ adopts a commensurately modulated $(2a \times 2a \times 3c)$ supercell at room temperature. Budkowski et al. [66] have shown that with a suitably transformed coordinate system a single $\mathbf{q}$-vector is sufficient to completely describe the superstructure of TaTe$_4$, however, here the modulation is decomposed into three components lying along simple crystallographic directions for ease of notation.

X-ray crystallographic analysis by several groups [13,48] has revealed that the superstructure consists of three energetically equivalent columns bearing CDW which are mutually $2\pi/3$ out of phase. With reference to Figure 5, the phasing along the columns are $\phi_I = \phi_{II} = 0$,

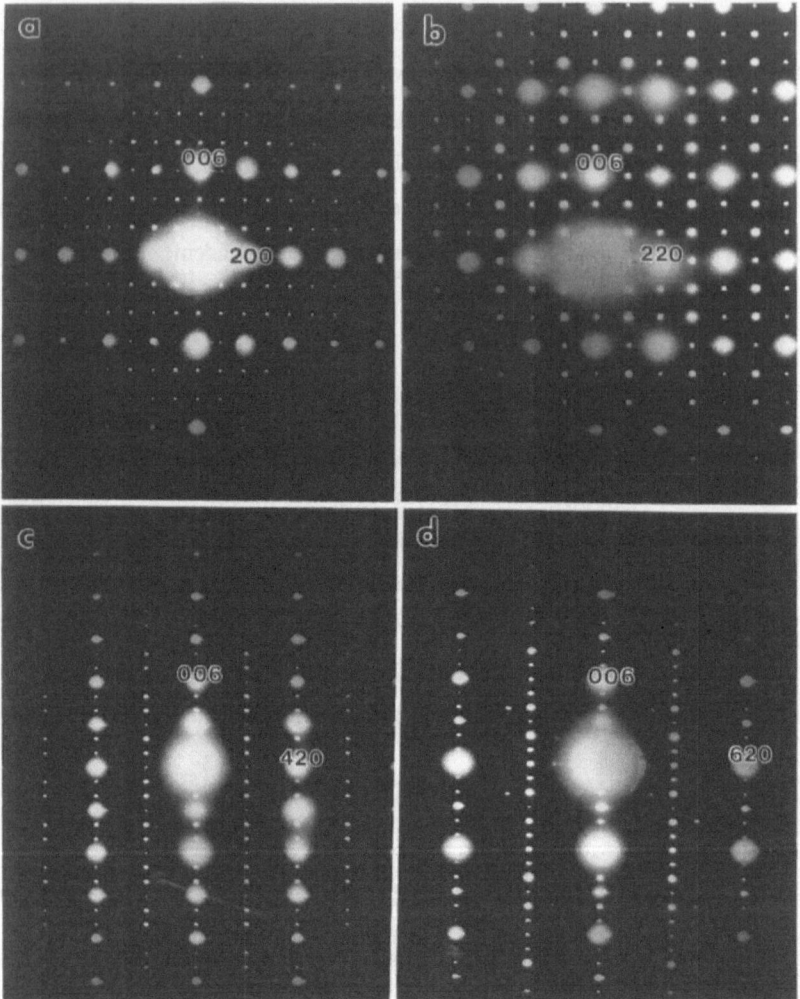

Fig. 13: Electron diffraction patterns from TaTe$_4$ at room temperature. The zone axes are labelled with respect to the $(2a \times 2a \times 3c)$ supercell. (a) [010] z.a., (b) [110] z.a., (c) [120] z.a. and (d) [130] z.a..

$\phi_{III} = 2\pi/3$ and $\phi_{IV} = 4\pi/3$. The symmetry of this structure has been determined to be P4/ncc [48] in agreement with the Landau free energy calculations of Walker [67]. Due to this phasing arrangement, the Ta atoms are displaced along the c-axis to form triplets on each of the columns. The displacement of the Ta atoms is accompanied by a distortion of the Te squares involving both a breathing mode and a weak longitudinal displacement while the CDW phasing is such that the intercolumn Te-Te bond lengths remain essentially unaffected. The main features of the modulation are thus very similar to those of NbTe$_4$.

As shown by Walker, the ground state of the CDW on each column has a discrete

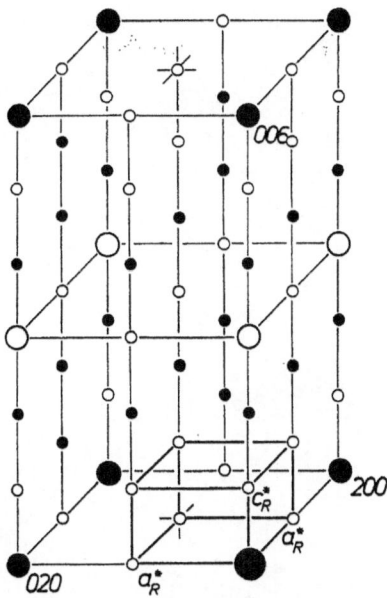

Fig.14: Reciprocal lattice of C phase of TaTe$_4$ indexed with respect to the (2a × 2a × 3c) supercell. Larger circles represent subcell reflections, smaller circles superlattice reflections. Systematically extinct reflections are indicated by open circles.

degeneracy since physically nonequivalent phasings have the same energy. Hence, the modulation may assume a certain phasing in one region while in the adjacent regions an alternate phasing may occur, generating a domain structure. Antiphase boundaries (APBs) occur where the various domains converge, corresponding essentially to "stacking faults" in the periodic arrangement of columns comprising the superstructure. From geometrical considerations, Boswell et al. [19] have determined the potential displacement vectors of the APBs as:

$$R_1 = \tfrac{1}{2}\{100\}, \quad R_2 = \tfrac{1}{2}\{110\}, \quad R_3 = \tfrac{1}{3}\{001\}, \quad R_4 = \tfrac{1}{6}\{302\}, \quad R_5 = \tfrac{1}{6}\{332\}.$$

SDF images of TaTe$_4$ obtained at room temperature typically reveal a three-dimensional network of APBs [18,19,21]. These defects are not visible in the corresponding BF or matrix DF images and are thus associated with the CDW modulation structure. CBED analysis [18] revealed no change in orientation or crystal symmetry occurred on crossing the position of the APBs, which indicates the boundaries are displacive and separate identical commensurate domains. From the observed diffraction contrast (Figure 15), two main types of APBs can be distinguished [19]. *Type-I* boundaries lie generally parallel to the c-axis direction and are associated with displacement vectors of the form $R_2 = \tfrac{1}{2}\{110\}$ while *type-II* boundaries are irregular and inclined to the c-axis with displacement vectors of the form $R_4 = \tfrac{1}{6}\{302\}$.

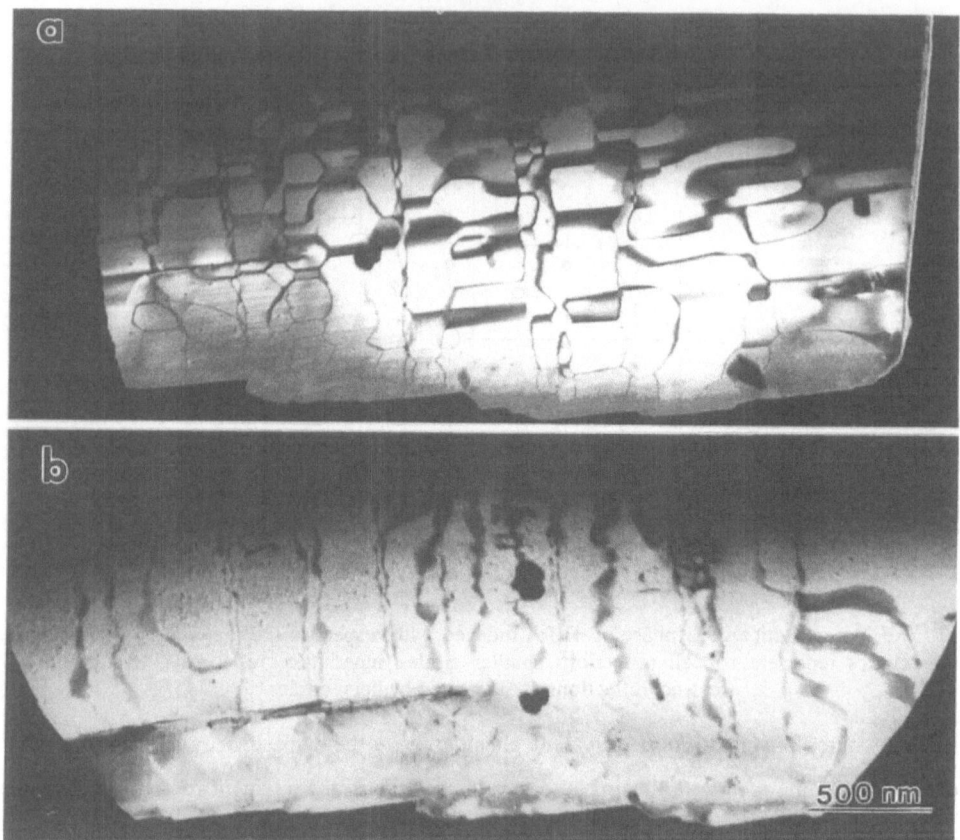

Fig. 15: SDF images of TaTe$_4$ at room temperature using (a) a q$_2$ satellite and (b) a q$_1$ satellite. Note the *type-I* APBs (arrows) are out of contrast in (b).

Although it is not possible to confirm the latter directly from the diffraction contrast, since the *type-II* boundaries show contrast in all of the available satellite reflections, lattice displacements observed in HREM images [19] which occur on crossing the boundaries are compatible with this hypothesis (Figure 16). Since APBs are translation interfaces, one would normally expect α fringe contrast at the position of the fault [68]. However, since the SDF images are obtained with superlattice reflections having quite long extinction distances, only a single fringe is typically formed within the specimen thickness. When viewed obliquely due to an inclination of the fault plane occupied by the boundary relative to the incident beam, this fringe may be considerably broadened (Figure 15). The observed sensitivity of the boundary contrast to slight deviations from the exact Bragg condition, i.e. for $s \neq 0$, is also expected for α fringes (Figure 17).

The two types of APBs are observed to interact with a *type-I* boundary frequently termin-

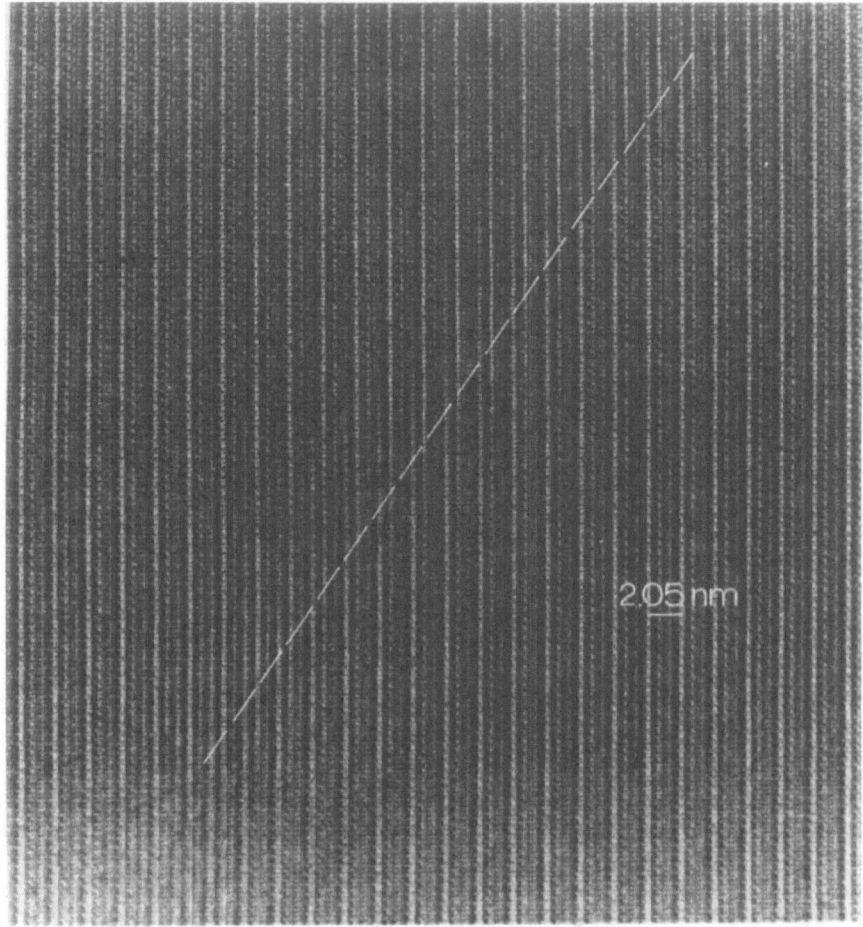

Fig. 16: [120] z.a HREM image of TaTe$_4$. The strong fringes are associated with the 3c modulation. The dashed line indicates the position of *type-II* APB (note the shift of one fringe spacing across the boundary).

ating at a junction with two *type-II* boundaries. At the junction, the following reaction consistent with the assigned APB R-vectors is proposed:

$$\tfrac{1}{2}[110] \rightarrow \tfrac{1}{6}[302] + \tfrac{1}{6}[03\bar{2}].$$

Similar to discommensuration walls, the c-component of the displacements associated with the *type-II* APBs produces phase slips in the CDW modulation of TaTe$_4$. However, the phase slips are presumably abrupt (occurring over at most a few Å) and insufficient in number to disturb the overall commensurate periodicity.

Ideally, APBs arising as a result of an ordering process are strain-free boundaries [69]. However, some features of the diffraction contrast suggest significant localized lattice strains are associated with the APBs in TaTe$_4$. Figure 18 shows that the image of a *type-I* boundary is

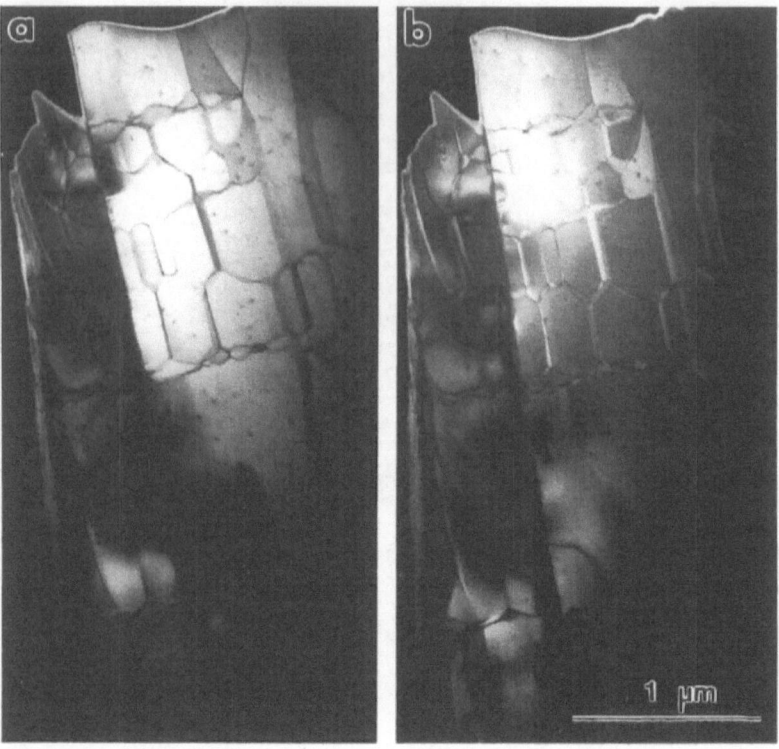

Fig. 17: SDF images of TaTe$_4$ at room temperature using a q$_1$ satellite showing the dependence of the boundary contrast on deviation parameter. (a) s = 0 and (b) s ≠ 0. Note the reversal of contrast between images (a) and (b).

slightly displaced on crossing the position of an extinction contour. In analogy with the diffraction contrast of edge dislocations, which are observed to undergo a similar displacement due to the effects of the associated strain field [70], this implies that local strains exist around the APBs. A source of this strain may be the strong Te-Te bonds coupling the CDW modulation on adjacent columns. For the room temperature modulation, x-ray diffraction studies [48,49] indicate that the Te squares are expanded relative where the Ta atoms shift towards one another along the chains and contracted where the Ta atoms move apart. In the unfaulted regions, intercolumn bonding couples Te atoms on expanded squares in one column with Te atoms on contracted squares in adjacent columns (and vice versa) while the presence of an APB would lead to coupling across the fault of, for example, Te atoms on expanded squares. This presumably produces lattice strain fields around the boundaries and accounts for the observed diffraction contrast. Detailed image calculations are necessary however to confirm this model.

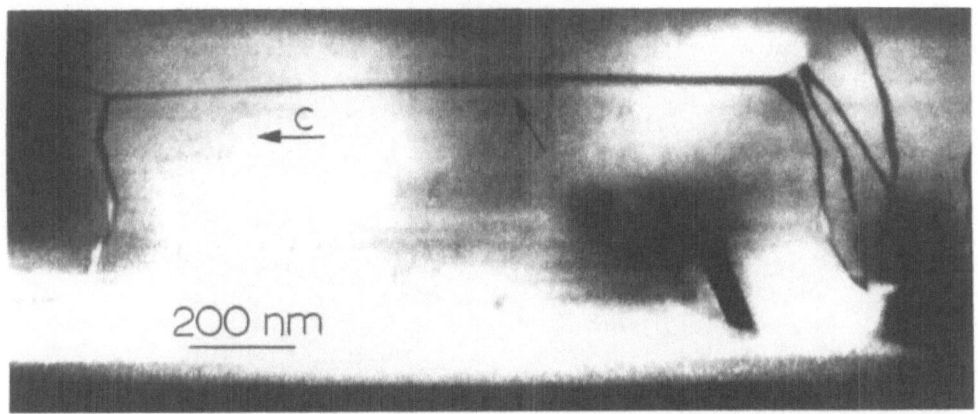

Fig. 18: SDF image of a displacement of a *type-I* boundary upon crossing an extinction contour.

3.3.2 The Commensurate to Commensurate Phase Transition

When heated above ~450 K, the satellites associated with wavevectors q_2 and q_3 are found to diminish in intensity and gradually disappear over a narrow temperature range (~30 K) while those associated with q_1 remain unchanged (Figure 19). These observations have been interpreted [19,21] in terms of a reversible transformation to a high temperature commensurately modulated phase based on a $\left(\sqrt{2}a \times \sqrt{2}a \times 3c\right)$ unit cell. The behaviour of the APBs upon heating to the commensurate to commensurate ($C \leftrightarrow C^*$) phase transition has been observed in detail using SDF microscopy. In SDF images formed using a q_2 or q_3 satellite, the APBs typically exhibit dark contrast (at s = 0) and appear fairly narrow at room temperature. On heating, the *type-II* boundaries become increasingly mobile which results in progressively shorter *type-I* boundaries until, very near the transition point, these latter are essentially eliminated. The disappearance of the *type-I* boundaries is accompanied by a noticeable increase in the width of the remaining of the remaining *type-II* boundaries. These boundaries are very unstable and, with a slight increase in temperature, rapidly expand to convert the entire specimen to the high temperature commensurate phase (i.e. to dark contrast coinciding with the disappearance of the q_2 and q_3 satellites). The effects described above are shown in Figure 20 and indicate that the $C \leftrightarrow C^*$ transition is of first-order transition and initiated at the APBs. Similar effects are observed in SDF images formed using q_1 satellites.

When a specimen is heated above the $C \leftrightarrow C^*$ transition temperature, the SDF images obtained using a q_2 or q_3 satellite show mostly dark contrast throughout the field of view. However, a few narrow regions of light contrast are invariably observed along the lines of intersection of expanding regions of the high temperature commensurate phase (Figure 21a).

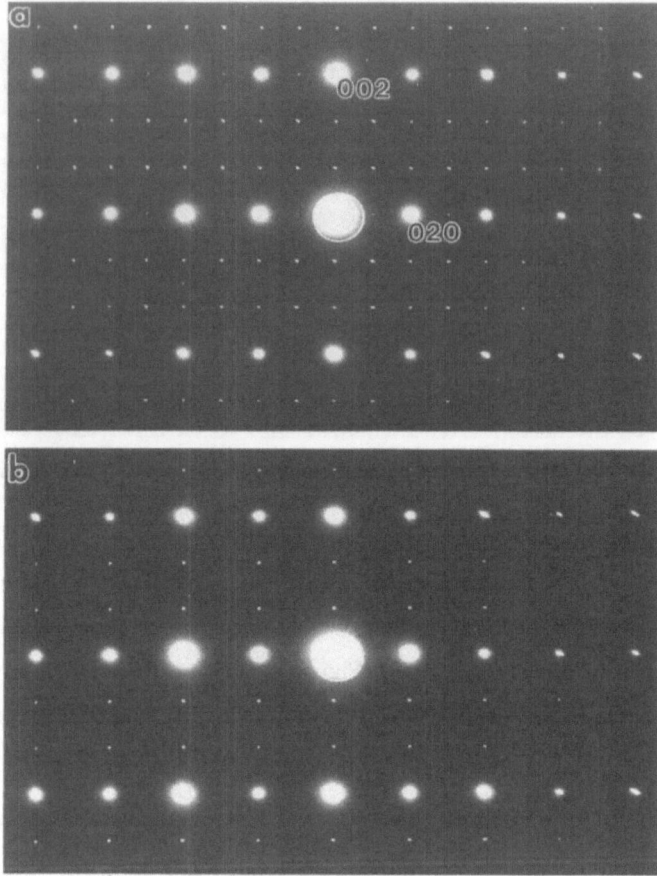

Fig. 19: [100] z.a. EDPs from TaTe$_4$ (a) at room temperature and (b) at ~450 K. Note the disappearance of the q$_2$/q$_3$ satellites in (b).

SDF images formed using $\mathbf{q}_1$ satellites clearly reveal these as domain boundaries (Figure 21b) which are apparently residual *type-II* APBs. The displacement vectors are thus of the $\mathbf{R}_4$ type which is equivalent to $\frac{1}{6}$[332] indexed in terms of the high temperature $\left(\sqrt{2}a \times \sqrt{2}a \times 3c\right)$ unit cell. Despite their occurrence at a temperature where APBs would be expected to be highly mobile, these high temperature boundaries are stable, an effect presumably related to strong pinning of the CDW.

SDF observations over several temperature cyclings through the C↔C* transition reveal a pronounced memory effect. Typically when a specimen is heated slightly above the transition point for a few seconds and then cooled, one observes the APBs to reappear in precisely the same configuration. This memory effect is lost if the specimen is held for too long at the higher temperature. The origin of the memory effect arises from the fact that nucleation of the

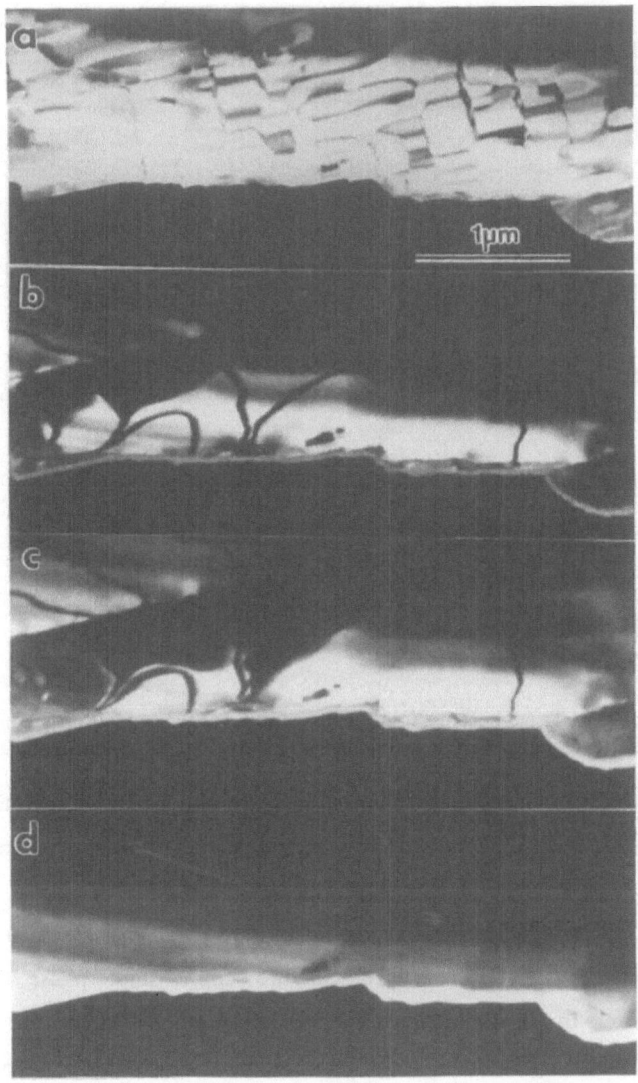

Fig. 20: SDF images of TaTe$_4$ formed with a q$_2$ satellite observed during heating to the C→C* transition at T$_c$ ~450K, (a) at room temperature, (b) heated nearly to T$_c$, (c) heated to T$_c$ and (d) heated slightly above T$_c$.

room temperature phase occurs at the domain boundaries in a process which appears to be the reverse of that observed on heating; i.e. the room temperature APBs are generated along the lines of intersection of expanding regions of the (2a × 2a × 3c) phase. In cases where a specimen was held above the transition temperature and then rapidly cooled, a new network of APBs was formed that was independent of the previous room temperature configuration. This "quenching in" of the APB network shows that the boundaries arise from phasing errors

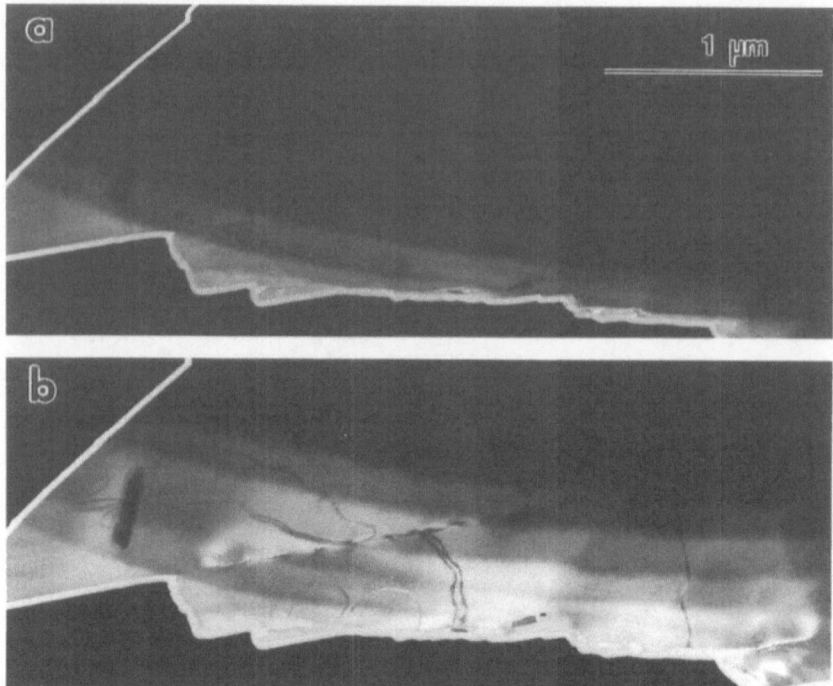

Fig. 21: SDF images of TaTe$_4$ heated slightly above 450 K (a) formed with a q$_2$ satellite and (b) formed with a q$_1$ satellite. Note the very faint boundaries (arrows) in (a) which are clearly visible in (b).

between CDW on neighbouring columns as discussed above. The observation that few APBs are observed in TaTe$_4$ crystals that are slow-cooled from the growth temperature also supports this hypothesis.

The APBs in the high temperature commensurate phase may be seen to originate in an analogous manner to those in the low temperature commensurate phase, i.e. they also result from stacking faults in the modulation superstructure. Here, any two of the three energetically equivalent CDW-bearing columns in the room temperature structure can be used to generate the high temperature superstructure and, as a result, several different configurations are possible. As revealed in the in-situ SDF images, at the transition, regions of high temperature phase nucleate at the existing APBs and expand throughout the crystal. If, during this process, two regions composed of different configurations of the high temperature structure impinge on one another, an APB with displacement vector $\mathbf{R} = \frac{1}{6}[332]$ would be expected (Figure 22). As discussed above, a strip of high temperature phase may be considered to lie along the domain boundaries in the room temperature phase. Nucleation of different configurations of the high temperature structure will be favoured on opposite sides of a room temperature APB, possibly accounting for the observed preference of the domains to expand in only one direction on

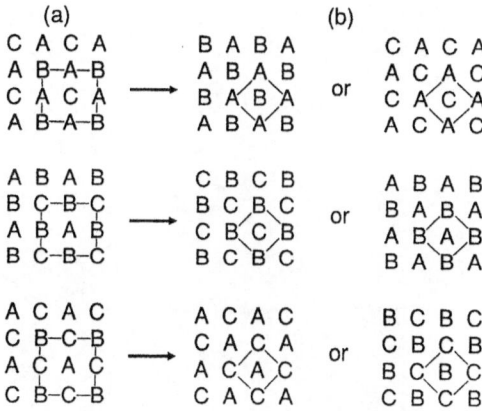

Fig. 22: Configuration of CDW domains for (a) C phase and (b) C* phase. Letters refer to CDW phasing along the columns; six different variants are possible for the C* phase.

heating. Conversely, on cooling, the nucleation of the room temperature structure is promoted at the APBs of the high temperature phase since in these regions the nearest neighbour columns are already shifted in phase by $2\pi/3$ as required while the next-nearest neighbour columns are in phase.

3.3.3   The Commensurate to Incommensurate Phase Transition

Additional changes occur in the EDPs of TaTe$_4$ upon heating above ~550 K (Figure 23) [21]. The remaining satellites shift to incommensurate positions defined by the modulation wave-vector $\mathbf{q}_i = (\frac{1}{2}\mathbf{a}^*, \frac{1}{2}\mathbf{b}^*, q_c \mathbf{c}^*)$ where q_c varies continuously from the commensurate value of 2/3 at 550 K to ~0.660 near 650 K where the crystals typically decompose. The electron diffraction observations reveal the development of an incommensurate phase of TaTe$_4$ having basal plane dimensions $(\sqrt{2}a \times \sqrt{2}a)$. This commensurate to incommensurate transition is reversible but exhibits considerable temperature hysteresis, occurring near 450 K on cooling.

SDF images reveal that the microstructure of TaTe$_4$ continues to evolve during the C*↔IC transition (Figure 24). Following the C↔C* transition at 450 K, relatively few APBs of *type-II* are visible in the SDF images (Figure 21b), however as the temperature increases, these defects gradually develop into multiple fine lines covering large regions of the crystal in an irregular fashion (Figure 24a). The defect density continues to increase with temperature until, near 550 K, a regular array of parallel "fringes" with an average separation of ~200 Å is observed, although considerable variability about the mean persists (Figure 24b). Electron diffraction reveals that the CDW modulation is incommensurate when TaTe$_4$ is in the

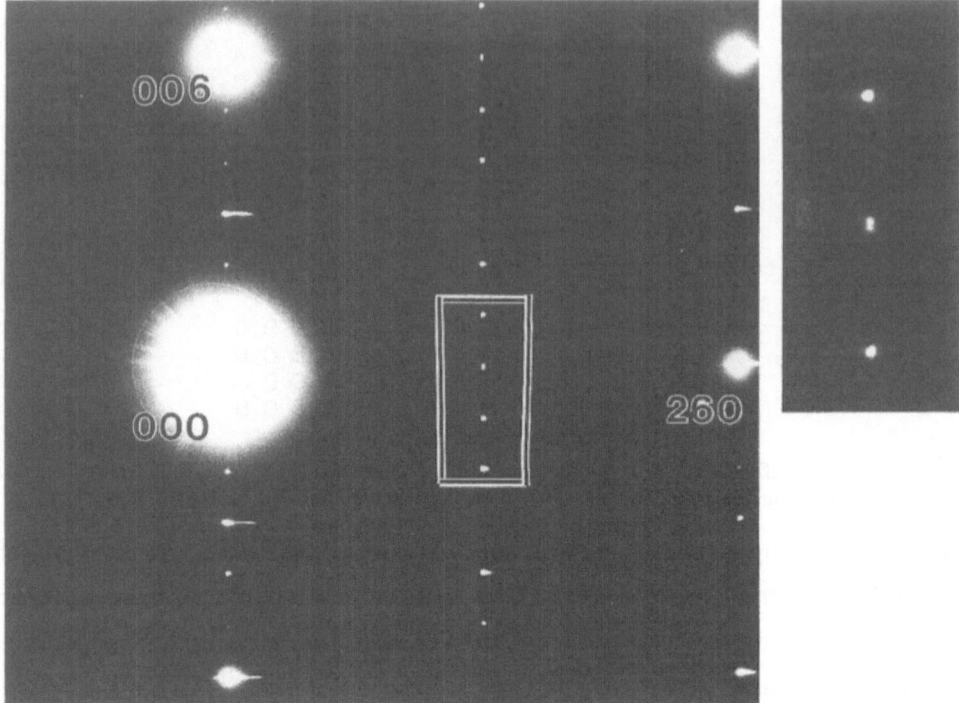

Fig. 23: [130] z.a. EDP from TaTe$_4$ at 550 K; the superlattice doublet revealing the incommensurate periodicity is enlarged in the inset.

"fringed" state. As discussed further below, these fringes may be interpreted as arising from the generation of discommensuration walls perpendicular to the c-axis. On heating, the mechanism for increasing the discommensuration wall density appears to involve the internal nucleation and motion of "discommensuration-dislocations", each consisting of a six-fold junction of discommensuration walls. These defects become increasingly mobile with temperature and new ones nucleate until eventually all of the commensurate domains contain a dense array of fringes. In the final IC state, the fringes are well-ordered and regularly spaced approximately 100 Å apart. The process is thus very similar to the discommensuration process occurring in NbTe$_4$. Similar phenomena have also been observed in a diverse collection of materials including 2H-TaSe$_2$ [55], Rb$_2$ZnCl$_4$ [71] and certain Ag-Mg alloys [72], all of which undergo a commensurate to incommensurate phase transition.

Once established, the discommensurate state of TaTe$_4$ persists on cooling to near 450 K. At this temperature, the crystal apparently transforms directly back to the $(2a \times 2a \times 3c)$ commensurate structure which is then retained down to room temperature. This is shown in SDF images formed with a q_1 satellite recorded during cooling from the incommensurate state

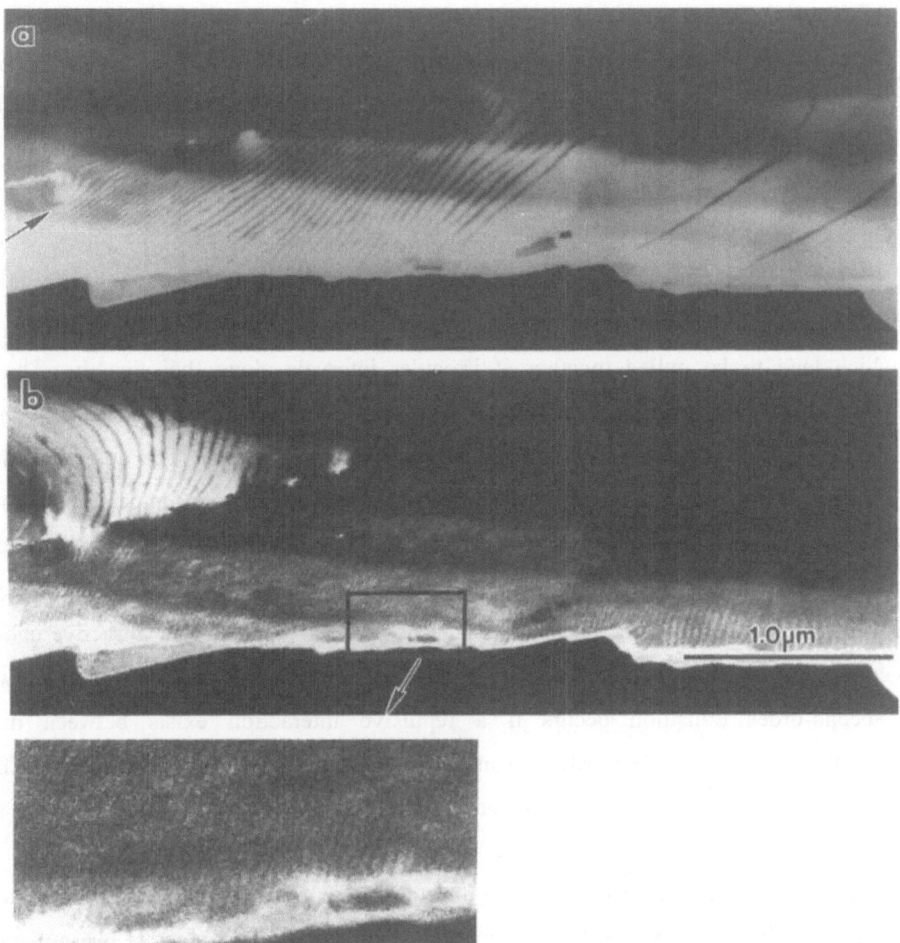

Fig. 24: SDF images using a q_1 satellite of TaTe$_4$ observed during heating to the C*→IC transition at ~550 K. The crystal has been heated slightly between (a) and (b). A "discommensuration-dislocation" is indicated by the arrow in (a) The inset shows the marked region at higher magnification.

(Figure 25). Near the transition at 450 K (Figure 25a), dark bands running perpendicular to the c-axis are observed co-existing with well-defined bands of parallel discommensuration walls. The dark bands exhibit bright contrast when imaged using a q_2 or q_3 satellite, indicating these are regions of room temperature commensurate phase (Figure 25b). These dark bands are, in fact, rarely continuous across the width of a crystal but are instead typically segmented by discommensuration walls running roughly parallel to the c-axis. Eventually, the commensurate bands drive forward and cover the entire crystal, leaving a relatively small number of APBs in the room temperature phase.

Since the phase transitions in TaTe$_4$ take place rapidly over a narrow temperature range which is difficult to control precisely in the electron microscope, it has not been possible to follow the evolution of the discommensuration array in detail. However, the SDF observations suggest the mechanisms involved in increasing the discommensuration wall density during heating are distinct from those which act to remove the discommensuration walls upon cooling. This may account for the marked thermal hysteresis associated with the transitions in $TaTe_4$. On cooling, the growth of the commensurate bands tends to lead to the formation of isolated triplets of discommensuration walls (Figure 25c) which, on further cooling, appear to coalesce along their lengths. In many cases, the result is a roughly parallel pair of *type-II* APBs which tend to be pulled together at irregular intervals along their lengths. These characteristic "jogs" of the discommensuration walls appear to limit the growth of the commensurate domains and represent an energetically stable configuration. The defects at the "jogs" are then precursors to the *type-I* boundaries observed in the room temperature phase. Interpreted in this light, the elimination of the discommensuration walls is thus mediated by the formation of the APBs. Hence, the SDF observations imply a strong mutual interaction between discommensuration walls and between APBs that may repulsive or attractive under various conditions. For the LT phase transition in $NbTe_4$, Morelli and Walker [62] have shown that a second-order transition occurs if a repulsive interaction exists between discommensuration walls and a first-order transition occurs if the interaction is attractive. The defects in $TaTe_4$, which have identical **R**-vectors, would be expected to behave similarly in this regard.

SDF images of $NbTe_4$ and $TaTe_4$ obtained during cooling from the incommensurate state exhibit a remarkable similarity. For both compounds, bands of a commensurate phase nucleate and grow leading to lock-in. However, the transition in $TaTe_4$ occurs without the formation of precursor phases analogous to the LT_1/LT_2 phases observed in $NbTe_4$. Mahy et al. [20] have proposed a mechanism for the generation of discommensurations in $NbTe_4$ which, while involving only small additional atomic displacements along the chains, requires a coordinated rearrangement of the CDW phasing on neighbouring chains. Presumably since the basal plane has already transformed from $2a \rightarrow \sqrt{2}a$ prior to the incommensurate tran-sition for $TaTe_4$, the discommensuration process is much more rapid. The increased thermal energies also likely promote the transition process.

It is interesting to compare the SDF observations of the discommensuration process in the tetrachalcogenides with those obtained for other CDW systems. Of these, the transition metal dichalcogenide $2H\text{-}TaSe_2$ has been the most extensively studied by TEM [41,55,73]. This

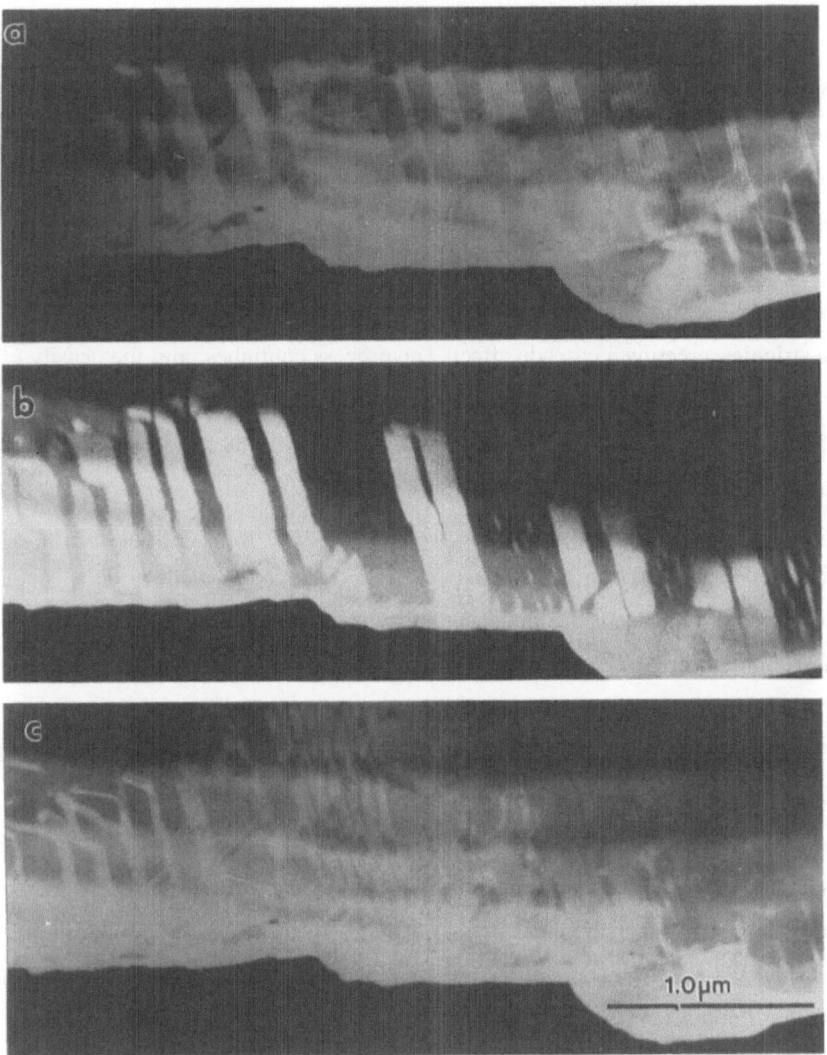

Fig. 25: SDF images of TaTe$_4$ obtained during lock-in using (a,c) a q$_2$ satellite and (b) a q$_1$ satellite. The temperature of the crystal has decreased slightly between (a) and (c).

compound undergoes a C↔IC transition at ~90 K. The discommensuration process has been shown by Fung et al. [55] to involve the nucleation of "stripples", each composed of three discommensuration walls, within commensurate domains. The motion of these stripples eventually generates an array of moiré-like fringes which are also interpreted in terms of discommensuration walls. The nucleation of the stripples in 2H-TaSe$_2$ was observed to occur at twin boundaries in the commensurate material; similarly, an enhanced nucleation of discommensurations occurs near the pre-existing APBs in the C* phase of TaTe$_4$. In addition, a stripple-like mechanism appears to act to increase the density of discommensuration walls in

TaTe$_4$. The stripple in this case consists of a junction of six discommensuration walls, the phase shift associated with each being $\pi/3$. Therefore it appears that the generation of a discommensurate state via the action of stripples is a characteristic feature of CDW compounds.

A microscopic model of a general commensurate $\rightarrow$ incommensurate phase transition has been developed by Parlinski and coworkers [74-76] using a molecular dynamics approach. Notably, the model predicts that thermal flucuations within the commensurate phase lead to the formation of a stripple and that the barrier for stripple nucleation is lowered near pre-existing stripples, creating a cascade effect. The process continues until the density of discommensurations walls in the system reaches an equilibrium condition. In general this picture is in good agreement with the SDF observations for TaTe$_4$, with the exception of the presence of pre-existing APBs in the commensurate phase which apparently accelerate the process by providing easy nucleation sites for the discommensurations. According to the molecular dynamics model, the addition of each new stripple changes the modulation $\mathbf{q}$-vector slightly and a temperature dependent incommensurability is expected. This is seen to be the case for TaTe$_4$ [77] but not for NbTe$_4$ where the periodicity of the LT phases is temperature independent [13]. This is presumably related to a "pinning" effect on the CDW due to the inter-column interactions for the LT phases of NbTe$_4$ which is absent in the case of TaTe$_4$, where the CDW phasing adjustments on neighbouring columns have already occurred prior to the $C^* \rightarrow IC$ transition.

In summary, the SDF observations of both TaTe$_4$ and NbTe$_4$ reveal that the transition from a commensurate to an incommensurate, or more specifically discommensurate, state is accomplished via the formation of various precursor phases which are thermodynamically stable within limited temperature ranges. In both cases structural defects associated with the CDW, namely discommensurations and APBs, play a central role in mediating the transitions. The rich diversity of phenomena observed for these nominally isostructural compounds dramatically illustrates the subtle balance of factors which determine the properties of CDW.

3.4 THE SOLID SOLUTION $(Ta_{1-x}Nb_x)Te_4$

In the initial study of mixed crystal series $(Ta_{1-x}Nb_x)Te_4$ with $0 \leq x \leq 1$, Boswell and Prodan [50] demonstrated the existence of a complete range of solid solutions and a variation in the CDW $\mathbf{q}$-vector as a function of composition. It was subsequently found using electron [51] and x-ray diffraction [77] that the periodicity remains commensurate until a threshold Nb concentration is exceeded and then jumps discontinuously to an incommensurate value.

Further jumps occur at progressively higher Nb contents [51] providing the first indication of a "Devil's staircase" [78,79] in this type of system. The mixed crystals are also observed to undergo CDW-driven phase transitions at various temperatures with the nature of the phase transition being dependent on the stoichiometry [51,77]. The contrasting behaviour of the various members of this series provides fascinating new insights into the nature of CDW in the tetrachalcogenides.

3.4.1   The Room Temperature Modulated Structures

The satellite positions in electron diffraction patterns reveal that a commensurate to incommensurate phase transition occurs in the $(Ta_{1-x}Nb_x)Te_4$ system for $x \geq \sim 0.3$ (Figure 26). Below the critical Nb concentration, the CDW modulation structure is identical to that of the room temperature phase of $TaTe_4$. The C→IC transition is marked by an abrupt change in the c-component of the q_1 wavevector (retaining for convenience the notation used for $TaTe_4$) while the q_2/q_3 satellites remain commensurate [51]. However, these latter satellites are broadened and become progressively more diffuse as the Nb concentration increases. The electron diffraction observations are consistent with the development of a new incommensurate modulation which retains the $(2a \times 2a)$ basal plane of the commensurate phase and is therefore analogous to the LT phases of $NbTe_4$.

Following the abrupt transition at $x \sim 0.3$, the incommensurability of the mixed crystals approaches that of $NbTe_4$ (Figure 27). The variation appears to be discontinuous, proceeding through a sequence of discrete jumps as the Nb content exceeds certain threshold values. Even for pure $NbTe_4$ these plateaus are not very sharply defined, presumably due to non-stoichiometry. The presence of these plateaux has been attributed to the existence of a series of long-period commensurate phases in the $(Ta_{1-x}Nb_x)Te_4$ system [51].

For Nb concentrations up to $x \sim 0.3$, SDF images reveal the presence of a fairly low density of APBs similar to those observed in $TaTe_4$. However, in the mixed crystals these APBs, predominately of *type-II*, occur even in crystals slow-cooled from the growth temperature and are probably associated with pinning of the CDW phases along the various columns by the Nb "impurities". These impurity pinning effects are characteristic of CDW systems and have been extensively studied in several di- and trichalcogenides [10,80], For Nb concentrations slightly above the commensurate threshold, all of the satellites retain sufficient intensity for SDF imaging. SDF images formed with q_2/q_3 satellites typically reveal a dense network of APBs (Figure 28a) while in the corresponding SDF images formed with q_1 satellites a patchwork of ill-defined, irregular domains are observed (Figure 28b). The appearance of the domains is a

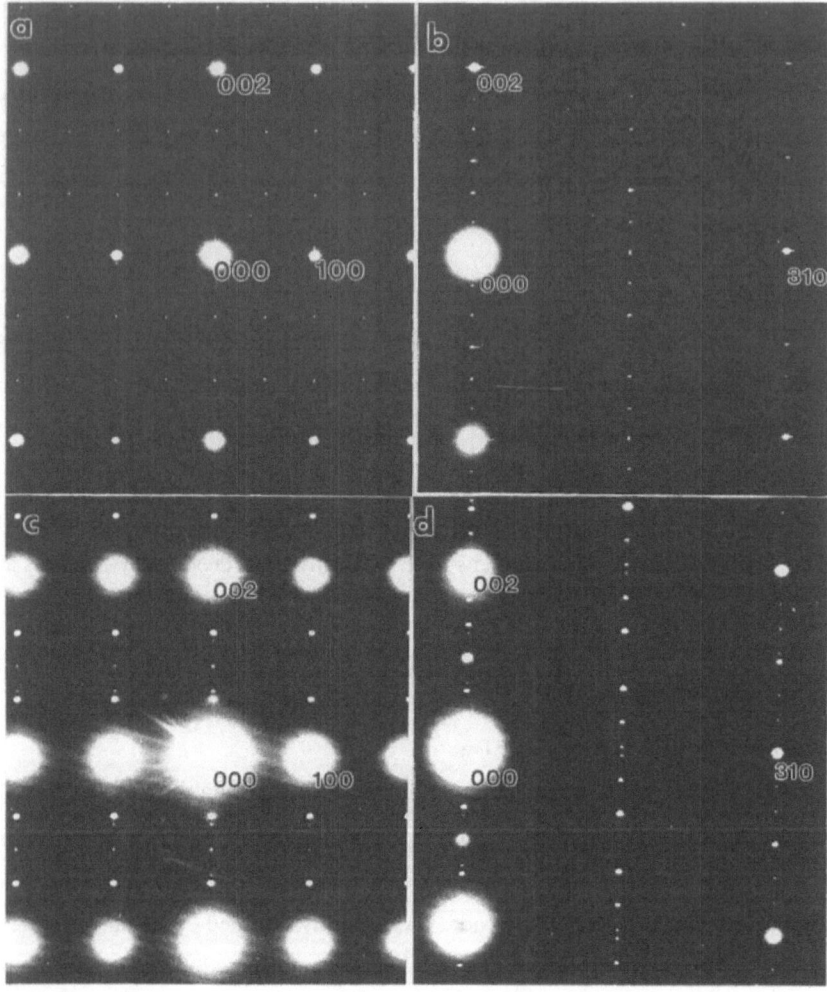

Fig. 26: EDPs from $(Ta_{1-x}Nb_x)Te_4$ at room temperature. (a) [010] z.a. and (b) [130] z.a. for x ~0.3; (c) [010] z.a. and (c) [130] z.a. for ~0.7.

sensitive function of the specimen tilt, suggesting lattice strains associated with the APBs effect the observed diffraction contrast. Superimposed on the domain contrast features, arrays of closely spaced fringes may also be observed (Figure 29). These fringes resemble those observed in the SDF images of the incommensurate phases of $NbTe_4$ and $TaTe_4$ described previously and again have been attributed to the formation of an array of discommensuration walls. The average spacing of the fringes varies from ~250 Å for compositions occupying the first plateau in Figure 27 to ~50 Å for Nb contents near x = 1. Unlike the very regular fringes observed for $NbTe_4$ and $TaTe_4$, the fringes in the mixed crystals exhibit significant variations in orientation and spacing, an effect again attributed to impurity pinning of the CDW. The

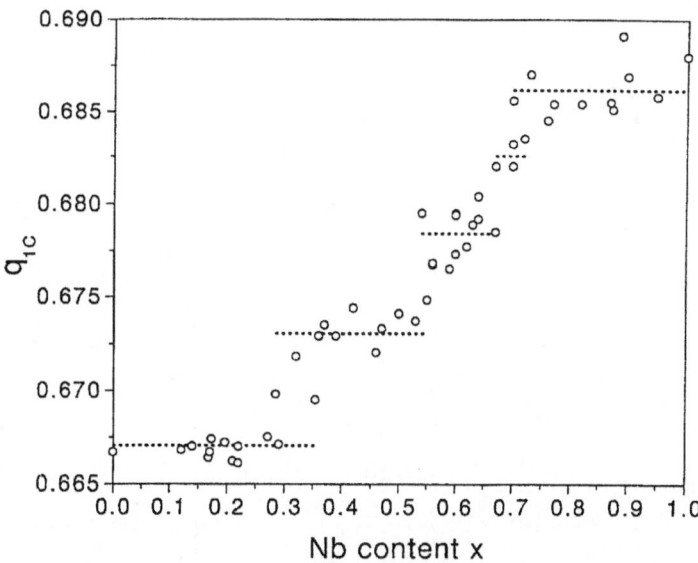

Fig. 27: Variation of the c*-component of the modulation **q**-vector with Nb content in $(Ta_{1-x}Nb_x)Te_4$. Open circles represent experimental measurements with compositions as determined from in situ EDS microsanalysis.

long-range coherence among CDW on adjacent columns rather than a direct effect of disorder in the discommensuration arrays.

3.4.2   CDW-Driven Phase Transitions

A variety of CDW-driven phase transitions have been observed to occur at elevated temperatures in the mixed crystals [51]. Crystals containing relatively small amounts of Nb up to x ~0.2 behave identically to $TaTe_4$, undergoing a transition from the C to C* phase at ~450 K. For the commensurate crystals containing near, but slightly less than, the critical Nb concentration of x ~0.3, two distinct phase transitions are observed: a C↔IC transition at ~350 K followed by an IC↔IC* transition at ~450 K. During the C↔IC transition, the modulation retains $(2a \times 2a)$ basal dimensions but the value of q_{1c} changes discontinuously to that of the first plateau in Figure 27. It would appear that smaller Nb concentrations are needed to stabilize the incommensurate, LT-like phase at elevated temperatures.

SDF images obtained during the C↔IC phase transition are shown in Figure 29. In the C state, a small number of APBs are typically observed in images formed using a q_2/q_3 satellite (Figure 29a). On heating into the IC state, a dense network of APBs forms throughout the crystal (Figure 29b). This process is usually completed within a few seconds and occurs over a

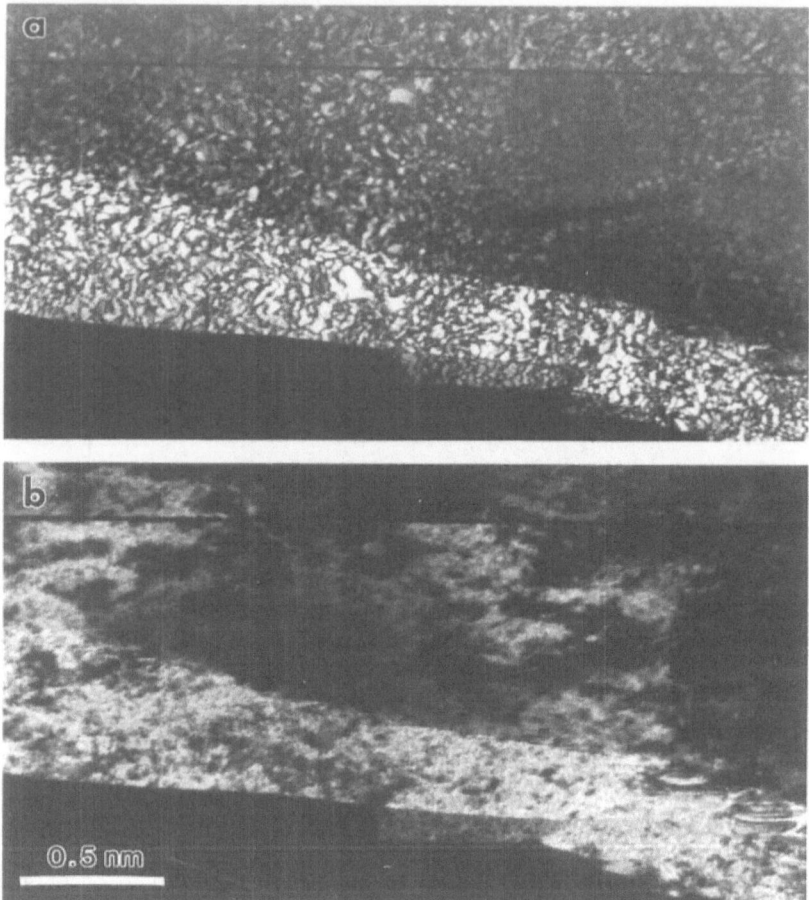

Fig. 28: SDF images of the same region of a $(Ta_{1-x}Nb_x)Te_4$, x ~0.3 crystal formed using a (a) q_2
satellite and (b) a q_1 satellite.

very narrow temperature range of about 5-10 K. The structural transformation appears to be
mediated by the nucleation of stripple-like defects composed of a junction of six APBs or
discommensurations. Once the transformation is complete, the SDF images are indistinguishable from those of crystals containing slightly higher Nb content that are incommensurate at
room temperature. The C↔IC transition is reversible, however there is again considerable
hysteresis between the warming and cooling cycles. On cooling, the IC state persists at room
temperature for a period of several hours, with the CDW defects being gradually eliminated by
diffusion to the crystal surfaces. On further heating these crystals, an abrupt reversible
IC↔IC* transition takes place at ~450 K. The disappearance of the q_2/q_3 satellites indicates a
transformation of the basal plane dimensions to $\left(\sqrt{2}a \times \sqrt{2}a\right)$. SDF images show that the

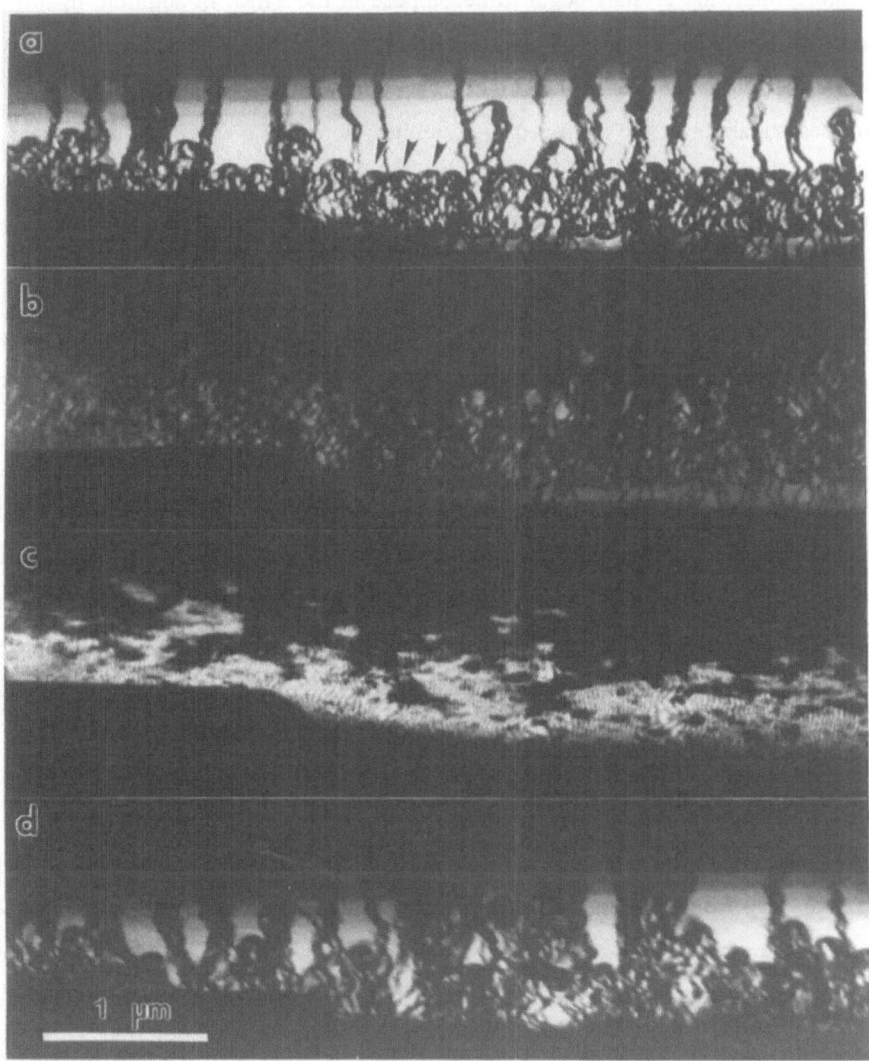

Fig. 29: SDF images of a $(Ta_{1-x}Nb_x)Te_4$, x ~0.2 crystal (a) slightly above room temperature (C phase), (b) (c) at ~350 K (LT phase), (d) after to cooling to room temperature for several hours. (a), (b), (d) are formed with a q_2 satellite, (c) is formed with a q_1 satellite.

IC$\leftrightarrow$IC* transition is initiated at the defects and is first-order while x-ray diffraction experiments show the degree of incommensurability is temperature dependent in the IC* state [77].

For the mixed crystals which are incommensurate at room temperature, an IC$\leftrightarrow$IC* phase transition also occurs at ~450 K. SDF images show the fringes associated with the discommensuration arrays are more uniform at higher temperatures, however the incommensurability appears to be temperature independent. The CDW modulations remain present at all higher temperatures until the crystals begin to decompose in the TEM. In addition, in-situ cooling experiments reveal that none of the mixed crystals that are incommensurate at room temperature undergo a lock-in transition. For the crystals having Nb contents $x > \sim 0.5$, pronounced streaking of the diffuse q_2/q_3 satellites along the c^*-direction does occur but no satellites analogous to those of the LT_1/LT_2 phases of $NbTe_4$ are resolvable. The presence of impurity atoms appears to inhibit the lock-in process, presumably due to phase pinning effects.

3.4.3   CDW Phase Relationships

The existence of a stepwise variation of the CDW modulation periodicity with composition for $(Ta_{1-x}Nb_x)Te_4$ system suggests the existence of a series of independent phases over the range of solid solubility. The structures of these phases have been modeled using a similar approach as that employed for the LT_1/LT_2 phases of $NbTe_4$ [58]. The analysis reveals that the various structures corresponding to the discrete plateaux in Figure 27 may be derived from a distortion of a common basic structure with the distortions being of the same type as seen for the LT phases. Thus the various mixed crystal phases occurring at room temperature can now be seen to form a series of closely related structures, which approximate to long-period commensurate, for which the LT phases of $NbTe_4$ constitute one end member. This model considerably clarifies the interpretation of the CDW phase relations in this "simple" system.

As discussed above, the LT_1/LT_2 distortions in $NbTe_4$ act to mediate the transition to the lock-in state. Thus, the presence of similar distortions in the mixed crystals implies a kind of precursor nature for these phases as well, although lock-in does not actually occur. Based on the available electron and x-ray diffraction data, a preliminary phase diagram can be constructed for the $(Ta_{1-x}Nb_x)Te_4$ system (Figure 30). At least four distinct modulated structures exist: a commensurate (C) phase having a $(2a \times 2a \times 3c)$ supercell, a commensurate (C*) phase with a $(\sqrt{2}a \times \sqrt{2}a \times 3c)$ supercell, an incommensurate (LT) phase with a $(2a \times 2a \times 1/q_{1c}c)$ unit cell, and an incommensurate (IC) phase with a $(\sqrt{2}a \times \sqrt{2}a \times 1/q_{1c}c)$ unit cell where q_{1c} varies from $\sim 0.660c^*$ for $TaTe_4$ to $0.691c^*$ for $NbTe_4$. A normal phase of $NbTe_4$, which is undistorted by a CDW, has also been identified at elevated temperature by Böhm [81]. The experimental phase diagram agrees remarkably well with the phase diagram that is derived from Landau free energy calculations by Walker and co-workers [32].

Several factors act to stabilize the various CDW phases occurring in the $(Ta_{1-x}Nb_x)Te_4$ system. For the C and C* phases, lock-in energy favours the formation of metal atom triplets while the intercolumn interactions dictate a phase difference for CDW on adjacent columns of

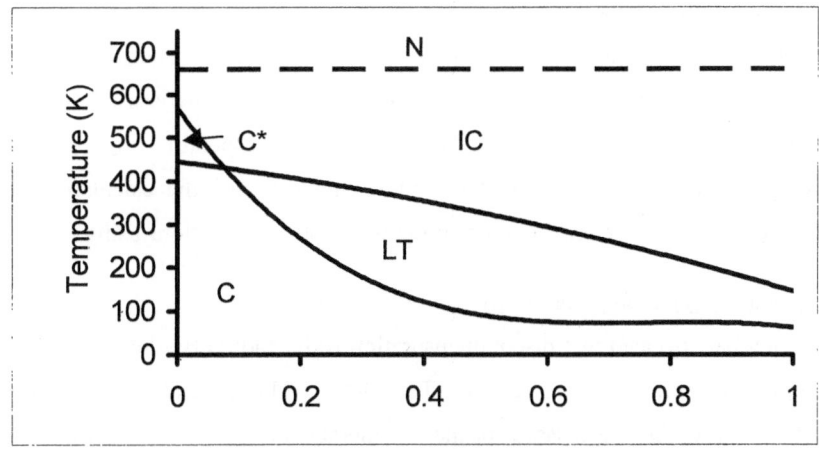

Fig. 30: Phase diagram of the $(Ta_{1-x}Nb_x)Te_4$ system. The following symbols are used: N = normal (unmodulated) phase, C = $2a \times 2a \times 3c$ commensurate phase, C* = $\sqrt{2}a \times \sqrt{2}a \times 3c$ commensurate phase, LT = $2a \times 2a \times \frac{1}{q_{1c}}c$ incommensurate phase, IC = $\sqrt{2}a \times \sqrt{2}a \times \frac{1}{q_{1c}}c$ incommensurate phase [after 77].

$2\pi/3$ and π, respectively. For the IC phase, electronic factors dominate and the largest possible phase shift of π is found between the columns. The LT phases incorporate aspects of both the C and IC structures. As for the IC phase, nearest neighbour columns are essentially antiphased but with an additional small phase shift ϕ associated with the LT_1 distortion. The small longitudinal displacements of the metal atoms, which effectively generate more triplets, result in a relative phasing of $(\pi - \phi)$ for one set of nearest neighbour columns and $(\pi + \phi)$ for the others. The possible values of ϕ are limited and this leads to the stepwise variation in the **q**-vector observed for the LT phases.

Considered as a series of long-period commensurate structures, the various LT phases represent an example of a so-called Devil's staircase. Dzailoshinsky [82] has concluded that for any **q**-vector lying close to some rational value, an infinitesimal energy contribution will cause the system to lock-in to that value. Therefore any variations resulting from changes in an external parameter, such as composition or temperature, should proceed via the formation of a infinity of long-period commensurate phases. Under certain conditions, the staircase may be "harmless", consisting of a finite series of steps [83]. Although certain phenomena in the electrical properties of o-TaS_3 have previously been interpreted in terms of a Devil's staircase [84], the stepwise variation in the c*-component of the modulation wavevector q_{1c} observed for the LT phases represents the first direct evidence for this type of behaviour in a CDW system. Pinning to a long-period commensurate structure would account for the lack of temperature dependence in the value of q_{1c} for the LT phases. The possibility of Devil's staircase behaviour is also relevant to the Landau theoretical analysis of the $(Ta_{1-x}Nb_x)Te_4$ system. Since the contributions responsible for lock-in correspond to high order terms in the free energy expansions, the observed behaviour suggests that the calculations need to be carried to high order for a complete description of the phase transitions using Landau theory.

3.4.4   Discommensurations and APBs

SDF imaging has revealed that discommensuration walls and APBs show extraordinarily rich and complex dynamics in the $(Ta_{1-x}Nb_x)Te_4$ system. The general characteristics of these defects have already been discussed for the end members of the series and would be expected to be similar for the mixed crystals. The fundamental difference between the defects lies primarily in the abruptness of the associated phase slip: gradual for discommensurations while for an APB the phase slip is by definition restricted to the immediate vicinity of the interface. For some compositions, both types of defects may be present simultaneously. In these cases, in analogy with the commensurate situation, the APBs are assumed to separate identical LT phase domains containing an appropriate density of discommensurations.

The C→LT transition in the mixed crystals is accompanied by a rapid increase in the number of APBs. According to the long-period commensurate models of the LT structures, the relative phasing of the CDW on nearest neighbour columns changes from $2\pi/3$ to $\pi \pm \phi$ as a result of the transition. Generation of large numbers of APBs implies that the atomic shifts needed to adjust the CDW phases cannot be easily accomplished, probably due to pinning by impurities. It also implies that the APBs are relatively low energy defects. This also supports the long-period commensurate interpretation of the LT structures since the energy costs of stacking disorder between columns with CDW phases $\pi + \phi$ or $\pi - \phi$ should be small.

The C→LT transition occurs very rapidly in the mixed crystals while the reverse process is sluggish, requiring several hours for completion at room temperature. Similar hysteresis effects are characteristic of both the C↔LT transition in $NbTe_4$ and the C^*↔IC transition in $TaTe_4$. The sluggishness of the lock-in transitions indicates that the elimination of discommensurations is a slow process independent of the temperature at which it occurs. It is interesting to note that both the subsequent LT↔IC transition in the mixed crystals and the C↔C^* transition in $TaTe_4$ occur rapidly, without any significant temperature hysteresis. Although substantial rearrangements of the CDW phasings along the columns are necessary for these transitions, neither involve the formation or elimination of discommensurations.

3.5 THE EFFECTS OF NON-ISOELECTRONIC SUBSTITUTIONS

It is evident that a complex interplay of factors controls the behaviour of CDW in the tetra-chalcogenides. Transition metals substitutions provide one avenue to explore these factors, particular those which are electronic in origin. As discussed above, isoelectronic substitutions in the $(Ta_{1-x}Nb_x)Te_4$ system have profound effects on the stability and periodicity of the various CDW phases. Non-isoelectronic substitutions in the $(A_{1-x}B_x)Te_4$ system with A = Nb,Ta and B = Ti,Zr,V have similar effects, generating changes in the modulation structures accompanied by the formation of characteristic lattice defects.

3.5.1 Modulated Structures and Defects

The substitution of non-isoelectronic transition metals into $TaTe_4$ and $NbTe_4$ produces changes in the modulation q-vector, as revealed by the satellite positions in the corresponding electron diffraction patterns (Figure 31). For the case of $(Ta_{1-x}B_x)Te_4$ with B = Ti or Zr, a commensurate to incommensurate (C→LT) transition occurs at a dopant concentration of ~5% with the c^*-component of the CDW modulation changing discontinuously from $2/3c^*$ to ~$0.660c^*$ [85]. The magnitude of the step in q_{1c}, i.e. ~$0.006c^*$, is approximately equal to that occurring in the $(Ta_{1-x}Nb_x)Te_4$ system but of opposite sign. No further changes occur in q_{1c} as the dopant concentration is increased to the estimated solubility limit of ~10 %, however the q_2/q_3 satellites grow progressively more diffuse while the q_1 satellites remain sharp. The behaviour is thus similar to that observed for the $(Ta_{1-x}Nb_x)Te_4$ system.

A decrease in the magnitude of q_{1c} has also been observed for the series $(Nb_{1-x}B_x)Te_4$ with B = Ti or Zr. In these crystals, q_{1c} varies strongly from values typical of pure $NbTe_4$, i.e. ~$0.691c^*$, to values less than that of pure $TaTe_4$, i.e. $2/3c^*$. *In situ* EDS microanalyses indicate that the solubility of Ti and Zr in $NbTe_4$ is limited to ~10 %, however due to uncertainties in stoichiometry it is not known if the q-vector varies in a stepwise fashion for these crystals. The intensity of satellites corresponding to second and higher harmonics of the q_1 wavevector is rapidly diminished by the dopant additions, implying the discommensurate character of the modulation is strongly suppressed. Surprisingly, substitutions of small amounts (< 1 %) of vanadium in $NbTe_4$ increase the value of q_{1c} to ~$0.697c^*$. Attempts to substitute V in $TaTe_4$ have not been successful, possibly due to the large size difference of these atoms.

SDF images of the $(A_{1-x}B_x)Te_4$ mixed crystals also reveal characteristic defect structures in

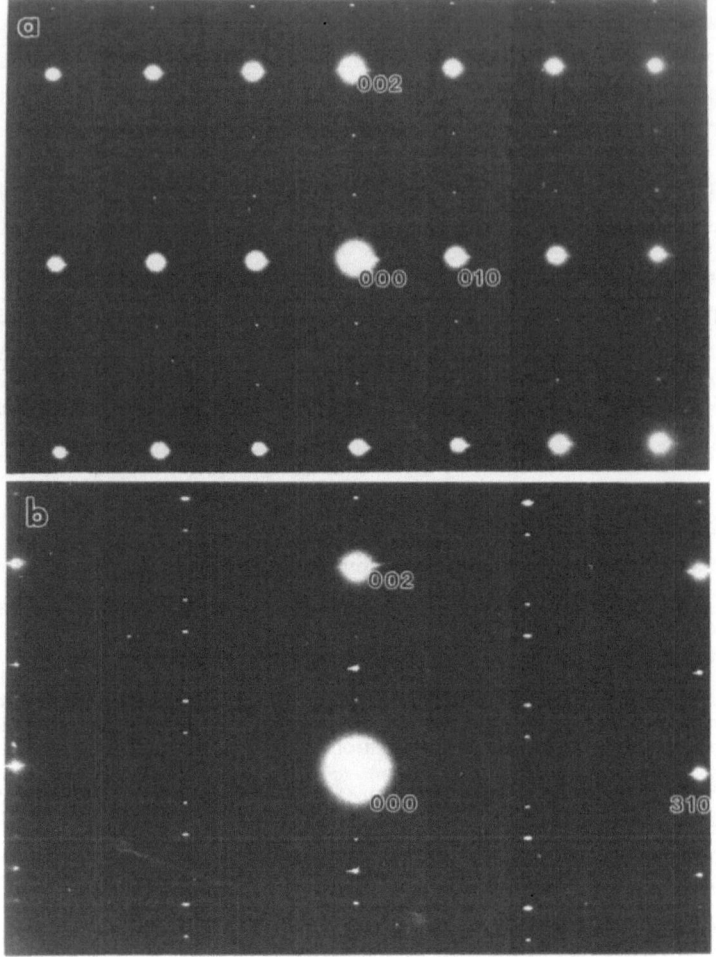

Fig. 31: EDPs from $(Ta_{1-x}Ti_x)Te_4$, x ~0.1. (a) [100] z.a. and [130] z.a. patterns.

the CDW modulation structure. For A = Ta and B = Ti or Zr, the SDF images show that doped, but still commensurate (i.e. x < ~0.05), crystals contain APBs in rough proportion to the level of substitution (Figure 32a). This behaviour is similar to that of the $(Ta_{1-x}Nb_x)Te_4$ system however the dopant concentration required to induce comparable densities of defects is much smaller for Ti or Zr doping. This indicates a much stronger interaction between the CDW and impurity potential in the case of these non-isoelectronic dopants. For dopant concentrations x > ~0.05, the incommensurate crystals typically contain a patchwork of irregular domains resembling those present in the LT phases of the $(Ta_{1-x}Nb_x)Te_4$ system (Figure 32b). Similar domains are observed in the $(Nb_{1-x}B_x)Te_4$ mixed crystals and appear to be characteristic of the incommensurate CDW. Interestingly, the fringe contrast previously super-

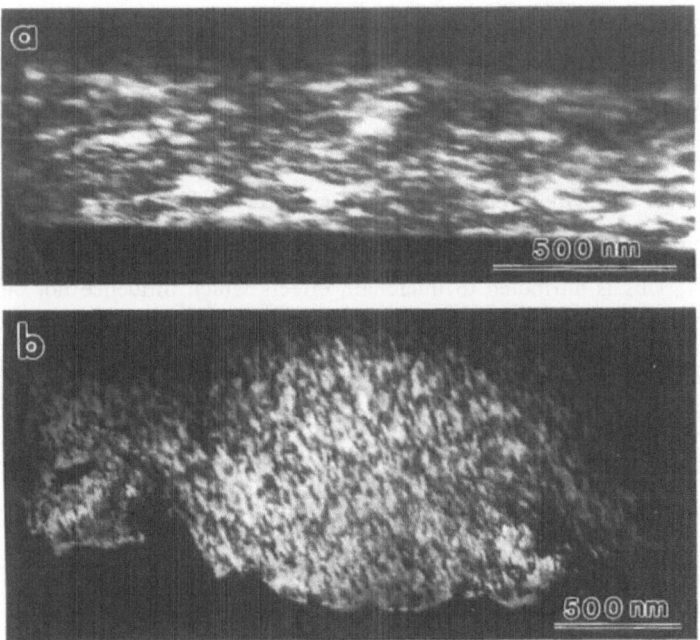

Fig. 32: SDF images of $(A_{1-x}B_x)Te_4$ formed using a q_1 satellite for (a) A = Ta, B = Ti, x = ~0.1 and (b) A = Nb, B = Ti, x ~0.1.

imposed on the domain contrast in SDF images from $(Ta_{1-x}Nb_x)Te_4$ and associated with discommensuration arrays, has not been observed in SDF images of the Ti, Zr or V doped compounds. In any case, the SDF images reveal that the non-isoelectronic substitutions introduce considerable disorder into the CDW modulation structure.

In summary, the SDF observations indicate that both isoelectronic and non-isoelectronic dopants act as pinning centers in the tetrachalcogenides. In the commensurate phases this is reflected by the presence of APBs. As expected lower concentrations of the non-isoelectronic dopants Ti and Zr, which interact more strongly with the CDW, are required to induce an equivalent density of APBs relative to the concentration of the isoelectronic dopant Nb. A similar condition applies for the dopant concentrations required to induce a C→LT transition. Impurities favour an incommensurate state since the CDW in this case is not pinned to the lattice by commensurability effects and can reduce its overall energy by adjusting its phase in response to randomly distributed impurity potentials. The behaviour of the tetrachalcogenides in this respect is similar to that observed for many other CDW systems [8].

3.5.2 Interchain vs. Intrachain Effects

In general, cationic substitutions produce two distinct effects: the introduction of randomness into the lattice potential and/or a change in the average conduction electron density ρ. For the case of an isoelectronic substitution, the **q**-vector of the CDW would be expected to remain constant independent of dopant concentration while for non-isoelectronic substitutions a variation in ρ, and hence **q**, should occur. This type of behaviour is in fact observed for transition metal substitutions in many CDW systems, for example 1T-TaS$_2$ [86]. As we have seen, the response of the CDW modulations in the tetrachalcogenides to transition metal substitutions is more complex. This is attributed to interchain effects which influence not only the relative phasing of CDW but also the modulation periodicity directly.

The calculation of the electronic band structure of the tetrachalcogenides by Whangbo and Gressier [52] provides a framework in which to interpret the experimental results. In the subcell, the shortest interchain Te bonds are much shorter than the bonds forming the Te squares and therefore each metal atom may be considered to be surrounded by eight $\left(Te_2^{2-}\right)$ dimers. Calculations based on this assumption show that the $p_{\pi*}$ orbitals of the dimers interact with all of the d orbitals of the metal M^{4+} ions with the exception of the d_{z^2} orbital. One electron is available per metal ion leading to a ½-filling of the d_{z^2} band. In this case, the CDW would be expected to generate metal atom pairing along the chains however triplets are in fact observed, indicating a $\frac{1}{3}$-filling for the conduction band. As noted by Whangbo [65], the reduced occupancy of the d_{z^2} band can be accounted for if the oxidation state of some of the interchain bonding units is $\left(Te^{2-}\right)_2$, rather than $\left(Te_2^{2-}\right)$, as a result of an increase in the bond length. This dissociation of the Te dimers draws electrons from the d_{z^2} band and thereby decreases the **q**-vector.

In the $(Ta_{1-x}Nb_x)Te_4$ system, the available structural data [47-49] indicates that the average interchain Te bonding distance decreases by ~0.3% as x increases from 0 to 1 while the magnitude of q_{1c} increases from $2/3c^*$ to $\sim0.690c^*$. This is consistent with the mechanism proposed by Whango and implies a higher fraction of the interchain Te bonds are dissociated in the case of TaTe$_4$. The perturbation of the interchain bonding is likely accentuated in the mixed crystals by CDW pinning effects associated with the dopant ions. To maintain nearly constant interchain Te bond lengths, a coupling of the CDW distortions on adjacent columns is required. This coupling would be disrupted by impurity pinning of the CDW phases at random sites along the columns, increasing the likelihood of dissociation.

The above arguments indicate dopant substitutions may effect the modulation periodicity directly, by changing the number of electrons contributed by the metal atoms in the case of non-isoelectronic dopants, or indirectly, by causing dissociation of the interchain Te dimers.

The observed stepwise variations in the **q**-vector with changing dopant concentrations indicate that commensurability effects also play an important role in determining the modulation periodicity. These factors are in fact interrelated since it is the necessity of maintaining interchain Te bond lengths as constant as possible that limits the possible phasing relationships for CDW on adjacent chains and results in a discontinuous series of long-period commensurate structures.

4. Concluding Remarks

In a crystalline material, the atoms are by definition arranged on a periodic lattice in three-dimensions. It has become increasingly evident that many "crystalline" materials fall outside the confines of this conventional definition. One example are modulated structures where physical processes with competing periodicities produce a structure in which no true translational symmetry exists along one or more directions. Over the last two decades, the challenge of providing a crystallographic description of modulated structures has stimulated intense theoretical and experimental activity [87-89]. Much of that activity has focussed on the study of one-dimensionally modulated materials, particularly displacively modulated structures arising in connection with charge density waves. Among these, the transition metal tetrachalcogenides provide an excellent case in point. Significant advances in superspace crystallography [47] and Landau theoretical analysis of phase transitions [32] have been achieved using these compounds as a model system.

As a prototypical system, a detailed characterization of the microstructure of $TaTe_4$, $NbTe_4$ and related compounds is of considerable importance. One finds a fascinating and diverse range of phenomena which nonetheless share some common characteristics. Strikingly, it is possible to unify the description of the various CDW phases using an approach whereby these are formally considered to constitute a series of long-period commensurate structures with phase transitions mediated by temperature dependent inter-column interaction strengths. Also arrays of discommensuration walls were found to be a prominent feature of all the incommensurate phases. However, counter to the predictions of the original Landau theory, these discommensurations are never sharply defined and have in all cases an appreciable spatial extension. On the other hand, sharply defined planar defects are commonly observed in these crystals which have been classified as APBs. The differentiation between these two types of defects is somewhat ill-defined, usually being based on the degree of abruptness of the associated phase slip. However, the pronounced difference in the behaviour observed *in situ* in the TEM during CDW-driven phase transitions, particularly in the mixed crystals where both defects may be present simultaneously, indicates a funda-mental distinction should be made. This arises from the fact that APBs are true translational defects while the discommensuration walls represent a projection of one-dimensional defects. A number of the details of the modulated structures, the phase transition processes, the interactions of APBs and discommensurations, and the processes by which these defects evolve remain puzzling. The search for the solutions will continue to make a unique contribution to the evolving definition of a "crystal".

Acknowledgements

The authors would like to thank the following for their kind permission to reproduce previously published figures: Institute of Physics (Figures 8, 10, 11 and 12) and Akademie-Verlag (Figure 16). This work received financial support from the National Science and Engineering Research Council of Canada.

References

1. R.E. Peierls, *Quantum Theory of Solids*, Oxford University Press, (1955).
2. R. Comes, S.M. Shapiro, G. Shirane, A. Garito and A.J. Heeger, *Phys. Rev. Lett.*, **35**, 1518, (1975).
3. *Low-Dimensional Electronic Properties of Molybdenum Bronzes and Oxides*, ed. C. Schlenker, Kluwer, (1989).
4. J.D. Axe, G. Grubel and G.H. Lander, *J. Alloys and Comp.*, **213**, 262 (1994).
5. S.M. Dubiel and S. Cieslak, Phys. Rev. B, 51, 9341 (1995).
6. C. M. Hwang and C.M. Wayman, *Scr. Met.*, **17**, 381 (1983).
7. C.M. Hwang, M. Meichle, M.B. Salamon, and C.M. Wayman, *Phil. Mag. A*, **47**, 31 (1983).
8. J.A. Wilson, F.J. DiSalvo and S. Mahajan, *Adv. Phys.*, **24**, 117 (1975).
9. R.L. Withers and J.A. Wilson, *J. Phys. C:Solid State Phys.*, **19**, 4809 (1986).
10. *Electronic Properties of Inorganic Quasi-One-Dimensional Compounds, Parts I and II*, ed. P. Monceau, D. Reidel, (1985).
11. A. Meerschaut and J. Rouxel, *J. Less Common Mat.*, **39**, 197 (1975).
12. J.L. Hodeau, M. Marezio, C. Roucau, R. Ayroles, A. Meerschaut, J. Rouxel and P. Monceau, *J. Phys. C:Solid State Phys.*, **11**, 4117 (1978).
13. F.W. Boswell, A. Prodan and J.K. Brandon, *J. Phys. C:Solid State Phys.*, **16**, 1067 (1983).
14. K. Suzuki, M. Ichihara, I. Nakada and Y. Ishihara, *Sol. State Comm.*, **52**, 743 (1984).
15. K. Suzuki, M. Ichihara, I. Nakada and Y. Ishihara, *Sol. State Comm.*, **59**, 291 (1986).
16. T. Sekine, Y. Kiuchi, E. Matsuura, K. Uchinokura and R. Yoshizaki, *Phys. Rev. B*, **36**, 3153 (1987).
17. K. Selte and A. Kjekshus, *Acta Chem. Scand.*, **18**, 690 (1964).
18. D.J. Eaglesham, D. Bird, R.L. Withers and J.W. Steeds, *J. Phys. C:Solid State Phys.*, 18, 11 (1985).
19. F.W. Boswell, A. Prodan, J.C. Bennett, J.M. Corbett and L.G. Hiltz, *Phys. Stat. Sol. A*, **102**, 207 (1987).
20. J. Mahy, J. van Landuyt, S. Amelinckx, K.D. Bronsema and S. van Smaalen, *J. Phys. C:Solid State Phys.*, **19**, 5049 (1986).
21. J.C. Bennett, F.W. Boswell, A. Prodan, J.M. Corbett and S. Ritchie, *J. Phys: Condens. Matter*, **3**, 6959 (1991).
22. A.J. Berlinsky, *Rep. Prog. Phys.*, **42**, 1245 (1979).
23. J. Bardeen, L.N. Cooper, and J.R. Schrieffer, *Phys. Rev.*, **108**, 1175 (1957).
24. T. Ekino and J. Akimitsu, *Jpn. J. Appl. Phys. 26 suppl.* **26-3**, 625 (1987).
25. P.M. de Wolff, T. Janssen and A. Janner , *Acta Cryst. A*, **37**, 625 (1981).
26. A. Janner, T. Janssen and P.M. de Wolff, *Acta Cryst. A*, **39**, 658 (1983).
27. A. Yamamoto, *Acta Cryst. A*, **38**, 87 (1982).
28. J.C. Toledano and P. Toledano, *The Landau Theory of Phase Transitions*, World Scientific Pub., (1987).

29. W.L. McMillan, *Phys. Rev. B*, **12**, 1187 (1975).
30. W.L. McMillan, *Phys. Rev. B*, **12**, 1197 (1975).
31. W.L. McMillan, *Phys. Rev. B*, **14**, 1496 (1976).
32. M.B. Walker, in *Nuclear Spectroscopy on Charge Density Wave Systems*, ed. T. Butz and A. Lerf, Kluwer, (1992).
33. M. Tanaka, *Acta Cryst. A*, **50**, 261 (1994).
34. M. Terauchi, M. Takahashi and M. Tanaka, *Acta Cryst. A*, **50**, 566 (1994).
35. A.W. Overhauser, *Phys. Rev. B*, **3**, 3137 (1971).
36. S. Amelinckx and D. van Dyck, in *Electron Diffraction Techniques Vol.2*, ed. J.M. Cowley, Oxford Science Pub., (1993).
37. J. van Landyut, R. De Ridder, R. Gevers and S. Amelinckx, *Mat. Res. Bull.*, **5**, 353 (1970).
38. R. De Ridder, J. van Landyut and S. Amelinckx, *Phys. Stat. Sol. A*, **9**, 551 (1972).
39. J.M. Corbett, L.G. Hiltz, F.W. Boswell, J.C. Bennett and A. Prodan, *Ultramic.*, **26**, 43 (1988).
40. J.W. Steeds, D.M. Bird, D.J. Eaglesham, S. McKernan, R. Vincent and R.L. Withers, *Ultramic.*, **18**, 97 (1985).
41. Y. Koyama and T. Onozuka, *Mat. Trans, JIM*, **31**, 636 (1990).
42. *Computer Simulation of Electron Microscope Diffraction and Images*, ed. W. Krakow and M. O'Keefe, The Minerals, Metals and Materials Society, (1989).
43. E. Bjerkelund and A. Kjekshus, *J. Less Common Met.*, **7**, 231 (1964).
44. J. Mahy, J. Van Landuyt, S. Amelinckx, Y. Uchida, K.D. Bronsema and S. Van Smaalen, *Phys. Rev. Lett.*, **55**, 11 (1985).
45. H. Böhm and H.G. Von Schnering, *Z. Kristallogr.* **162**, 26 (1983).
46. H. Böhm and H.G. Von Schnering, *Z. Kristallogr.* **171**, 41 (1985).
47. S. van Smaalen, K.D. Bronsema and J. Mahy, *Acta Cryst. B*, **42**, 43 (1986).
48. K.D. Bronsema, S. van Smaalen, J. de Boer, G.A. Wiegers and F. Jellinek, *Acta Cryst. B*, **43**, 305 (1987).
49. D. Kucharczyk, A. Budkowski, F.W. Boswell, A. Prodan and V. Marinkovic, *Acta Cryst. B*, 46, 587 (1990).
50. F.W. Boswell and A. Prodan, *Mat. Res. Bull.*, **19**, 93 (1984).
51. J.C. Bennett, S. Ritchie, A. Prodan, F.W. Boswell and J.M. Corbett, *J. Phys: Condens. Matter*, **4**, 2155 (1992).
52. M.H. Whangbo and P. Gressier, *Inorg. Chem.*, **23**, 1228 (1984).
53. J.A. Wilson, in *Physics and Chemistry of Electrons and Ions in Condensed Matter*, ed. J.V. Acrivos, D. Reidel, (1984).
54. S. Tadaki, N. Hino, T. Sambongi, K. Nomura and F. Levy, *Syn. Metals*, **38**, 227 (1990).
55. K.K. Fung, S. McKernan, J.W. Steeds and J.A. Wilson, *J. Phys. C:Sol. State Phys.*, **14**, 5417 (1981).
56. J. Kusz and H. Bohm, *Z. Kristallogr.*, **208**, 187 (1993).
57. F.W. Boswell and A. Prodan, *Phys. Rev. B*, **34**, 2979 (1986).
58. A. Prodan, F.W. Boswell, J.C. Bennett, J.M. Corbett, T. Vidmar, V. Marinkovic and A. Budkowski, *Acta Cryst. B*, **46**, 587 (1990).
59. M.B. Walker, *Can. J. Phys.*, **63**, 46 (1985).
60. R. Morelli, D. Sahu and M.B. Walker, *Phys. Rev. B*, **33**, 4843 (1986).
61. M.B. Walker and R. Morelli, *Phys. Rev. B*, **38**, 4836 (1988).
62. R. Morelli and M.B. Walker, *Phys. Rev. B*, **40**, 7542 (1990).
63. M.B. Walker and R. Morelli, *Phys. Rev. Lett.*, **62**, 1520 (1989).
64. Z.Y. Chen and M.B. Walker, *Phys. Rev. B*, **40**, 8983 (1989).
65. M.H. Whangbo, in *Crystal Chemistry and Properties of Materials with Quasi-One-Dimensional Structures*, D. Reidel, (1986).
66. A. Budkowski, A. Prodan, V. Mainkovic, D. Kucharczyk, I. Uszynski and F.W. Boswell, *Acta Cryst. B*, **45**, 529 (1989).
67. D. Sahu and M.B. Walker, *Phys. Rev. B*, **32**, 1643 (1985).
68. S. Amelinckx and J. van Landyut, in *Diffraction and Imaging Techniques in Materials Science*, ed. S. Amelinckx, R. Gevers and J. van Landyut, North Holland Pub., (1987).
69. M.J. Marcinkowski, in *Electron Microscopy and Strength of Crystals*, ed. G. Thomas, J. Wiley and Sons, (1963).
70. G. Thomas, *Transmission Electron Microscopy of Metals*, J. Wiley and Sons, (1962).
71. Y. Yamada, I. Shibuya and S. Hoshino, *J. Phys. Soc. Jap.*, **18**, 15934 (1963).
72. D. Broddin, G. van Tendeloo and S. Amelinckx, *J. Phys. Cond. Matter*, **2**, 3459 (1990).
73. D.M. Bird, S. McKernan and J.W. Steeds, *J. Phys. C:Sol. State Phys.*, **18**, 499 (1985).
74. K. Parlinski, *Phys. Rev. B*, **35**, 8680 (1987).
75. K. Parlinski, *Phys. Rev. B*, **39**, 12154 (1989).
76. K. Parlinski and F. Denoyer, *Phys. Rev. B*, **41**, 11428 (1990).

77. J. Kusz, H. Böhm and J.C. Bennett, *J. Phys. Cond. Matter*, **7**, 2775 (1995).
78. P. Bak, *Rep. Prog. Phys.*, **45**, 587 (1982).
79. S. Aubry, F. Axel and F. Vallet, *J. Phys. C:Sol. State Phys.*, **18**, 753 (1985).
80. *Charge Density Waves in Solids, Modern Problems in Condensed Matter Sciences*, ed. L.P. Gor'kov and G. Grüner, North Holland Pub., (1989).
81. H. Böhm, *Z. Kristallogr.* **110**, 113 (1987).
82. I.E. Dzailoshinski, *Zh. Eksp. And Teor. Fiz*, **46**, 1420 (1964).
83. J. Villain and M. Gordon, *J. Phys. C*, **13**, 3117 (1980).
84. D.V. Borodin, F. Ya Nad, S.Y. Savitskaya and S.V. Zaitsev-Zotov, *Physica B*, **143**, 73 (1986).
85. J.C. Bennett, F.W. Boswell, A. Prodan, J.M. Corbett and S. Ritchie, *Aust. J. Chem.*, **45**, 1363 (1374).
86. F.J. DiSalvo, J.A. Wilson, B.G. Bagley and J.V. Waszczak, *Phys. Rev. B*, **12**, 2220 (1975).
87. *Methods of Structural Analysis of Modulated Structures and Quasicrystals*, eds. J.M. Pérez-Mato, F.J. Zúñiga and G. Madariaga, World Scientific Pub (1991).
88. *Quasicrystals and Incommensurate Structures in Condensed Matter*, eds. J.M. Yacaman, D. Romeu, V. Castano, World Sci. Pub., (1990).
89. S. van Smaalen, *Cryst. Rev.*, **4**, 79 (1995).

TRANSMISSION ELECTRON MICROSCOPY OF CDW-MODULATED TRANSITION METAL CHALCOGENIDES

JAMES M. CORBETT

Guelph-Waterloo Program for Graduate Work In Physics
Waterloo Campus, Waterloo, Ontario, Canada N2L 3G1

1. Introduction

The transmission electron microscope (TEM) has proved to be an extremely useful tool for investigating the periodic structural distortions associated with CDW in low-dimensional materials. With this instrument, numerous studies of the modulated phases and transitions between phases have been carried out on the family of two- and three-dimensionally modulated transition metal chalcogenides. The principal advantage of the TEM for this work is that both imaging and diffraction may be performed on the same area of specimen. Because the atomic scattering factor for electrons is the order of a thousand times greater than that for x-rays, electron diffraction patterns may be recorded from very small volumes, having cross-sections as small as a few nanometers in diameter. This is particularly useful for elucidating orientation variants in specimens in which the CDW form a microdomain structure. Diffraction patterns may be recorded with essentially parallel illumination (selected area diffraction) or with finely focused illumination (convergent beam diffraction). All diffraction patterns display superlattice or satellite reflections associated with the modulations. Diffraction contrast in images formed using these satellite reflections (satellite dark field imaging) may be used to reveal details of the microstructures associated with the modulations, including antiphase domains, domain boundaries and discommensuration walls. In some studies, lattices and superlattices have been imaged directly at very high magnification (high resolution electron microscopy).

Specimen holders are available for heating or cooling samples in the electron microscope, permitting modulated phases and phase transitions to be studied over a range of approximately 15 to 1500 K.

The various diffraction and imaging techniques noted above will be discussed in detail, with selected examples, in subsequent sections of this chapter.

The TEM has already been employed to study a broad range of materials with compositions mainly of the form MX_2 [1-31], MX_3 [32-39], MX_4 [40-55], M_2X_3 [56] and M_3X_4 [57,58], where M represents a transition metal, pure or partly substituted with a second metal, and X

121

F. W. Boswell and J.C. Bennett (eds.),
Advances in the Crystallographic and Microstructural Analysis of Charge Density Wave Modulated Crystals, 121–151.
© 1999 *Kluwer Academic Publishers.*

represents a chalcogen (S, Se, Te). Within this large group of materials, two systems have been studied in considerable detail, viz. 2H-TaSe$_2$ [14-26] and tetragonal Nb$_x$Ta$_{1-x}$Te$_4$, with $0 \leq x \leq 1$ [42-55]. These particular systems display an intriguing variety of modulated phases and microstructures. For this reason, the majority of the examples employed in subsequent sections of this chapter have been drawn from the investigations of these two systems.

2. Selected Area Diffraction

In the TEM, the back focal plane of the objective lens contains a diffraction pattern formed by electrons transmitted through the specimen. If a suitable diaphragm is introduced into the objective image plane, the electrons contributing to the diffraction pattern may be limited to those which originate only from a small selected region of the specimen. The selected-area diffraction (SAD) patterns formed in this way may be limited to cross-sections of the specimen as small as a few nanometers in diameter. If the TEM condenser lens system is adjusted for nearly parallel illumination, the SAD pattern recorded from a single crystal consists of a cross-grating of very fine spots, for example as shown in Figure 1. It is important to note that in the TEM, diffraction patterns display several orders of Bragg reflection, owing to a relaxation of the Bragg condition in the very thin specimens employed, and to the relatively large radius of Ewald's sphere for high energy electrons (e.g. 400 nm^{-1} at 200 kV). For this reason, a typical SAD pattern from a single crystal may be considered to be a map of a plane section of the reciprocal lattice of the crystal. The positions of the diffraction spots in the SAD pattern from a CDW-modulated crystal may be adequately described by kinematical diffraction theory. In this approximation, it has been shown [2] that for small distortions, the diffraction amplitude may be written

$$A(g) = \sum_H v_H \, \delta(g-H) - i \sum_H \sum_G v_H W_{H,G} \, \delta[g-(H+G)], \tag{1}$$

where g is a position vector in reciprocal space, v_H the Fourier coefficient of the crystal potential, H a reciprocal lattice vector for the undistorted crystal and $w_{H,G}$ the coefficients describing the periodic structural distortions. The diffracted intensity will therefore have peaks at $g = H$, the points in the basic lattice or subcell and at $g = H + G$, points which define a superlattice of satellite reflections.

It may additionally be concluded that the satellite spots will be very weak compared with subcell spots and that the intensity of the spot at $H + G$ becomes larger for larger values of H Although the diffracted intensities should strictly be treated by dynamical theory, the trend in satellite intensities predicted by the simple kinematical theory is often observed.

The specimen holder in a TEM provides for adjustment of the sample orientation about one or two axes over a range that is typically ±30° and may be as large as ±60°. This allows SAD patterns to be recorded from a sufficiently large number of zone-axis orientations that a complete mapping of the superlattice reflections in three dimensions may be accomplished.

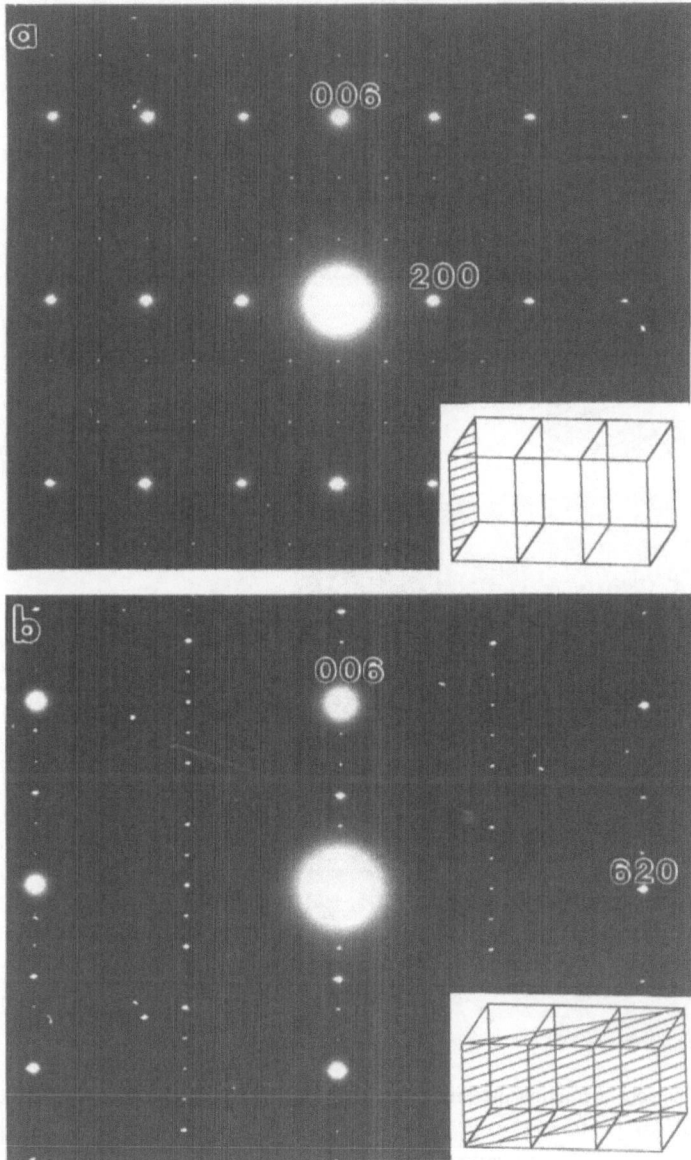

Fig. 1. Electron diffraction patterns from TaTe$_4$ at room temperature. (a) [100], (b) [1$\bar{3}$0], (c) [1$\bar{2}$0] and (d) [1$\bar{1}$0] zone axes. The insets indicate the orientations of the corresponding planes sampled in the reciprocal lattice (e).

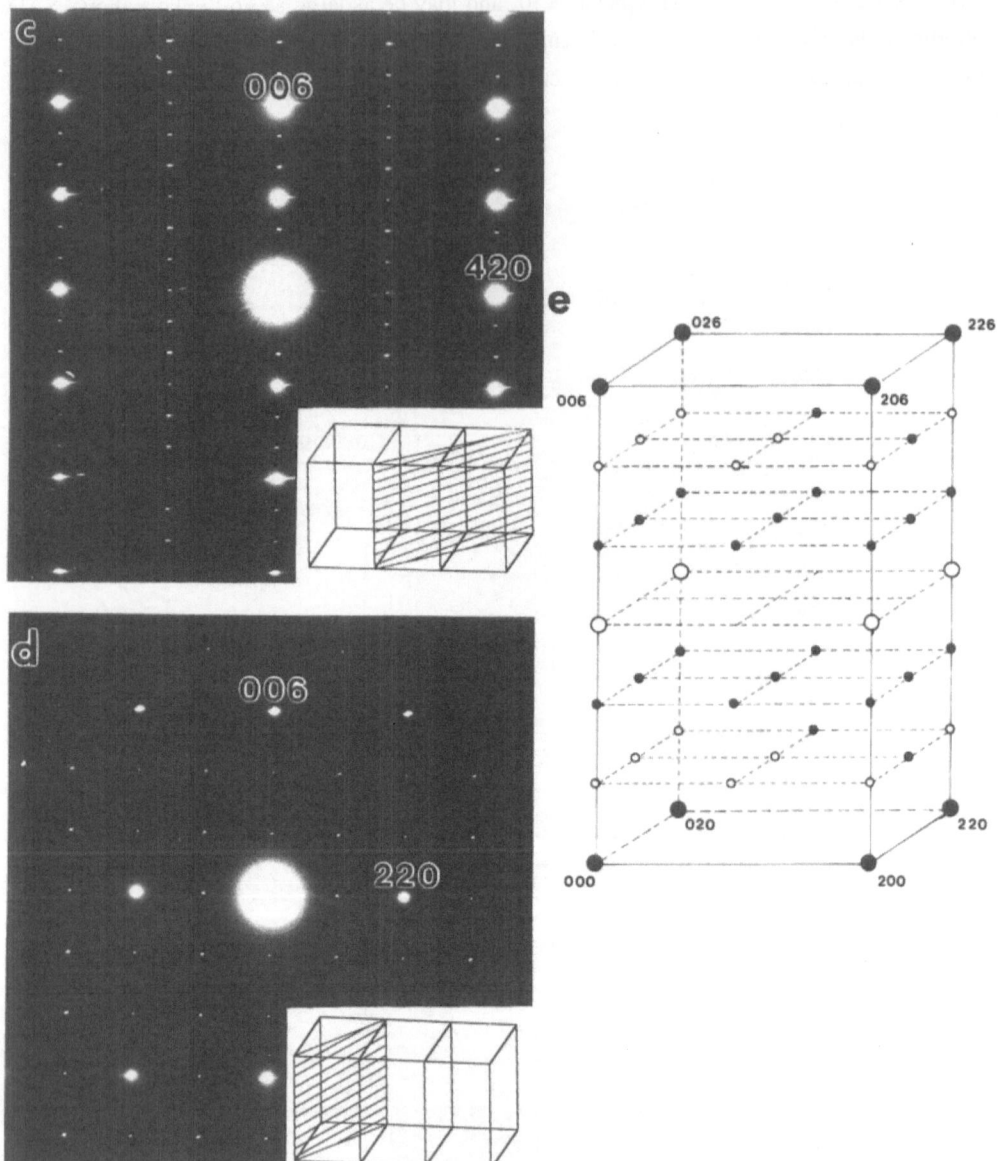

Fig. 1 (con't). Electron diffraction patterns from TaTe$_4$ at room temperature. (a) [010], (b) [1$\bar{3}$0], (c) [1$\bar{2}$0] and (d) [1$\bar{1}$0] zone axes. The insets indicate the orientations of the corresponding planes sampled in the reciprocal lattice (e).

Examples are the room-temperature structures of the tetragonal compounds TaTe$_4$ and NbTe$_4$. Figure 1 shows SAD patterns recorded from TaTe$_4$ with a) [010], b) [$\bar{1}$30], c) [$\bar{1}$20] and d) [$\bar{1}$10] axes parallel to the incident electron beam, together with the fully constructed reciprocal lattice derived from these patterns. Measurements of the spacings of the diffraction spots in a properly exposed pattern allow the satellite reciprocal lattice points to be located with a precision of the order of 0.1%.

TEM specimen holders are available for heating or cooling the specimen in the microscope. Both images and diffraction patterns may be monitored as functions of specimen temperature. In this way, high- and low-temperature modulated phases may be elucidated. Results for the system $Nb_xTa_{1-x}Te_4$ are described in detail in Chapters 1 and 3. For example, in $TaTe_4$ the modulated structure forms a $2a_o \times 2a_o \times 3c_o$ superlattice at room temperature. At about 450 K the superstructure transforms to $\sqrt{2}a_o \times \sqrt{2}a_o \times 3c_o$, as revealed in the [010] SAD patterns shown in Figure 2. In this figure, pattern (a) was recorded at room temperature and pattern (b) at 450 K. Note that columns of superlattice (SL) spots observable between the columns containing both subcell and SL spots in pattern (a) are absent in pattern (b). A further change at about 550 K is revealed in Figure 3. This is a $[\bar{3}10]$ SAD pattern in which spots in the SL columns between the columns containing subcell spots have become split in the c^*-direction, indicating that the SL is no longer commensurate (C) with a period of $3c_o$, but rather is incommensurate (IC) with respect to the subcell.

Diffraction patterns from CDW compounds often display precursor effects associated with an IC to C transition. A notable example of this behaviour is the transition in $NbTe_4$. The

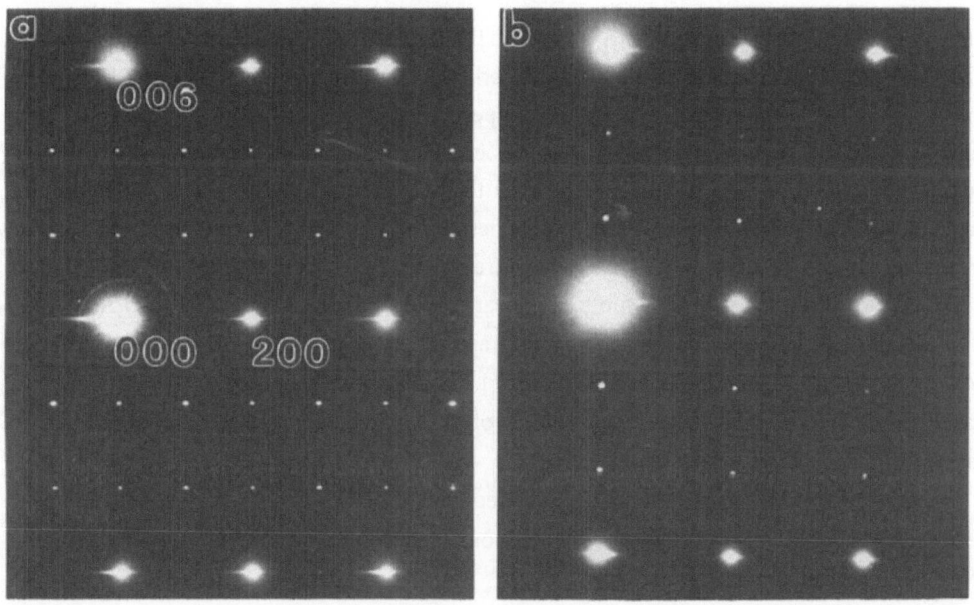

Fig. 2. Electron diffraction patterns from $TaTe_4$ in [010] zone-axis orientation (a) at room temperature and (b) at 450 K. Note that the columns of SL spots indicated by arrows in (a) are absent in (b).

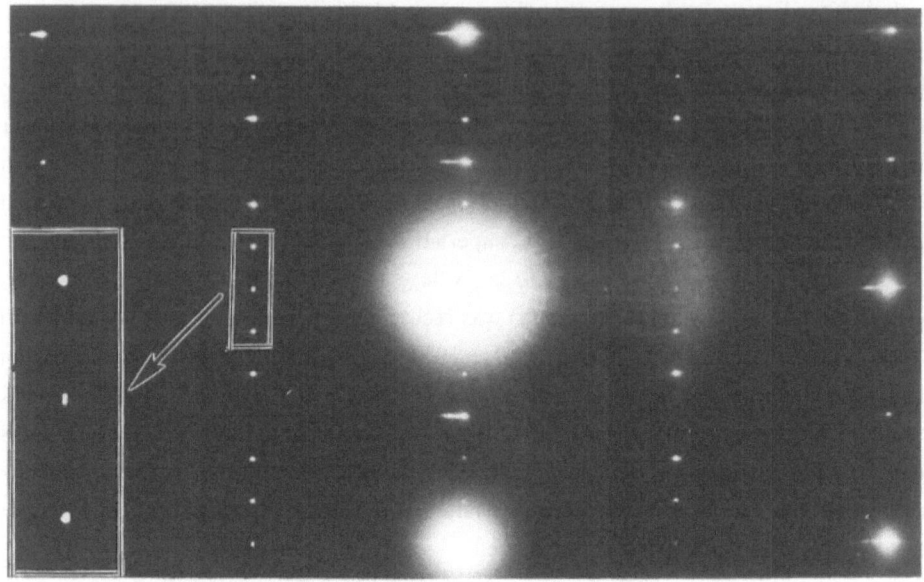

Fig. 3. Electron diffraction pattern from TaTe$_4$ in [$\bar{3}$10] zone-axis orientation at 550 K. Note the splitting in the c-direction of spots indicated by arrows.

subcell of NbTe$_4$ is isostructural with that of TaTe$_4$, but at room temperature NbTe$_4$ displays an IC $\sqrt{2}a_o$ x $\sqrt{2}a_o$ x 3c$_o$ modulation with superlattice spots located at q_2 = (0,0,2/3-2δ) and q_3 = (0,0,2/3+4δ). In addition, the room temperature patterns display very faint spherical regions of diffuse intensity at (1/2,0,2/3). On cooling below about 200 K, these spherical regions are replaced by sharp streaks parallel to the c*-direction. With further cooling to the range 100-50 K, two sets of well-defined spots designated LT$_1$ and LT$_2$ may develop along the streaks, as shown in Figure 4. The LT$_1$ spots are located at (1/2,0,2/3-2δ), (1/2,0,2/3+δ) and (1/2,0,2/3+4δ), etc. and the LT$_2$ spots are located at (1/2,0,2/3-δ/2), (1/2,0,2/3+5δ/2), etc. Both the LT$_1$ and LT$_2$ sets have spacings of 3δ and the spots in the LT$_2$ lie mid-way between the LT$_1$ spots, but no spot in either set falls on (1/2,0,2/3), corresponding to the lock-in which develops below 50K. Detailed explanations of these precursor effects may be found in Chapter 1.

Identification of the superstructures associated with CDW by investigating the locations of satellite spots in SAD patterns is an essential component of every TEM study conducted on CDW-modulated materials.

Fig. 4. Electron diffraction pattern from $NbTe_4$ in [010] zone-axis orientation at 77 K. The LT_1 and LT_2 spots are indicated by arrows.

3. Convergent Beam Diffraction

An alternative means of forming a diffraction pattern in the TEM is to adjust the condenser lens system to form a small-diameter focused probe on the specimen. Area-selection is then performed by locating the probe on the desired region of the specimen. The pattern obtained under these conditions is known as a convergent beam diffraction (CBD) pattern.

If the region of specimen illuminated by the small probe is a single crystal, then the CBD pattern will consist of an array of discs rather than the sharp spots obtained with parallel illumination in SAD. In CBD, the diameter of the discs is governed by the diameter of a diaphragm placed in the condenser system, and it is often desirable to choose the diameter of this diaphragm to allow adjacent discs to nearly touch one another. Since the illumination is in the form of a cone, each disc displays a two-dimensional "rocking curve" of diffracted intensity.

An example CBD pattern recorded by K.K. Fung et al. from 2H-TaSe$_2$ [15] is shown in Figure 5.

If a zone axis of the crystal is carefully aligned with the axis of illumination, the CBD discs will display dynamical diffraction information which may be used to determine the symmetry of the crystal. A pattern recorded at a short effective projection length may display diffraction information from upper Laue zones which can be combined with information from the zero-order zone to yield symmetry information in three dimensions. Under favourable circumstances it is possible to use CBD to determine the crystallographic point group or even the space group of the crystal [59-62]. For this reason, CBD has proved to be an extremely valuable tool for elucidating CDW-modulated structures [63]. Several excellent books concerning CBD are now available [64,65], and the reader may find the experimental details together with the methods for interpreting CBD patterns described therein.

One important example of the application of CBD in the study of CDW is the determination of broken hexagonal symmetry in the commensurate phase in 2H-TaSe$_2$ [15]. Early studies of

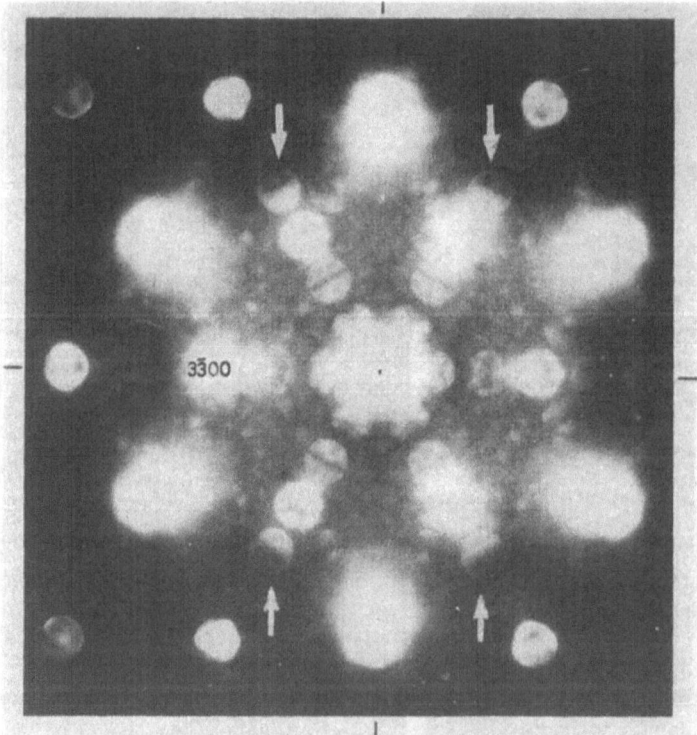

Fig.5. Convergent beam electron diffraction pattern from 2H-TaSe$_2$ recorded below the C lock-in temperature [15]. The $\pm(40\bar{4}0)$ and $\pm(0\bar{4}40)$ reflections, indicated by arrows, are much stronger than the $\pm(4\bar{4}00)$ reflections, thus breaking the hexagonal symmetry.

the modulations in this polytype of TaSe$_2$ by SAD and neutron diffraction had revealed an apparently hexagonal 3a$_0$ x 3a$_0$ x c$_0$ superlattice below a lock-in temperature of about 90 K [4]. The [001] CBD pattern from 2H-TaSe$_2$ shown in Figure 5 was recorded with a fine probe of diameter 40 nm and the specimen held well below the C lock-in temperature. In this pattern the indices are referred to a 3a$_0$ hexagonal supercell. It is clear that the $\pm 4\overline{4}00$ reflections are much weaker than the $\pm 40\overline{4}0$ and $\pm 0\overline{4}40$ reflections, thus breaking the hexagonal symmetry. Within the limit (0.1%) of detection however, no difference was observed in the interplanar spacings. Fung and co-workers [15] have concluded from the details of their CBD data that the supercell is orthorhombic 3a$_0 \times 3\sqrt{3}$a$_0 \times$c$_0$, belonging to space group number 63 (Cmcm). Satellite dark field microscopy, discussed in the next section of this chapter, revealed a microstructure of $\cong 1 \mu$m orthorhombic domains oriented in three equivalent directions relative to the hexagonal P6$_3$/mmc subcell. A CBD pattern recorded in a TEM may be taken from a single domain, revealing the orthorhombic symmetry. Because of the relatively large probe size, a neutron or x-ray diffraction pattern must represent an average over many domains in the three equivalent orientations, thus yielding apparently hexagonal symmetry.

Following this discovery of orthorhombic domains in commensurately-modulated 2H-TaSe$_2$, D.M. Bird and co-workers [18,19] determined the displacements of the Ta and Se atoms from the subcell positions by analysing higher-order Laue zone (HOLZ) diffraction in CBD patterns. The basis of the method is the application of Bloch wave analysis to relate the intensities of the fine structure lines in HOLZ reflections to structure factors of the modulated crystal. These structure factors are very sensitive to small changes in atomic position.

Figure 6 shows one quadrant of a [001] CBD pattern. The inner ring of first-order Laue zone (FOLZ) reflections and the outer ring of second-order Laue zone (SOLZ) reflections are circled. In the analysis it was demonstrated that the Ta atoms make no contribution to the [001] FOLZ diffraction in the parent crystal. The intensities of the reflections in the FOLZ ring may therefore be used to study the Se atoms in isolation. Similarly, the Se contributions to the intensities of reflections in the SOLZ ring nearly vanish, with the result that the SOLZ intensities are dominated by the Ta atoms. In the FOLZ ring some reflections, labelled a in Figure 6, have very little intensity, whilst others typically show a double line corresponding to Bloch waves associated with branches (2) and (6) of the dispersion surface. In the modulated structure, the lines associated with branch (2) were predicted to be unaffected by the Ta displacements. From the observation that the intensity ratio $I_{subcell}/I_{satellite} \cong 100$ in the FOLZ branch (2) lines, Bird et al. deduced that the Se displacements were < 0.015 Å in-plane and < 0.04 Å out-of-plane. The Ta in-plane displacements were estimated in be $\cong 0.05$ Å from the observation that $I_{subcell}/I_{satellite} \cong 5$ in the SOLZ branch (6) singlets. These magnitudes compare favourably with the results of neutron diffraction [66], which of course are averages over many orthorhombic domains.

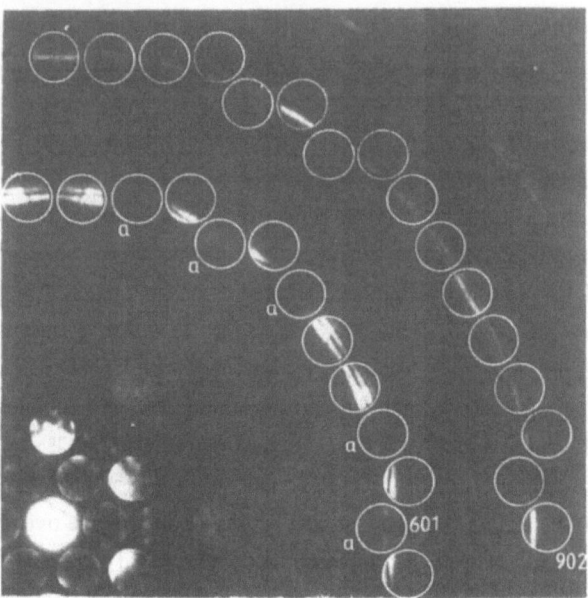

Fig. 6. A quadrant of the convergent beam diffraction pattern from C-modulated 2H-TaSe$_2$ [18]. The FOLZ and SOLZ reflections are indicated by circles. The reflections labelled (a) in the FOLZ display very little intensity.

Figure 7 shows the displacements in one plane of the superstructure, as predicted by Bird et al [19] after fitting about 100 HOLZ reflections recorded in the range 80-120 kV. It should be emphasized that, owing to the small size of the orthorhombic domains, it is unlikely that the broken hexagonal symmetry could have been discovered by any other method.

4. Satellite Dark Field Microscopy

The most powerful feature of the TEM as an instrument for the characterization of materials is the ability to form images and diffraction patterns from the same volume of a specimen. In the usual mode of operation for imaging, a small aperture is placed in the back focal plane of the objective lens. This aperture may be centred on the directly transmitted beam, blocking out all of the diffracted beams. The image formed in this way is known as the bright field image. Such images display diffraction contrast, since any region of the sample which is diffracting strongly will appear dark. Diffraction contrast is used to reveal lattice defects, for example dislocations or stacking faults. Bright field images recorded from materials modulated by CDW display the usual contrast associated with lattice defects in the subcell, but no feature attributable to the CDW. This is because the weak diffraction effects associated with the modulations are masked by the very strong diffraction by the subcell.

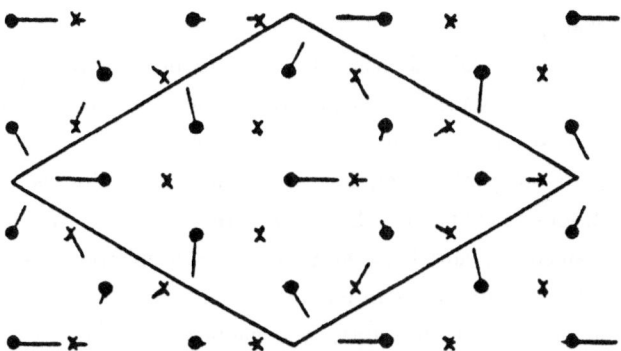

Fig.7. The displacement pattern deduced for one layer of 2H-TaSe$_2$ [19]. Circles denote Ta atoms with their displacements magnified by 20X. Crosses denote Se atoms with their displacements magnified 50X.

An alternative way of operating the TEM to form an image is to displace the small objective aperture to allow only the electrons in a particular diffracted beam to contribute to the image. The image formed in this way is known as a dark field image. To minimize the effects of spherical aberration, a dark field image is often formed by leaving the objective aperture centred on the optic axis and tilting the incident electron beam to cause a particular diffracted beam to fall on the aperture, rather than by mechanically displacing the aperture assembly. If the diffracted beam passing through the aperture is a superlattice reflection, as described in the previous section on SAD, the image is known as a satellite dark field (SDF) image. Contrast features associated with structural modulations may be revealed by SDF microscopy. It should be noted that, depending on the details of the modulations, the contrast may vary according to the particular satellite reflection employed and that SDF images formed with some satellites may reveal no contrast features of interest.

SDF microscopy has been used to reveal complex microstructures in both C and IC phases in a variety of materials. Studies of Nb$_x$Ta$_{1-x}$Te$_4$ [44,46,47,49,53,54] and 2H-TaSe$_2$ [14-16,20,21] have been particularly interesting because of the rich variety of structures revealed. A detailed description of the microstructure in the modulated phases of Nb$_x$Ta$_{1-x}$Te$_4$, including observations on the evolution of these microstructures during phase transitions is given in chapter 3 of this volume. The results of the studies on 2H-TaSe$_2$ will be reviewed in this section, followed by a discussion of some extremely interesting features observed in NbSe$_3$, and finally by a discussion of radiation-induced effects in 1T-TaS$_2$ and 1T-TaSe$_2$.

The polytype 2H-TaSe$_2$ is remarkable for the variety of CDW states and for the large thermal hysteresis effects it exhibits. The incommensurate-commensurate (IC) transition temperature occurs at 84 K on cooling and 92 K on warming. Between 92 and 112 K on warming a 2I x 1C intermediate phase develops. A first order transition to a triply incommensurate "hexagonal" phase occurs at $T \cong 112$ K, as revealed by x-ray diffraction [67]. Extensive TEM studies of these modulated phases have been reported by K.K. Fung et al. [15] and by C.H. Chen et al. [16]. As discussed in the previous section, the former group used CBD to detect broken hexagonal symmetry in the low temperature C phase. Both groups have employed SDF imaging to study the evolution of the microstructure as a function of temperature. Below the C lock-in temperature SDF images reveal a microstructure of light and dark domains $\cong 1\mu$m in diameter. Figure 8 shows a series of SDF micrographs recorded from the same region of specimen, but formed with different satellite reflections, each lying in the direction of one of the three modulation q-vectors, i.e. along one of the three hexagonal axes of the subcell. It may readily be seen that the dark domains cover the entire area, but that a particular domain appears dark in the SDF image corresponding to only one q-vector and bright in the images corresponding to the other two q-vectors. The interpretation is that each domain has orthorhombic symmetry, as determined by CBD, and that the direction of the orthorhombic distortion differs by $2\pi/3$ between adjacent domains. One unusual contrast feature is revealed in Figure 8(d), in which the contrast is reversed compared with Figure 8(a), although the satellite reflections used to form these two images lie along the same q-direction. A very small change in the specimen orientation ($\cong 0.2°$) in the TEM may also induce a contrast reversal. The origins of these contrast reversals remain to be determined and may require dynamical calculations of the diffracted intensity for elucidation. A similar effect was observed in the contrast from domain boundaries in the C phase of $TaTe_4$.

On warming from the low temperature 3C phase regime, the transition to the 2I x 1C phase proceeds by the nucleation of stripes within the orthorhombic domains. Figure 9 shows the evolution of the stripe phase with temperature. Note that the density of stripes increases with temperature and that above the transition at 112 K to the 3I state, the fringes have rotated through $\pi/6$ from a $<10\overline{1}0>$ orientation to a $<11\overline{2}0>$ orientation, perpendicular to the imaging q-direction. In the stripe-phase regime between 92 K and 112 K, the contrast in a bright domain alternates between bright and dark stripes and displays nodes at which three dark stripes join together. Examples of this contrast are shown in Figure 10. The interpretation of the contrast is that within each dark and bright stripe the modulated structure is orthorhombic, but the orthorhombic distortion direction alternates between adjacent stripes, e.g. the white stripes may be distorted in the q_1-direction and the black stripes in the q_2-direction. The boundaries between the stripes would then be q_1/q_2 discommensurations and the nodes at which three black stripes terminate are designated as CDW dislocations. The double-layer crystal structure of $2H$-$TaSe_2$ must be taken into account to understand the phase relationships between the stripes. It turns out that three stripes of the same type and three pairs of discommensurations are required to form a CDW dislocation such that the total phase change along any path around the dislocation is 2π.

The interpretation of the contrast in the 3I phase above 112 K is not entirely straightforward. The initial proposal by K.K. Fung and co-workers [15] was that the fringes were associated with a double-honeycomb array of discommensurations. According to this picture, each dark region of the fringes is a chain of one type of C domain which does not contribute to the SDF imaging reflection. In a later study, T. Onozuka and co-workers [20] discovered that the SL spots in the SAD patterns from 2H-TaSe$_2$, recorded in the 3I temperature regime on cooling, were finely split as shown schematically in Figure 11. In this diagram, the splitting is exaggerated for clarity of presentation. The separation of the spots 3δ

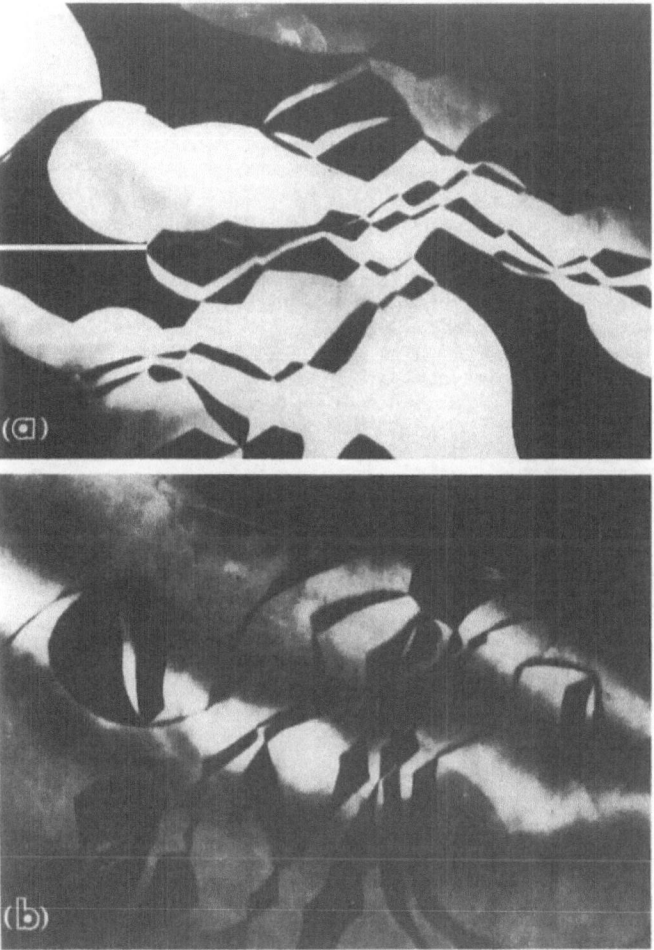

Fig. 8. Satellite dark field images recorded from the same area of the C-phase in 2H-TaSe$_2$ at 33 K [16], (a)-(c) with different q-vectors, (d) with a different SL reflection than (a), but in the same imaging direction. (e) A line diagram illustrating the combined domain patterns of (a)-(c).

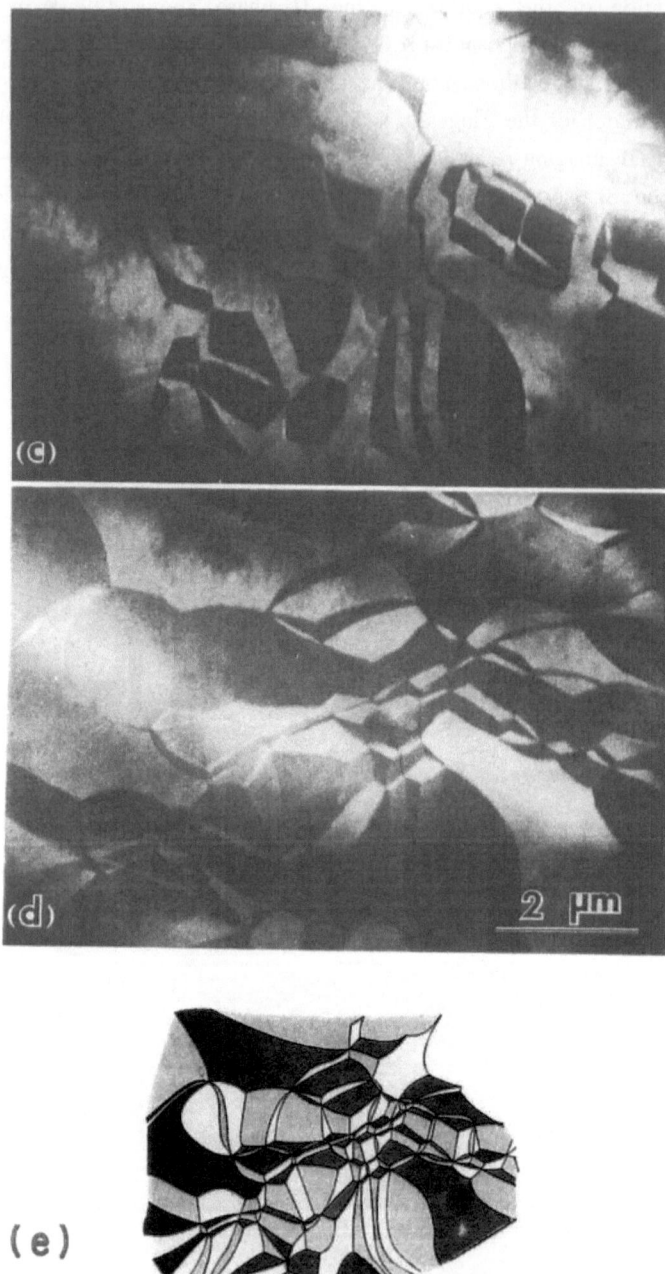

Fig. 8   Satellite dark field images recorded from the same area of the C-phase in 2H-TaSe$_2$ at 33 K
(con't.) [16], (a)-(c) with different q-vectors, (d) with a different SL reflection than (a), but in the
               same imaging direction. (e) A line diagram illustrating the combined domain patterns of (a)-(c).

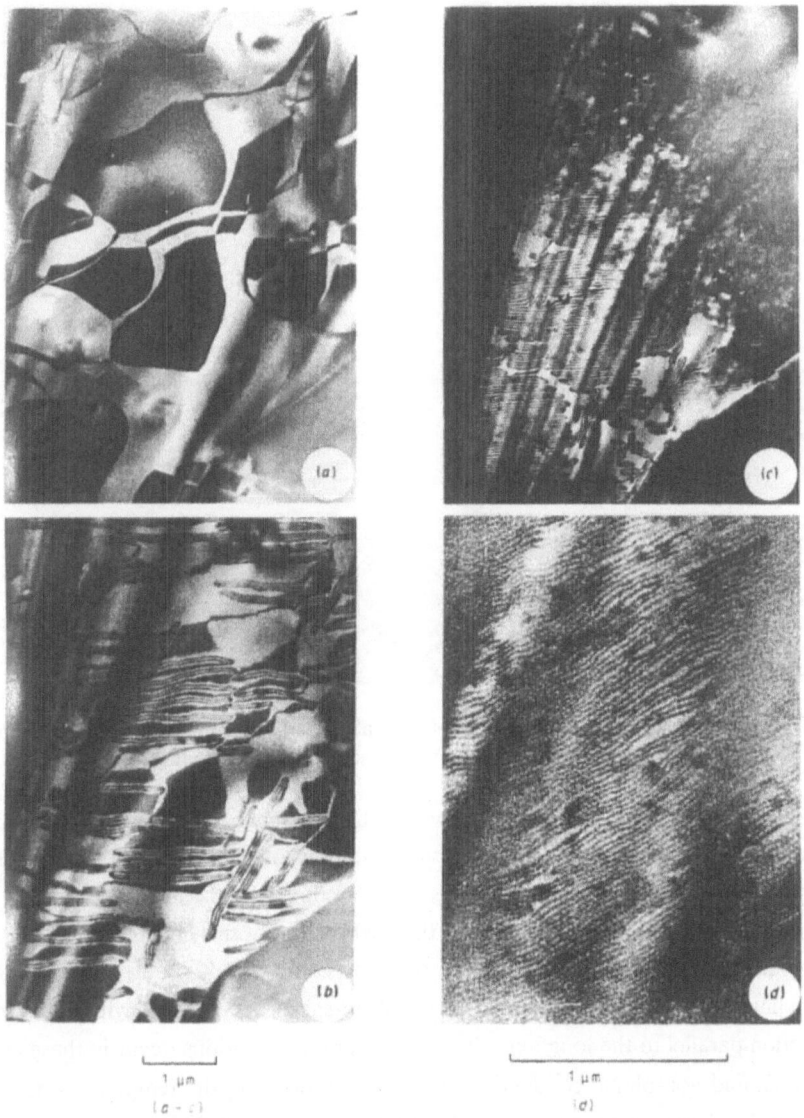

Fig. 9. Satellite dark field images recorded from the same area of 2H-TaSe$_2$ on warming [15]. Recorded (a) below the C lock-in temperature, (b) slightly above 90 K, (c) at about 100 K and (d) above 112 K. Note that the fringes have rotated through $2\pi/12$ in (d) and become very fine thus a 3X higher magnification is employed.

corresponded to about $(300\ \text{Å})^{-1}$ at 97 K. This splitting had not been previously reported in any of the TEM studies, although similar splitting was observed by neutron diffraction [65] and ascribed to first and second order reflections associated with the lattice modulations. The fringes observed in the SDF images were then identified as resulting from interference between

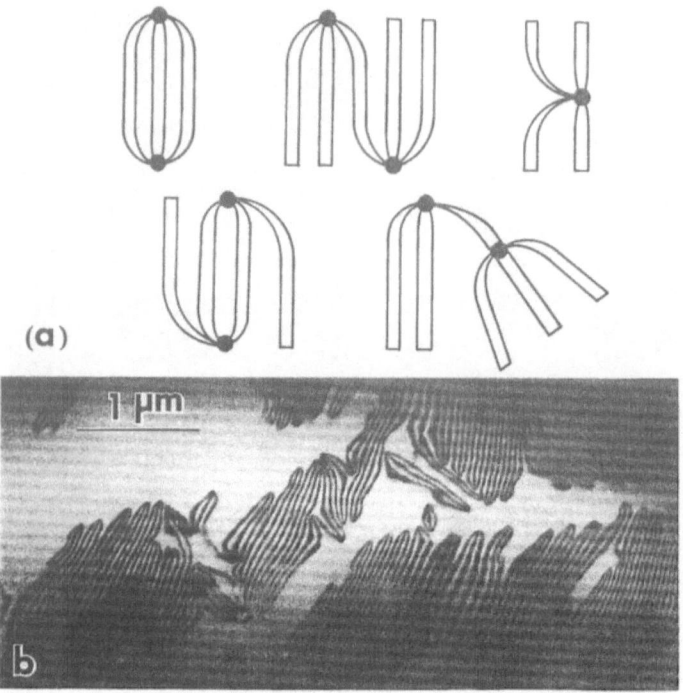

(a)

Fig. 10. CDW dislocations in the stripe phase of 2H-TaSe$_2$ [16]. Line diagrams (a) illustrating the configurations which may be observed in an SDF image (b).

the primary (P) and secondary (S) spots shown in the diagram, and it was shown that the directions and spacings of the fringes were consistent with this interpretation. In the same study, high resolution imaging was employed to complete the description of the 3I microstructure. The results of these experiments will be discussed in the final section of this chapter.

Another remarkable type of microstructure was observed by K.K. Fung and J.W. Steeds [32] in monoclinic NbSe$_3$. This compound forms blade-shaped crystals or whiskers with the unique b direction parallel to the long axis. Two CDW phase transitions occur in these crystals at 144 and 59 K, and non-ohmic conductivity has been observed in the temperature regime in which the CDW has been established [68]. The SDF images reveal two types of contrast, as shown in Figure 12. The whiskers appear to be broken up into dark and light strands approximately parallel to the whisker-axis, but with deviations up to 15°. The images sometimes reveal modulations along the strands. Both of the micrographs in Figure 12 display this effect. An unusual feature of these images is that they were found to be time-dependent with different rates of evolution above and below the lower transition temperature.

On cooling below 40 K the images were reported to be constantly moving with individual strands lighting up, darkening or branching to give a twinkling appearance. Initially it was

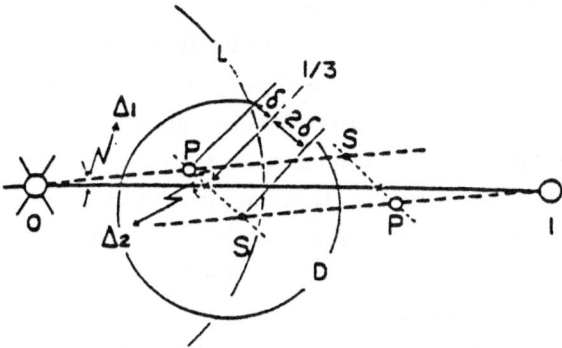

Fig. 11.   Schematic diagram of the diffraction pattern of 2H-TaSe$_2$ along one of the subcell hexagonal axes [20]. The symbols 0 and 1 represent two subcell spots, whilst P and S represent the primary and secondary SL spots, respectively.

thought that the twinkling appearance might result from the CDW sliding with respect to the crystal lattice, however later experiments by R.M. Fleming et al [34] showed that there were no observable changes in the appearance of these contrast effects on application of an electric field sufficient to produce conduction along the whisker.

One other unusual feature of the contrast is the appearance of the contours associated with the slight bending present in all TEM specimens. These are continuous in a bright field or subcell dark field image but appear to be broken and shifted between strands in the satellite images from these crystals.

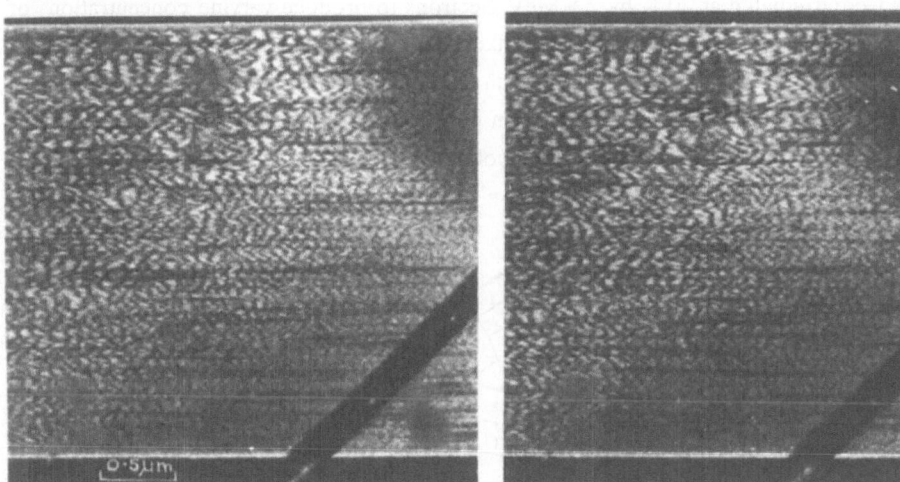

Fig. 12.   Satellite dark field images recorded from NbSe$_3$ with two-second exposures at a time interval of approximately one minute between exposures [32], revealing strands parallel to the whisker axis. The dark band is the shadow of another whisker.

A plausible interpretation of these various contrast effects is as follows [33]. The strands correspond to domains of dimensions $\cong 1$ μm x 20 nm. Within each domain the modulation q-vector is constant. Between adjacent domains the q-vector is slightly rotated, most probably in the narrow direction of the blade-like whiskers, as illustrated in Figure 13. The q-vectors in alternative domains are expected to deviate in opposite senses from the b-direction, in order to minimize strain energy. The time variation or twinkling effect arises from variations in q over a given region of the specimen as a result of electron bombardment in the TEM. Specimen cooling is very rapid in the TEM and the CDW may well be in a metastable state before the specimen is irradiated by the electron beam.

The broad modulations or fringes are likely to be a Moiré effect arising from overlapping domains in the direction of the electron beam. Note that the sense of the deviation of q from the b-axis should be opposite between adjacent domains in both the narrow and broad directions of the blade-like whiskers, and that there are likely to be two domains present in the narrow direction in a typical TEM specimen.

In the sharpest SDF images, weak sets of fine fringes with spacings in the range 10-20 nm and approximately normal to the whisker axis were also observed. Definitive experiments to deduce the origins of these fringes have not been possible because they display extremely weak contrast.

Another interesting area of application of SAD methods combined with SDF imaging is the study by H. Mutka and co-workers of the effects of irradiation-induced lattice defects on the modulated structures of 1T-TaS$_2$ and 1T-TaSe$_2$ [23-25]. In the case of 1T-TaS$_2$, the specimens were irradiated at 20 K by 2.5 MeV electrons to produce varying concentrations of imperfections, most likely Ta defects of the Frenkel type.

Several distinct phases have been observed in non-irradiated samples of 1T-TaS$_2$ [23]. An IC phase denoted 1T$_1$ exists above 350 K. On cooling this IC phase transforms abruptly at ~350 K to a nearly C phase, 1T$_2$. On further cooling the 1T$_2$ phase transforms at ~200 K by a

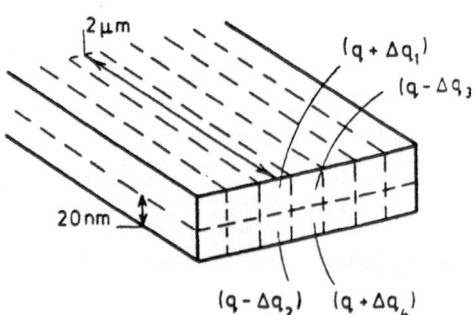

Fig. 13. Schematic diagram of a model for the stranded structure in CDW modulated NbSe$_3$ [32].

gradual rotation of the SL cell to a commensurate phase, $1T_3$. Figure 14 shows SDF images recorded at 100 K and 350 K from the non-irradiated material. The contrast visible in the $1T_3$ image at 100 K is typical of a domain-structure. The image recorded at 350 K shows a two-phase region in which the IC $1T_1$ phase and the nearly C $1T_2$ phase are clearly separated by a sharp phase boundary. Also visible in the 350 K image are defects displaying black-white contrast and

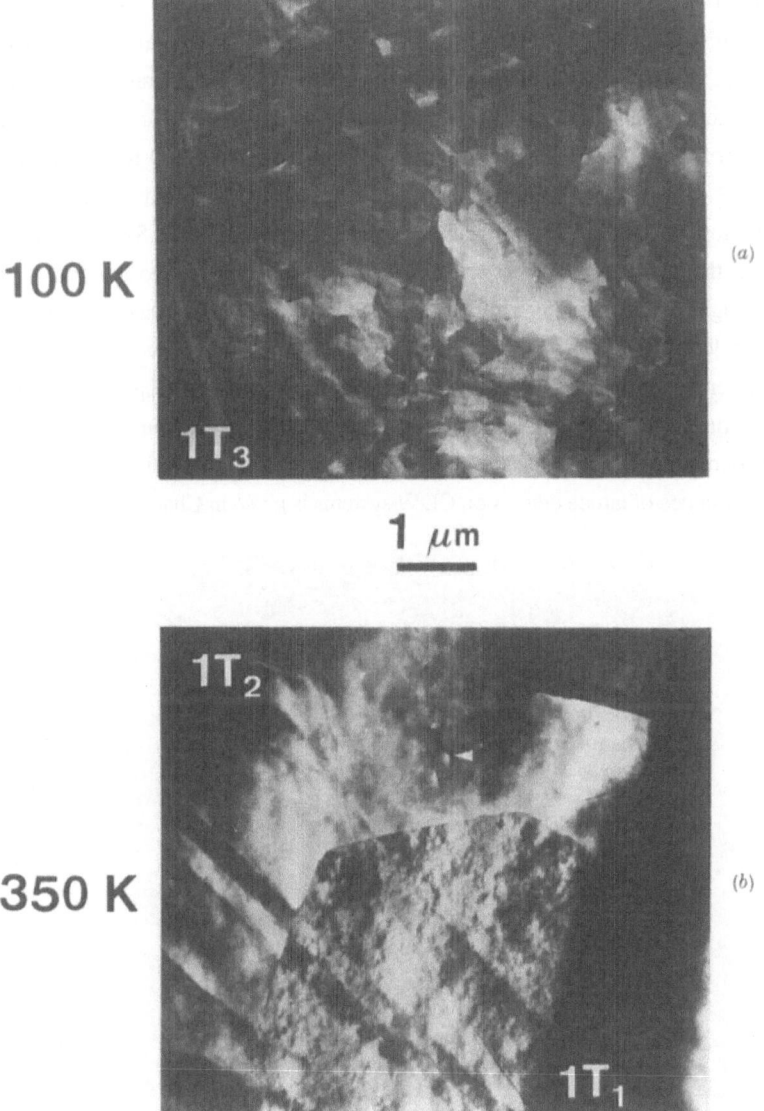

Fig. 14. Satellite dark field images showing the three distinct CDW phases in $1T$-TaS_2 [23]. (a) $1T_3$ recorded at 100 K. (b) $1T_2$ and $1T_1$ recorded at 350 K. The arrow points to a black-white contrast feature associated with a CDW defect.

several subcell dislocations. The former features were later identified by high resolution imaging to be CDW dislocations [25]. These defects are very mobile. Their density increases as the $1T_2$ - $1T_3$ transition is approached and ultimately they generate the commensurate domains on further cooling.

Figures 15-17 show the effects on the SL structure due to varying concentrations of irradiation-induced defects. At 10^{-4} dpTa (displacements per Ta atom) the C domains are clearly visible in the $1T_3$ image recorded at 100 K, but they are considerably smaller than those observed in the virgin material. A slowing of the $1T_2$ -$1T_3$ transition kinetics was also observed. After irradiation to 10^{-3} dpTa, the $1T_3$ phase could no longer be observed by SAD, even to a temperature of 7 K [23]. At this dosage the density of CDW dislocations had increased considerably and at 5×10^{-3} dpTa a very high concentration of these defects was visible in the $1T_2$ phase at both 100 K and 300 K (Figure 16). After 1×10^{-2} dpTa, the SDF images recorded at 100 K and 300 K displayed a similar fine grainy contrast (Figure 17), although subtle differences were observed in the SAD patterns. At the lower temperature, the SL spots were split into doublets corresponding to two distortion-variant $1T_2$ domains. SDF imaging with either member of the doublet revealed the domain structure. At the highest dosage employed, 3×10^{-2} dpTa, very little contrast could be observed in the SDF images recorded at either temperature. This series of experiments shows that the response of CDW to the introduction of irradiation-induced defects into the $1T$-TaS_2 lattice is not homogeneous, but rather the displacement damage was accompanied by CDW dislocations and domains. A comprehensive discussion of the influence of lattice defects on CDW systems is given in Chapter 5 of this volume.

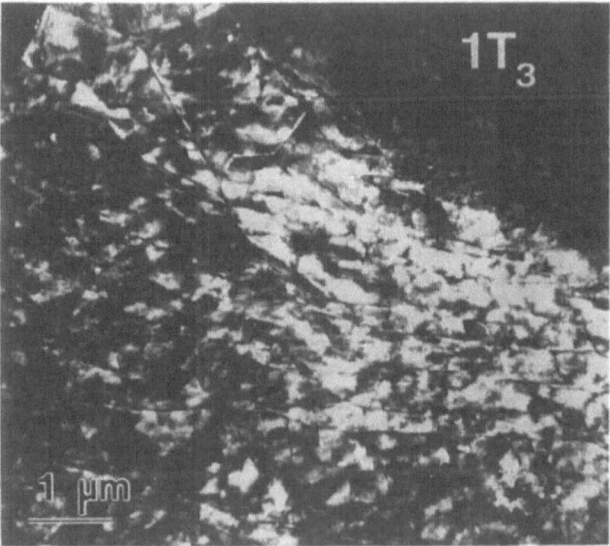

Fig. 15. Satellite dark field image recorded at 100 K from $1T$-TaS_2 irradiated to 10^{-4} displacements per Ta atom [23]. The C domains are considerably smaller than observed in the virgin material (Figure 14a).

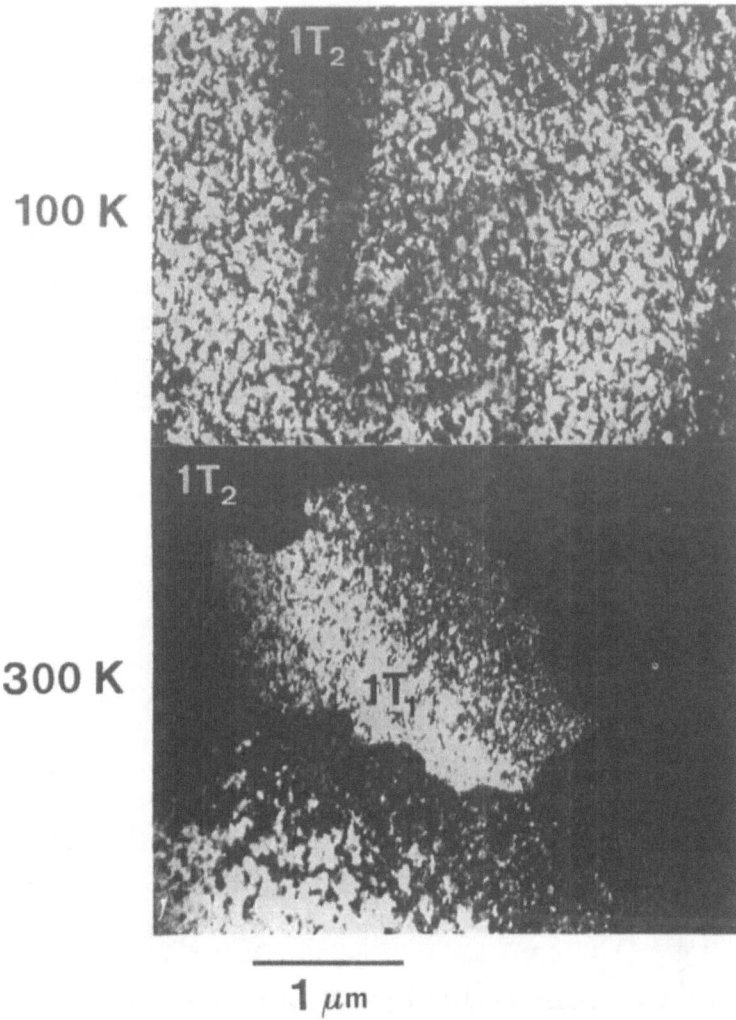

Fig. 16. Satellite dark field image recorded at (a) 100 K and (b) 300 K from 1T-TaS$_2$ irradiated to 5 x 10^{-3} displacements per Ta atom [23].

5. High Resolution Electron Microscopy

All commercial TEM manufacturers produce instruments which are capable of performing high resolution electron microscopy (HREM), by which it is meant that images recorded at high magnification may display structural details at least as small as 0.3 nm. The principal requirements for HREM are a properly aligned, astigmatism-compensated optical system, highly stable lens currents and accelerating voltage and a high degree of mechanical stability,

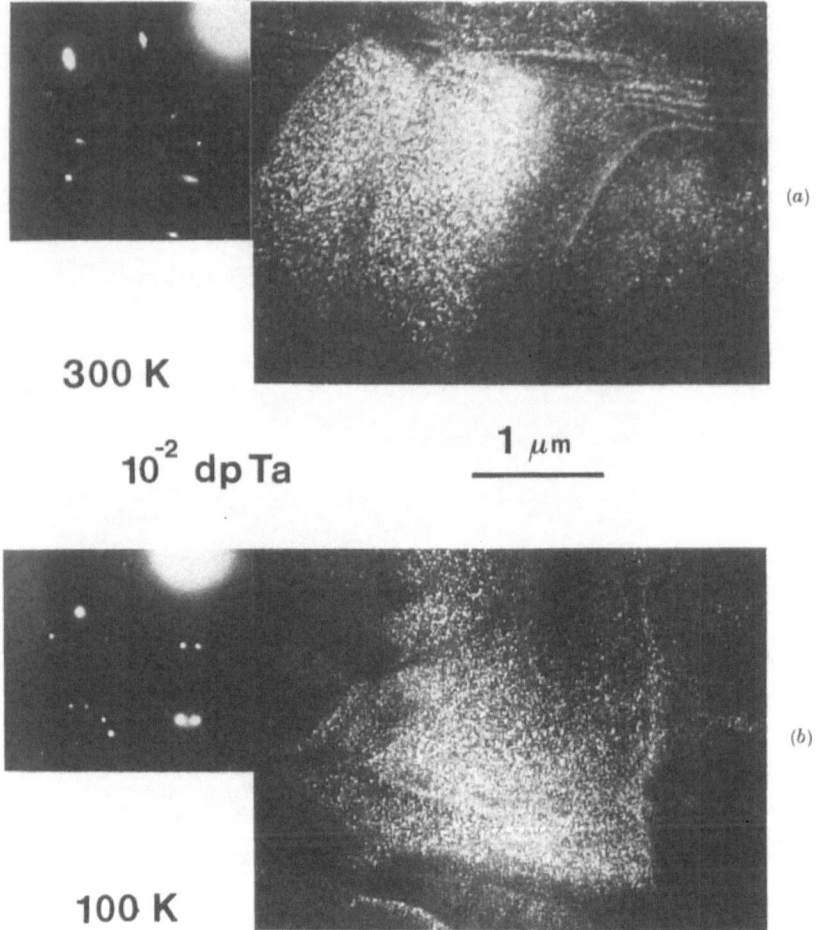

300 K

10^{-2} dp Ta

1 μm

100 K

Fig. 17. Satellite dark field images recorded at (a) 100 K and (b) 300 K from 1T-TaS$_2$ irradiated to 10^{-2} displacements per Ta atom [23].

particularly in the specimen stage. If a crystalline specimen is viewed with axial illumination and oriented with a prominent zone-axis accurately aligned parallel to the optic axis, then a high magnification image will display a cross grating of lattice fringes, providing that a number of Bragg-diffracted beams are allowed to pass through the objective aperture to interfere in the image plane. The interpretation of a lattice image formed in this way, however, is not entirely straight-forward. The image may be regarded as a map of the projected potential of the crystal only under very limited conditions, namely that the objective lens is operated at optimum defocus and that the crystal is sufficiently thin that it may be regarded as a weak phase object. These two conditions are seldom both met in practice. At present, the standard way of interpreting a high resolution image is to compare it with computer simulations based on assumed models for the atomic positions and calculated for a range of values of objective lens defocus and crystal thickness encompassing the nominal values assigned to the experimental micrograph.   Ideally, one would prefer to have the capability to perform electron crystallography, that is to solve the structure without any prior knowledge of the possible atomic positions beyond the constraints set by diffraction pattern symmetries. While significant progress has been made along these lines, much remains to be done before it will be possible to completely determine an unknown structure by HREM in anything approaching a routine fashion. There are currently available several excellent references on the theory and practice of the standard methods for performing HREM, for example the comprehensive book by J.C.H. Spence [69].   A summary of recent advances towards electron crystallography by focus-variation reconstruction may be found in a review by D. van Dyck et al [70].

The methods of HREM have found limited applications in the study of CDW, principally because most specimen holders capable of reaching the low temperatures required to yield transitions to CDW phases are subject to thermally induced mechanical instability.   Notable exceptions are applications, which will be discussed in this section, to the studies of modulated structures in 2H-TaSe$_2$, NbTe$_4$ and TaTe$_4$.

HREM images of the 3C phase in 2H-TaSe$_2$ were recorded at ~60 K by J.M. Gibson and co-workers [17].   These images revealed further evidence for the orthorhombic distortion which had previously been deduced by CBD [15]. Figure 18 is an image recorded from a single domain, showing a hexagonal array of bright dots corresponding to the intersection of three sets of <10$\bar{1}$0> lattice planes. A remarkable feature of this image is a 3a$_0$ modulation of the intensities of the bright dots, predominantly in the direction labelled q$_1$ on the micrograph. Optical density traces revealed that indeed there were intensity modulations in all three q-directions, but that the amplitude was much greater parallel to the orthorhombic direction, q$_1$. Simulations of these images based on model structures, however, have not been reported.

In a later investigation of 2H-TaSe$_2$ [20], T. Onozuka et al. reported HREM imaging of the 3I phase obtained on cooling to ~100 K, the lowest temperature which could be reached in their instrument. As discussed in the previous section, these authors detected a previously unreported splitting of the SL reflections recorded in their SAD patterns, and they have therefore interpreted the Moiré-like contrast commonly observed in SDF images of the 3I phase, e.g. as shown in Figure 9d, as arising from interference between first-and second-order diffracted beams from the IC structure rather than diffraction-contrast involving domains in the commensurate structure. This interpretation was confirmed by detailed observations of lattice images recorded using SL reflections at ~100 K. In order to enhance the contrast of the lattice fringes, they tilted their crystals slightly from the zone-axis orientation in the direction of one of the three equivalent <10$\bar{1}$0> diffraction vectors. The resulting images thus displayed only a

Fig. 18. High resolution image of 2H-TaSe$_2$ recorded at ~60 K [17], showing a strong CDW modulation with wave vector in the indicated direction, q_1.

single set of parallel fringes with a spacing of ~0.9 nm, representing the period of the IC phase. Regions of weaker contrast of width ~10 nm and period ~30 nm corresponding to the spacing of the SDF Moiré-like fringes were also observed. In all of the images recorded in this way, the lattice fringes were visible over the entire field and showed no evidence for existence of domains having different orientations. On the other hand the spacing of the lattice fringes was observed to vary slightly, larger by ~0.3 nm over 10 nm in the regions of weaker contrast, as shown in Figure 19. This resulted in a phase shift of ~$2\pi/3$ across the regions of weaker contrast. Moreover, there appeared to be a shift in the location of the fringes across the bands of weaker contrast in images in which the lattice fringes and contrast bands were not parallel. Again, this corresponded to a phase slip of $2\pi/3$. These observations are illustrated schematically in Figure 20. The 3I phase is thus seen to be intrinsically incommensurate and the dark Moiré-like fringes observed in SDF images may be interpreted as discommensurations or regions of phase slip.

The pseudo-one-dimensional crystals NbTe$_4$ and TaTe$_4$ present a distinct advantage for HREM in that they have simple structures and both sustain CDW modulations at room temperature.

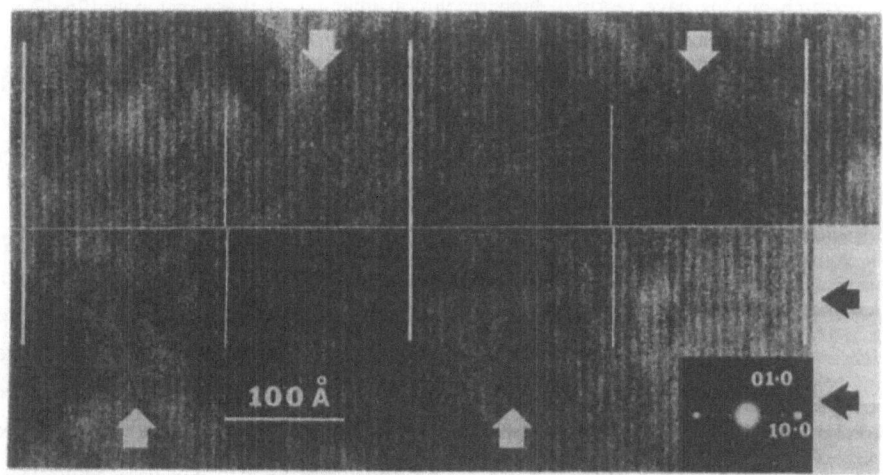

Fig. 19   Lattice fringe image of 2H-TaSe$_2$ recorded at ~100 K [20], showing periodic change of the CDW lattice spacing associated with phase slip. The lower half of the image is shifted to the left by ~150 Å to allow comparison of the lattice fringe spacing in the region of phase-slip with the spacing in the other region.

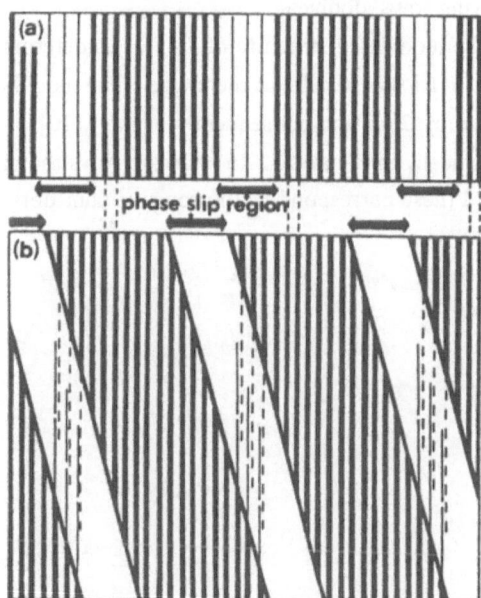

Fig. 20. Schematic diagram of the CDW lattice fringes viewed with respect to the phase-slip regions [20]. In (a) the lattice fringes are shown parallel to the regions of phase slip, as observed in Figure 19. In (b) the fringes are inclined ~15° with respect to the phase slip region. The phase slip is ~2π/3 in either case.

Several research groups have reported high resolution imaging of these materials [26,47,50,51].
J. Mahy and co-workers [47] have recorded [110] zone axis images from NbTe$_4$ at room
temperature, as shown in Figure 21. The principal features observed in this image are a D = 16/11 c$_o$
Moiré fringe associated with the IC modulation as well as an L = 8c$_o$ pattern, which the authors
have interpreted as representing an array of discommensurations. No image simulations have
been presented to support this interpretation.

High resolution [110] images from TaTe$_4$ at room temperature have also been reported by
Mahy et al [50]. An example is shown in Figure 22. The principal feature of this image is a
rectangular lattice of bright dots interspersed by alternating rows of weaker dots, denoted A
and B on the micrograph. The brightness of the weaker rows varies periodically as indicated in
the lower right corner. In TaTe$_4$, the CDW modulations form a commensurate 2a$_o$ x 2a$_o$ x 3c$_o$
structure, as discussed in Chapter 3. The atomic positions have been determined by an x-ray
structure refinement in space group P4/ncc [71]. Mahy and co-workers have employed this
data to prepare computer simulations of [110] images for a range of values of crystal thickness
and objective lens defocus. On the basis of comparing the computed and experimental images,
they have interpreted the brightest dots in Figure 22 as relating to the Te sublattice. The
weaker dots are related to the arrangement of Ta atoms. In alternating rows these dots form
triplets and in the interleaving rows, doublets.

A notable feature of the modulated structure of TaTe$_4$, as determined by x-ray diffraction,
is that the Ta atoms group in triplets parallel to the c-axis. The modulations on adjacent chains
of Ta atoms differ in phase by $2\pi/3$ and consequently in the [110] projection, the triplets are
arranged as shown schematically in Figure 23. In alternate rows the projected positions of the
triplets overlap exactly and these correspond to the rows of faint dots labelled A in Figure 22.

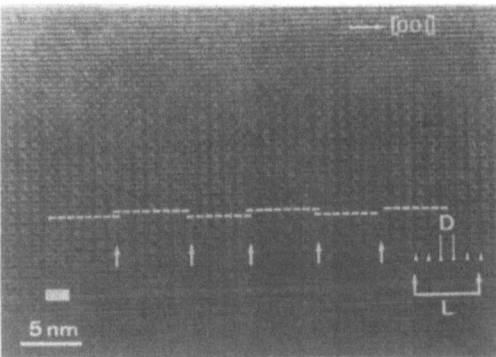

Fig. 21.   High resolution [110] zone-axis image recorded from NbTe$_4$ at room temperature [47]. A
2a$_o$ x 3c$_o$ cell is marked by a white rectangle. The authors have interpreted the spacing D as
a 16/11 c$_o$ Moiré fringe associated with the CDW deformation modulation and the spacing
L = 8c$_o$ as representing the spacing of a discommensuration array.

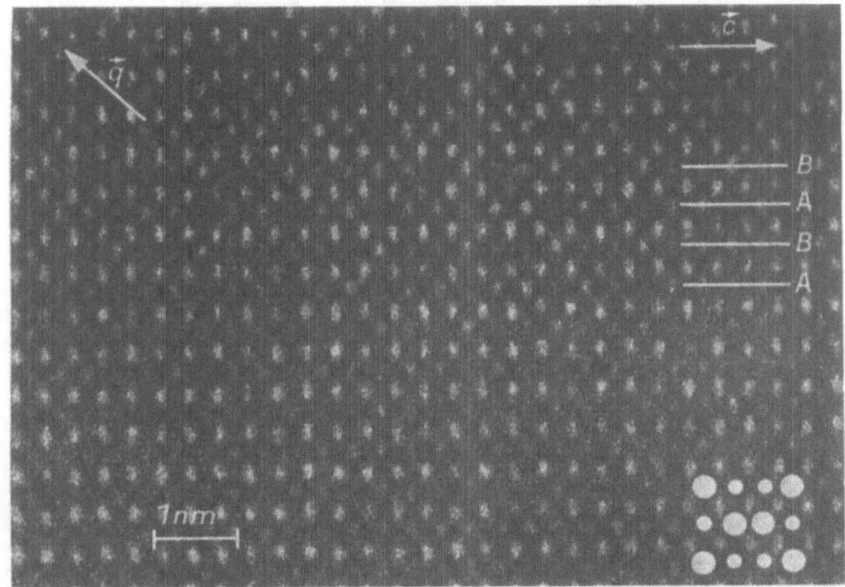

Fig. 22. High resolution [110] zone-axis image recorded from $TaTe_4$ at room temperature [50]. The brightest dots are related to the Te sublattice, whilst the weaker dots are associated with the Ta atoms. The brightness of the weaker rows varies periodically as indicated in the lower right corner. Two interleaved types of weaker rows are designated A and B.

In the interleaving rows the triplets project with their centres separated by one Ta-atom spacing, creating doublets. The corresponding rows of faint dots are labelled B in Figure 22. The experimental high resolution micrographs are therefore in good agreement with the P4/ncc structure deduced by x-ray crystallography.

In an independent study [51], high resolution micrographs of $TaTe_4$ were recorded in [210] projection. In this orientation, the modulations were observed to have a more pronounced effect on the image than in either the [100] or [110] projections. Figure 24 shows an example [210] micrograph, in which the $C_R = 3c_0$ modulations are clearly seen. Computer simulations based on the x-ray structural refinement in space group P4/ncc [71] and on an alternative acceptable refinement in P4/mcc were compared to the experimental image in two regions of differing specimen thickness, labelled a and b in Figure 24. The results are shown in Figure 25. The computed images based on the P4/ncc model are clearly a good match at both thicknesses. Note that in both the simulated and experimental images, the n-glide produces a zig-zag arrangement of dots perpendicular to the c-axis, as opposed to the mirrored arrangement predicted in P4/mcc. These results, together with the studies of [110] images discussed above, provide ample confirmation of the structural model deduced by x-ray crystallography.

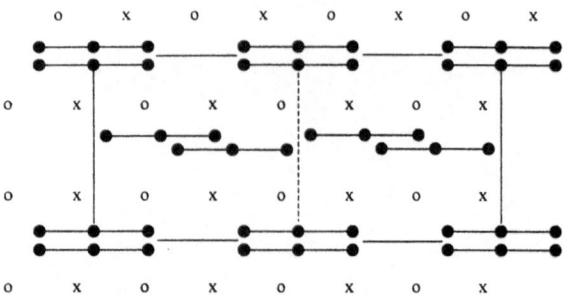

Fig. 23. Schematic diagram of the structure of TaTe$_4$ viewed along the [110] zone [50]. The Ta atoms form triplets which overlap in this projection. For clarity they have been drawn shifted sideways. The crosses and open circles represent Te atoms arranged in two kinds of squares, forming anti-prisms.

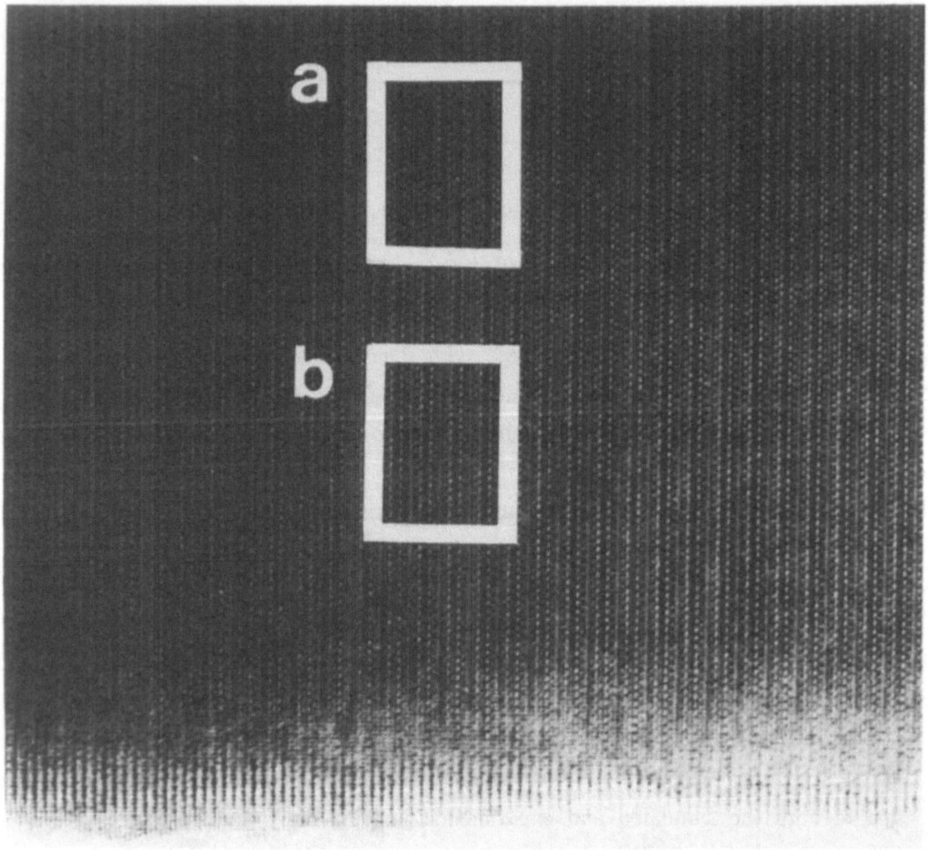

Fig. 24. High resolution [210] zone-axis micrograph of TaTe$_4$ recorded at room temperature. The c-axis runs parallel to the edge of the crystal and modulations with the $c_R = 3c_0$ spacing of the SL are clearly visible. The areas outlined in white borders were chosen for image matching (Figure 25).

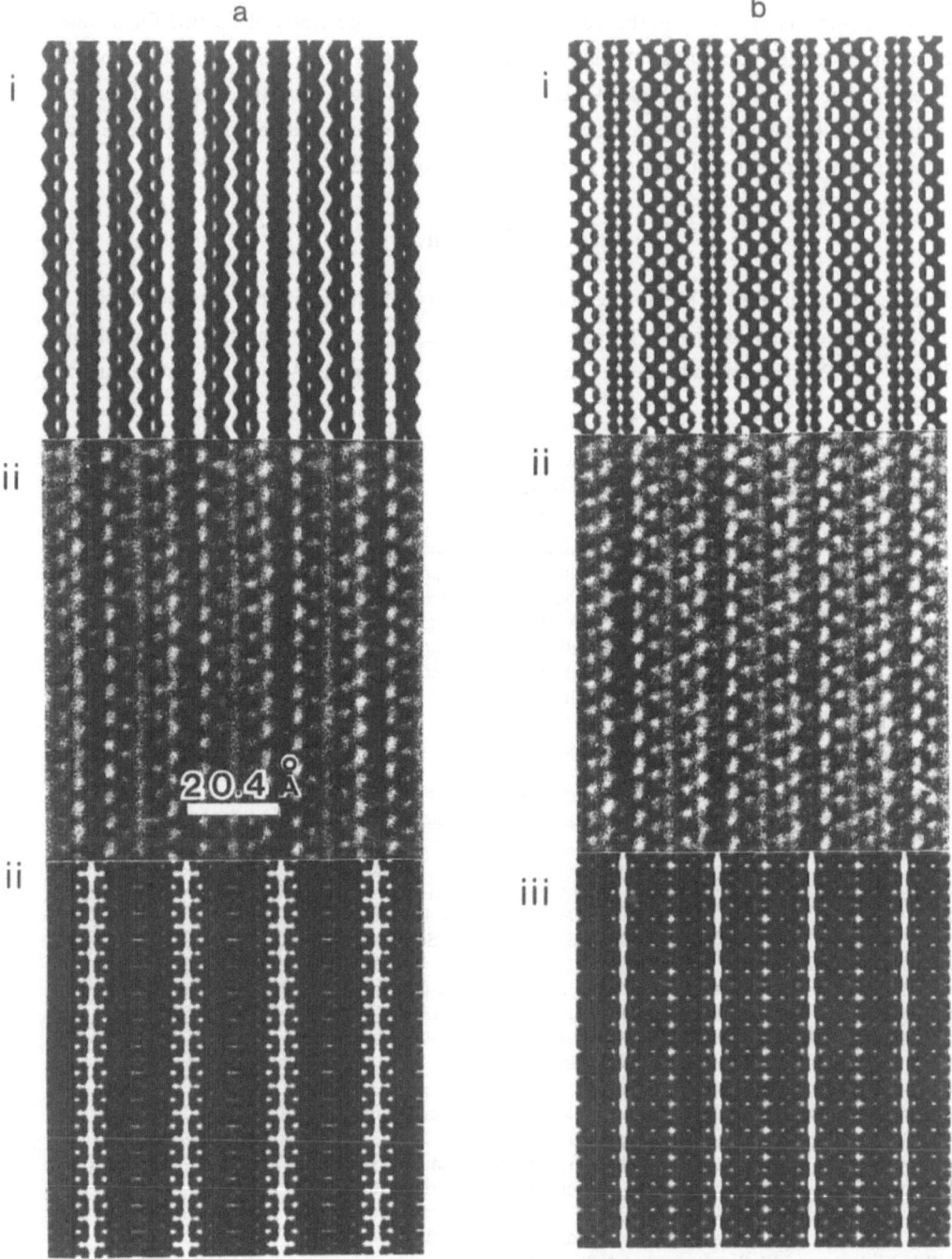

Fig. 25.   Comparison of [210] zone-axis images of TaTe$_4$. (i) computed using a P4/ncc model, (ii) experimental and (iii) computed using a P4/mcc model. Crystal thicknesses employed are (a) 1515 Å and (b) 1573 Å. Microscope parameters are those of Philips EM 430 T operated at 250 kV and near optimum defocus.

The experiments described in this section demonstrate the potential of HREM for detailed analysis of modulated structures. As digital image-recording systems based on charge coupled devices become routinely mounted on TEM columns, it is likely that the characteristic drift associated with heating and cooling specimen holders will be corrected on-line [72] and HREM will find even wider applications in the study of CDW modulations.

Acknowledgments

All figures which have been previously published were reproduced with the kind permission of the American Institute of Physics (Figures 8,9,10,11,12,13,18,19,20), the Institute of Physics (Figures 5,6,7,21), Taylor and Francis (Figures 14,1516,17) and the Akademie-Verlag (Figures 22, 23).

References

1. J. Van Landuyt, G. Van Tendeloo and S. Amelinckx, *Phys. Stat. Sol. (a)* **26**, 359 (1974).
2. J. Van Landuyt, G. Van Tendeloo and S. Amelinckx, *Phys. Stat. Sol. (a)* **26**, K9 (1974).
3. P.M. Williams, G.S. Parry and C.B. Scruby, *Phil. Mag.* **29**, 695 (1974).
4. J.A. Wilson, F.J. DiSalvo and S. Mahajan, *Adv. Phys.* **24**, 117 (1975).
5. J. Van Landuyt, G. Van Tendeloo and S. Amelinckx, *Phys. Stat. Sol. (a)* **29**, K11 (1975).
6. J. Van Landuyt, G. Van Tendeloo and S. Amelinckx, *Phys. Stat. Sol. (a)* **30**, 299 (1975).
7. J. Van Landuyt, G. Van Tendeloo and S. Amelinckx, *Phys. Stat. Sol. (a)* **36**, 757 (1976).
8. W.M. Stobbs, *Phil. Mag.* **35**, 1001 (1977).
9. J. Van Landuyt, G. Wiegers and S. Amelinckx, *Phys. Stat. Sol. (a)* **46**, 479 (1978).
10. J. Van Landuyt, *Physica* **99B**, 12 (1980).
11. K.K. Fung, J.W. Steeds and J.A. Eades, *Physica* **99B**, 47 (1980).
12. G. Van Tendeloo, J. Van Landuyt and S. Amelinckx, *Phys. Stat. Sol. (a)* **64**, K105 (1981).
13. J. Mahy, J. Van Landuyt and S. Amelinckx, *Phys. Stat. Sol. (a)* **74**, K89 (1982).
14. C.H. Chen, J.M. Gibson and R.M. Fleming, *Phys Rev Lett* **47**, 723 (1981).
15. K.K. Fung, S. McKernan, J.W. Steeds and J.A. Wilson, *J. Phys. C* **14**, 5417 (1981).
16. C.H. Chen, J.M. Gibson and R.M. Fleming, *Phys. Rev. B* **26**, 184 (1982).
17. J.M. Gibson, C.H. Chen and M.L. McDonald, *Phys Rev Lett* **50**, 1403 (1983).
18. D.M. Bird, *J. Phys. C* **18**, 481 (1985).
19. D.M. Bird, S. McKernan and J.W. Steeds, *J. Phys. C* **18**, 499 (1985).
20. T. Onozuka, N. Otsuka and H. Sato, *Phys. Rev. B* **34**, 3303 (1986).
21. Y. Koyama, Z.P. Zhang and H. Sato, *Phys Rev B* **36**, 3701 (1987).
22. D.J. Eaglesham, S. McKernan and J.W. Steeds, *J. Phys. C* **18**, L27 (1985).
23. H. Mutka and N. Housseau, *Phil. Mag. A* **47**, 797 (1983).
24. H. Mutka, N. Housseau, J. Pelissier, R. Ayroles and C. Roucau, *Solid State Commun.* **50**, 161 (1984).
25. G. Salvetti, C. Roucau, R. Ayroles, H. Mutka and P. Molinié, *J. Physique Lett.* **46**, L507 (1985).
26. D.M. Bird, D.J. Eaglesham, R.L. Withers, S. McKernan and J.W. Steeds, *CDW in Solids*, Proc. Int. Conf., Budapest 1984, Ed. Gy. Hutiray and J. Solyom, Springer-Verlag 1985.
27. K.K. Fung, *J. Phys. C* **18**, L489 (1985).
28. D.J. Eaglesham, R.L. Withers and D.M. Bird, *J. Phys. C* **19**, 359 (1986).
29. R.L. Withers and J.W. Steeds, *J. Phys. C* **20**, 4019 (1987).
30. J.A. Wilson, *J Phys: Condens. Matter* **2**, 1683 (1990).
31. T. Ishiguro and H. Sato, *Electron Microscopy* 1990. Proc 12th Int. Cong., San Francisco **4**, 168 (1990).
32. K.K. Fung and J.W. Steeds, *Phys. Rev. Lett.* **45**, 1696 (1980).

33. J.W. Steeds, K.K. Fung and S. McKernan, *J. Physique* **44**, C3-1623 (1983).
34. R.M. Fleming, C.H. Chen and D.E. Moncton, *J. Physique* **44**, C3-1651 (1983).
35. C.H. Chen and R.M. Fleming, *Solid State Commun.* **48**, 777 (1983).
36. C. Roucau, *J. Physique* **44**, C3-1725 (1983).
37. H. Mutka, N. Housseau, L. Zuppiroli and J. Pelissier, *Phil. Mag. B* **45**, 361 (1982).
38. D.J. Eaglesham, J.W. Steeds and J.A. Wilson, *J. Phys. C* **17**, L697 (1984).
39. R. Ayroles and C. Roucau, *Electron Microscopy* 1986. Proc. 11th Int. Cong., Kyoto. J. Electron Microscopy **35**, Supplement 1235 (1986).
40. C. Roucau, R. Ayroles, P. Gressier and A. Meerschaut, *J. Phys. C* **17**, 2993 (1984).
41. C. Roucau and R. Ayroles, *CDW in Solids*. Proc. Int. Conf., Budapest 1984, Ed. Gy. Hutiray and J. Solyom, Springer-Verlag 1985 pp 65-70.
42. F.W. Boswell, A. Prodan and J.K. Brandon, *J. Phys. C* **16**, 1067 (1983).
43. J. Mahy, J. Van Landuyt and S. Amelinckx, *Phys. Stat. Sol. (a)* **77**, K1 (1983).
44. D.J. Eaglesham, D. Bird, R.L. Withers and J.W. Steeds, *J. Phys. C* **18**, 1 (1985).
45. F.W. Boswell and A. Prodan, *Mat. Res. Bull.* **19**, 93 (1984).
46. J. Mahy, J. Van Landuyt, S. Amelinckx, Y. Uchida, K.D. Bronsema and S. Van Smaalen, *Phys. Rev. Lett* **55**, 1188 (1985).
47. J. Mahy, J. Van Landuyt, S. Amelinckx, K.D. Bronsema and S. Van Smaalen, *J. Phys. C* **19**, 5049 (1986).
48. F.W. Boswell and A. Prodan, *Phys. Rev. B* **34**, 2979 (1986).
49. F.W. Boswell, A. Prodan, J.C. Bennett, J.M. Corbett and L.G. Hiltz, *Phys. Stat. Sol. (a)* **102**, 207 (1987).
50. J. Mahy, J. Van Landuyt and S. Amelinckx, *Phys. Stat. Sol. (a)* **102**, 609 (1987).
51. J.M. Corbett, L.G. Hiltz, F.W. Boswell, J.C. Bennett and A. Prodan, *Ultramicroscopy* **26**, 43 (1988).
52. A. Prodan, F.W. Boswell, J.C. Bennett, J.M. Corbett, T. Vidmar, V. Marinkovic and A. Budkowski, *Acta Cryst. B* **46**, 587 (1990).
53. J.C. Bennett, F.W. Boswell, A. Prodan, J.M. Corbett and S. Ritchie, *J. Phys: Condens. Matter* **3**, 6959 (1991).
54. J.C. Bennett, S. Ritchie, A. Prodan, F.W. Boswell and J.M. Corbett, *J. Phys: Condens. Matter* **4**, 2155 (1992).
55. J.C. Bennett, F.W. Boswell, A. Prodan, J.M. Corbett and S. Ritchie, *Australian J. Chem.* **45**, 1363 (1992).
56. J. Van Landuyt, G. Van Tendeloo and S. Amelinckx, *Phys. Stat. Sol. (a)* **30**, 299 (1975).
57. K. Suzuki, M. Ichihara, I. Nakada and Y. Ishihara, *Solid State Commun.* **52**, 743 (1984).
58. K. Suzuki, M. Ichihara, I. Nakada and Y. Ishihara, *Solid State Commun* **59**, 291 (1986).
59. B.F. Buxton, J.A. Eades, J.W. Steeds and G.M. Rackham, *Phil. Trans. Roy. Soc.* **281**, 171 (1976).
60. M. Tanaka, R. Saito and H. Sekii, *Acta Cryst.* **A39**, 357 (1983).
61. J. Gjønnes and A.F. Moodie, *Acta Cryst.* **19**, 65 (1965).
62. M. Tanaka, H. Sekii and T. Nagasawa, *Acta Cryst.* **A39**, 825 (1983).
63. J.W. Steeds, D.M. Bird, D.J. Eaglesham, S. McKernan, R. Vincent and R.L. Withers, *Ultramicroscopy* **18**, 97 (1985).
64. M. Tanaka and M. Terauchi, *Convergent-Beam Electron Diffraction*, JEOL, Tokyo, 1985.
65. J.C.H. Spence and J. Zuo, *Electron Microdiffraction*, Plenum, New York, 1992.
66. D.E. Moncton, J.D. Axe and F.J. DiSalvo, *Phys. Rev. B* **16**, 801 (1977).
67. R.M. Fleming, D.E. Moncton, D.B. McWhan and F.J. DiSalvo, *Phys. Rev. Lett.* **45**, 576 (1980).
68. N.P. Ong and P. Monceau, *Phys. Rev. B* **16**, 3443 (1977).
69. J.C.H. Spence, *Experimental High Resolution Electron Microscopy*, Oxford University Press, Oxford, 2nd Ed 1988.
70. D. Van Dyck, M. Op de Beeck and W.M.J. Coene, *Microscopy Soc America Bull.* **24**, 426 (1994).
71. K.D. Bronsema, S. Van Smallen, J.L. de Boer, G.A. Wiegers and F. Jellinck, *Acta Cryst.* **B43**, 305 (1987).
72. E.D. Boyes, L. Hanna and P.L. Gai, *Proc 50th Annual Meeting of the Electron Microscopy Soc America*, San Francisco Press, San Francisco 1992, pp 98-99.

INFLUENCE OF DEFECTS AND IMPURITIES ON CHARGE DENSITY WAVE SYSTEMS

HANNU MUTKA

*Institut Max von Laue - Paul Langevin,
B.P. 156, F-38042 Grenoble Cedex 09, France*

1. Introduction

Charge density-waves (CDW) have been the subject of intensive research for about twenty years. Most often they are found in synthetic compounds issuing from the assiduous efforts of chemists. These fascinating materials, inorganic and organic quasi-one-dimensional and quasi-two-dimensional conductors, have metallic properties that are unusual in many ways [1-4]. The key feature is the unstable metallic state, apt to spontaneously form a charge-density modulation $\rho_{CDW}(r)$ and an associated lattice distortion. The modulation period is determined by the conduction electron density, i.e. related to the Fermi wave vector k_F,

$$\rho_{CDW} = \rho_o sin(qr + \phi) = \rho_o sin(2k_F r + \phi)$$ (1)

in 1D presentation. As a consequence of this instability there is a rich variety of low tempera-ture phase transitions driven by the temperature dependence of the CDW amplitude ρ_o and the interaction of its phase ϕ with the underlying lattice. One can find metal-insulator transitions, incommensurate and commensurate modulated structures for example. A central and most intriguing ingredient is the collective sliding mode conductivity [5-9] where the CDW conden-sate moves as a whole under an applied field. In practice this sliding is hindered by defects, since the phase ϕ of the modulation will have preferred values at defect sites, opposing the ideally free choice of position of the modulated charge density. This strong connection to defects influences the whole physics of charge density-waves with consequences that are manifest in a wide space and time scale from microscopic to macroscopic. Nevertheless the initial effects are found at the spatial scale of the defects and of the CDW wavelength. This is the reason why microstructural characterization of the CDW is of primary importance for understanding the CDW physics and of course not only with regard to the sliding CDW but also in the full scope of CDW related features.

Defects in charge density-wave compounds are inevitable and their influence on the charge-density wave phenomena is more a rule than an exception. Various descriptions of point defects, borrowed from semiconductor and metal physics are often useful as a first approach to

153

F. W. Boswell and J.C. Bennett (eds.),
Advances in the Crystallographic and Microstructural Analysis of Charge Density Wave Modulated Crystals, 153–184.
© 1999 *Kluwer Academic Publishers.*

understand the defect induced effects. As in a conventional semiconductor a foreign atom, by donating or accepting electrons, may change the average conduction electron density, i.e. the band filling. Indeed, such a global effect does not pass unnoticed in a system where the charge-density controls the periodicity of the modulated structure. On the microscopic scale the foreign potential of a defect can act as a scattering center, but rarely simply as a point scatterer as in most metals where individual collisions and, in consequence, a scattering time (or mean free path) approach can explain the effects of defects. In a charge-density wave system the collective response to a local potential is much stronger because the screening of the local charge is less complete than in normal metals. Locally, a high amplitude oscillating response (the Friedel oscillation) forms with the same periodicity ($2k_F$) as that which character-izes the CDW itself. This local charge oscillation interacts with the CDW and the phase of the CDW becomes pinned. Thus the defects have long range effects explaining why their presence is of determinate importance for the properties of the CDW. The pinning influences all CDW phenomena including the phase transitions, the structural characteristics, and the sliding of CDW. Due to strains in the pinned state, the ideal periodic order of the CDW may well be perturbed also by defects of the CDW modulation: CDW dislocations can occur and, as a particular feature, discommensurations, defects of the CDW akin to the antiphase boundaries of ordered alloys, can appear, either as intrinsic phenomenon or in association with the atomic lattice defects.

When treating the influence of defects on CDW one wishes to find out the genuinely "universal" defect phenomena and distinguish them from particular material properties, defect chemistry or defect physics. This is a very difficult practical problem. Our aim is to try to follow what has been observed and search for unifying features. In this context it is good to remember that while the materials we are examining are quasi-one-dimensional and quasi-two-dimensional, the CDW pinning takes place in three dimensions. From this point of view many of the problems are similar and independent of the systems examined.

Even though not much is known about the atomic scale structure of the defects, we try to briefly examine the expected varieties of point defects in section 2. After this the effects of defects on the formation of CDW and the appearance of the 3D ordered phase are introduced in the section 3. Note that this transition must pass through a characteristic 3D fluctuating regime and similar effects can be expected both for quasi-1D and -2D systems. In this section the fundamental reasons leading to the existence of disordered, pinned CDW are introduced. Section 4 is devoted to the structural variations in the CDW phase; the way defects affect the incommensurate-commensurate transition for example. A close connection between micro-scopic and macroscopic features is found. Phenomena such as the hysteresis of bulk properties due to variation of the CDW configuration, are seen to be amplified by defects. Some of these phenomena may have a connection with mobile defects and are explained within the concept of the Defect Density Wave, i.e. a defect configuration that mimics the CDW periodicity. To complete the two themes first treated, we then discuss the local defects of the CDW and their connection to the structural characteristics in section 5. Section 6 approaches the problem of evaluating the strength of the defect-CDW interaction. This is a central theme in the field because it is not evident under which conditions the effects of individual defects are simply additive and when they combine in a more complicated collective perturbation of the CDW. Under all these themes some cases of particular effects or specific compounds are examined to give more insight into the defect properties.

2. Point Defects in CDW Materials

2.1 GENERAL REMARKS

In this review we are mainly concerned with the point defects of the atomic lattice and interested in their influence on the CDW. Considerable literature exists on defect properties of even rather complicated materials [10-13] but the CDW systems are not yet well documented from the point of view of atomic defects. Of course the physical and chemical nature of the point defects is very variable in the great variety of compounds that show CDW phenomena. One may expect atomic scale defects due to imperfect crystallization or non-stoichiometry, and impurities in the starting materials. Impurities can be intentionally added: doping, substitution or even alloying can be used for changing the defect concentration. Irradiation with energetic particles (electrons, neutrons, high energy photons) has been used for defect production on several occasions. The general problem of all these approaches is the characterization and quantification of the defects. A vast majority of CDW literature makes reference to defect associated phenomena and there is a general agreement on their importance. Nevertheless, most often one finds only qualitative statements and meaningful quantitative features are harder to extract. In fact, in the absence of detailed thermodynamic data on defects, the assumption of a random quenched distribution considered in theoretical treatments is usually taken for granted. Even the basic defect properties, which have been extensively investigated in insulators, semiconductors and metals, are still to be assessed in the CDW materials.

Due to the multitude of effects the study of the defect associated phenomena calls for a wide range of experimental methods, both for characterizing the defects themselves and for investigating the perturbed CDW. The choice of methods is of course dictated partly by the applicability to a specific compound. The evaluation of the microstructural and structural properties needs the full power of modern scattering techniques, x-rays also with synchrotron sources, neutrons, electrons. Imaging with the help of electron microscopy and the powerful new tools, scanning tunneling and atomic force microscopies are the methods that approach the defect scale in space resolution. Electronic transport properties in a very wide spectral range, from dc to infrared are of interest. The magnetic resonance spectroscopies, ESR and NMR are very useful also as classical tools for defect physics. When it finally comes to examining the influence of defects on the CDW the ideal would of course be a method that is sensitive simultaneously to the CDW and to the defects, giving a signature, a spectrum, or an image of the two together.

2.2. SUBSTITUTIONAL IMPURITIES

Substitutional impurities in CDW compounds have been examined on several occasions. Both isoelectronic and non-isoelectronic substitutions have been done in the transition metal chalcogenides with chain and layer structures [1, 14-16], in blue bronzes [17-19] and in organic CDW materials. Isoelectronic substitution, for example can easily lead to alloying because some of the systems are isostructural and stable in the full concentration range. Non-isoelectronic impurities often have a rather low solubility and it is not always clear if they enter as true substitution or as interstitials in the structure. The general problem of substitution is that at very low concentrations ($c < 10^{-2}$ atomic fraction) the control of the impurity level, as well as of the homogeneity of the impurity distribution is very difficult. Also it is not evident how random is the spatial distribution of the impurities which are probably mobile and interacting at the crystal growth temperatures. The control of the growth conditions may be critical for obtaining the desired result as shown only very recently in the case of Ta defects in NbSe$_3$ where the oxygen content of the growth atmosphere has a drastic effect on the obtained concentration [20].

In certain cases impurity specific local probes have helped to characterize the substitutional defects. ESR has been used to investigate the V-impurity in NbSe$_3$ [21] and Mössbauer spectroscopy has been used to look for Fe substitution in the blue bronze [22].

2.3. IRRADIATION DEFECTS

2.3(a) *Production*

The production of point defects by irradiation with fast neutrons, electrons or gamma rays is a very practical method that offers at least two advantages. First, the defect production is linear with irradiation dose at low doses and the relative concentration of defects can be controlled with precision over a broad range covering several orders of magnitude below 10^{-2} atomic fraction. Second, at low temperature where the defects are immobile, the irradiation produces a truly random frozen in defect population whose effects can be examined *in-situ* with convenient methods. The inconvenience of irradiation is that defect production mechanisms and production rates can depend on the systems studied and have to be examined case by case.

In inorganic CDW crystals vacant lattice sites and interstitial atoms are produced by electron irradiation for example. Detailed case specific studies are necessary if one wants to know the production rate and characterize further the stable defects at more elevated temperatures. Some work along these lines has been done in transition metal chalcogenides. The defect production rate under electron irradiation was determined for $1T$-TaS_2 and TaS_3 in terms of the fraction of displacements in the Ta sublattice [23]. In the case of the blue bronze $K_{0.30}$ MoO_3, the ESR signal associated with irradiation defects helped to quantify the defect concentration [24]and later studies on the related red bronze $K_{0.33}MoO_3$, a semiconductor without CDW, gave further details on the possible defects and confirmed the order of magnitude of the damage rate [25].

In organic, molecular CDW materials the defect production by electronic excitations dominates [26-29]. The end product, a highly damaged molecule is a "point" defect in the molecular lattice.

2.3(b) *A Defect Model*

In the layered dichalcogenides a particular defect site is the interlayer van der Waals gap in which many sorts of intercalates can enter. Also self-intercalation of extra transition metal atoms in this space is plausible, and definitely occurs in $1T$-VSe_2 [30]. This defect process can be investigated by X-ray diffraction through change of lattice parameters and by magnetic susceptibility since the intercalated V-atom is a magnetic defect [30,31]. Comparison of the self intercalated $V_{1+x}Se_2$ and irradiated VSe_2 showed very interesting results that helped to specify the role of irradiation induced vacancies in the VSe_2 layers [32]. From the variation of lattice parameters it was possible to determine the formation volumes of both vacancies and interstitials: $-V/V_o = -0.54$ and 0.14, respectively, where V_o is the unit cell volume. The strong volume change associated with the vacancy type defect points out its force as a local electronic perturbation. Figure 1 shows these model defects in the layered structure.

2.3(c) *Prelude - the Effect of Defects on the Average Charge-Density*

Before getting involved with the intricacies of the CDW-defect interaction it is instructive to show results that nicely illustrate the band-filling or average charge-density effect, Figure 2. These results were and are of fundamental importance for the CDW physics. Alloying the $1T$ polytype of TaS_2 with Ti (d^0) which has one d-electron less than tantalum (d^1) decreases the d-type conduction band filling in direct proportion to Ti-concentration. The decrease of conduction electron density n is directly observable by electron diffraction [1] and by scanning tunneling microscopy [33] as an increase of the periodicity of the CDW due to the variation of the Fermi wave vector k_F. For a rigid 2D band, $2k_F/a^* = n^{1/2}/\pi$ which is in good agreement with the low-density region. These results demonstrate a high defect concentration effect that can be seen due to a fortunate circumstance: solid solutions of the $Ta_{1-x}Ti_xS_2$ system exist for the whole range of x. This example of the microstructural characterization of the defect influence on CDW shows the interest of electron diffraction and scanning tunneling microscopy,

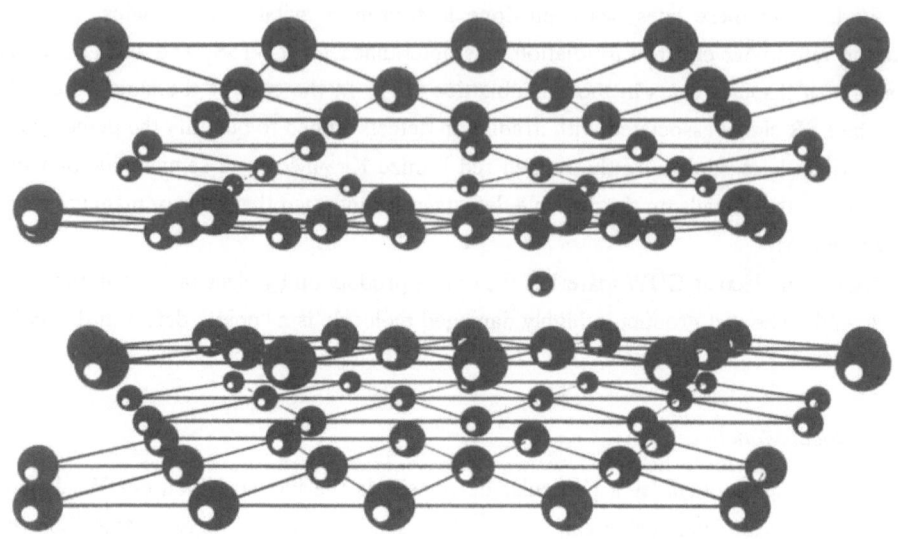

Fig.1 A defect model for the layered MX_2-compounds. Evidence for interstitials defects in the van der Waals gap and for vacancies in the layers has been obtained from magnetic susceptibility and X-ray diffraction experiments.

that are both presented in the articles of this volume.

3. CDW in the Presence of Defects

3.1. ONSET OF THE STATIC CDW

CDW do exist in strongly perturbed materials as shown by the above example of the Ta-Ti disulphide alloy for which CDW-modulation are found practically in the whole concentration range. Note that in a certain concentration range there must be on average several impurities in every structural unit of the CDW which have dimensions defined by the CDW wavelength. Clearly this implies that the alloying is a rather weak perturbation that cannot significantly influence the conditions of existence of the CDW-state. However, the long-range order is definitely perturbed. Early studies have suggested that long-range order of the CDW does not exist at all in the defect containing material [34,35]. Given that defect free materials do not exist, and that experimentally well-defined critical temperatures can be found, this latter conjecture is on first sight somewhat difficult to understand.

How do defects influence the stability of the CDW? To answer this question we must first know what controls the CDW formation at the limit of very pure material. In the

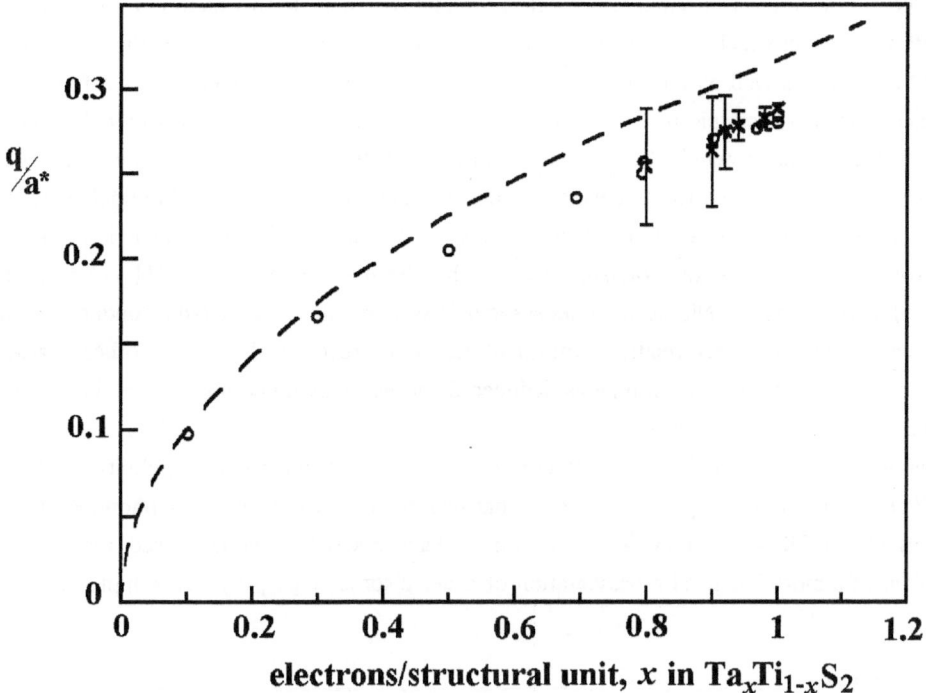

Fig.2 Defects that control the charge density can change the period of the CDW as shown by these results in Ti-doped $1T$-TaS$_2$. The modulation wave vector q/a^*, normalized to the reciprocal unit, is plotted versus the number of electrons per structural unit added by doping. The small circles represent data from electron diffraction (after [1]) and the crosses/error bars represent data from scanning tunnelling microscopy images (after [33]). The dashed line represents the variation that would be expected for an ideal 2D metal, $q/a^* = \sqrt{n}/\pi$, supposing that the electron density is equal to the Ta concentration $n = x$.

traditional approach it is easy to understand why the CDW phase is stable: there is gain of kinetic energy when the conduction electrons are condensed below the gap associated with the periodic lattice distortion that creates new "Brillouin zones". In this idealized model the CDW-formation is a consequence of the diverging electronic susceptibility $\chi(2k_F)$. The defects attenuate this divergence by decreasing the life-time τ of the electronic states, $\tau \propto 1/c$ [35-37]. A rapid decrease of the critical temperature of the CDW onset, initially equal to $T_O=T_{Oi}$, reaches absolute zero at a critical defect concentration c_{cr}, i.e. for a scattering rate

$$c_{cr} \propto \tau_{cr}^{-1} = \frac{\pi}{3.56\hbar} k_B T_{Oi} \tag{2}$$

that is of the order of $\tau_{cr} = 5\infty10^{-13}$ s for $T_{Oi}= 100$ K. Of course in this theory the low-temperature order parameter, the CDW modulation amplitude, decreases in proportion with the decrease of the onset temperature. It is difficult to reconcile this picture with the typical CDW materials in which the conduction electron collision times are short - on the order of $\tau = 10^{-15}$ seconds, as well as the mean free path - of the order of nanometers, i.e. a few lattice spacings, or only a couple of CDW wavelengths. The only case where this model seems to work is the layered system NbSe$_2$ in which a universal decrease of the CDW onset temperature occurs and can be related to the conduction electron lifetime [38-40], i.e. to the defect associated residual resistivity, as depicted in the Figure 3 where data for residual impurities in as-grown samples, and for irradiated samples has been compiled [38,39,41]. Among the layered dichalcogenides NbSe$_2$ is the most metallic system with good metallic conductivity and by consequence the above relations are quantitatively correct. In the case of other quasi-2D systems the electronic collision time as deduced from resistivity could not explain the decrease of the critical temperature [40].

Now, one might argue that the description in terms of electronic susceptibility controlled CDW formation is incomplete, that additional degrees of freedom such as phonons play an important role [42-44]. The CDW transition mechanism should then be revised, maybe up to the limit of a more localized representation of order-disorder type [45]. This approach might

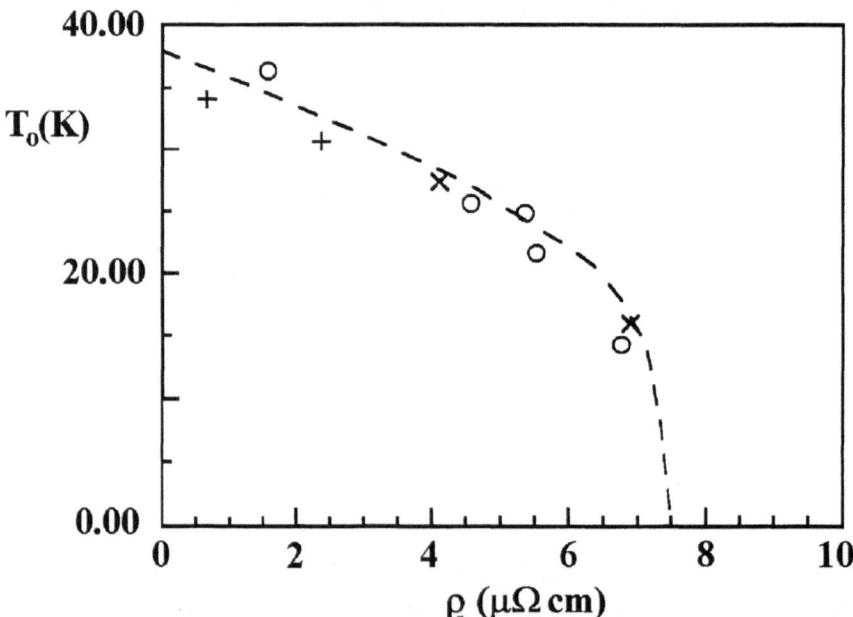

Fig.3 The charge density wave onset temperature T_0 versus the defect associated resistivity ρ_r in NbSe$_2$. A universal dependence, independent of defect type is observed. It follows the theoretical curve (dashed line, see [35-37]) that explains the decrease of the CDW onset temperature by the increased electron scattering rate. Data points are for residual impurities in as-grown material and for intentionally doped and irradiated samples.(o: as grown from [38], +: as grown from [41], x: electron irradiated from [39])

help to resolve the quantitative problem of the short life-time of the electronic states. In such cases, instead of the electronic susceptibility, another generalized susceptibility should be defined, and of course that could be influenced by defects in a qualitatively similar fashion. One might expect a renormalisation of the rate of decrease of the onset temperature and of the critical defect concentration.

In most CDW materials the critical temperature of the onset transition decreases when defects are introduced but the decrease definitely does not follow a simple universal dependence. This is demonstrated in Figure 4 which shows how the critical temperatures decrease in some representative quasi-1D CDW with irradiation defects. In search of the universal behaviour we have scaled the defect concentration with the initial onset temperature, see equation (2). The resulting initial slopes for the three systems become quite similar but this coincidence may be fortuitous. The important point is that the functional form is strikingly different. The common features are that the initial decrease quickly levels off at a rather elevated, system dependent, value and that the CDW onset transition disappears.

To clarify the loss of the phase transition one can examine the rather complete experimental characterization of the defect phenomena coming from scattering experiments that probe directly the fluctuations associated with the CDW transition, i.e. the dynamical susceptibility. X-ray and neutron scattering are suitable and have been used in some cases to examine the influence of defects. For example in the quasi-1D TMTSF-DMTCNQ the decrease of the critical temperature [46] and the loss of coherence [47] of the subsisting CDW was observed as a function of irradiation induced defect concentration. The important observation, found also in many other CDW systems is that even though defects suppress the phase transition they cannot totally extinguish the dynamical susceptibility. The disappearance of the phase transition leaves behind a disordered CDW, with ill-defined spatial coherence but considerable intensity.

The existence of such defect induced disordered CDW can be understood as a consequence of the local charge-density oscillations associated with defects. These enhanced Friedel oscillations are fixed on the defects, and already present above the phase transition temperature. Experimental evidence for their existence has been found by NMR both in quasi-1D and -2D systems (see section 3.3). More recently scanning tunneling microscopy images have been suggested to contain such features in the normal metallic state of quasi-2D CDW compounds.

The picture of defect-induced local CDW distortions approaches the treatment used in the general context of displacive phase transitions in the presence of defects. The phenomenon of the central peak associated with symmetry breaking defects has been treated by several authors over the years. Quite recent work on this subject has produced a rather appealing scenario in which the local distortions associated with defects produce a small static order parameter throughout the bulk of the material already well above the critical temperature [48,49]. The latter, at least for small defect concentrations, still persists, but the critical features are smeared

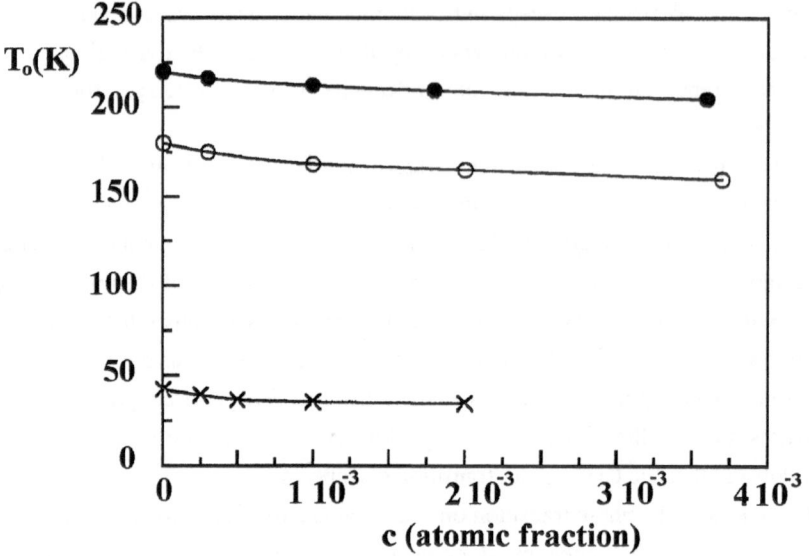

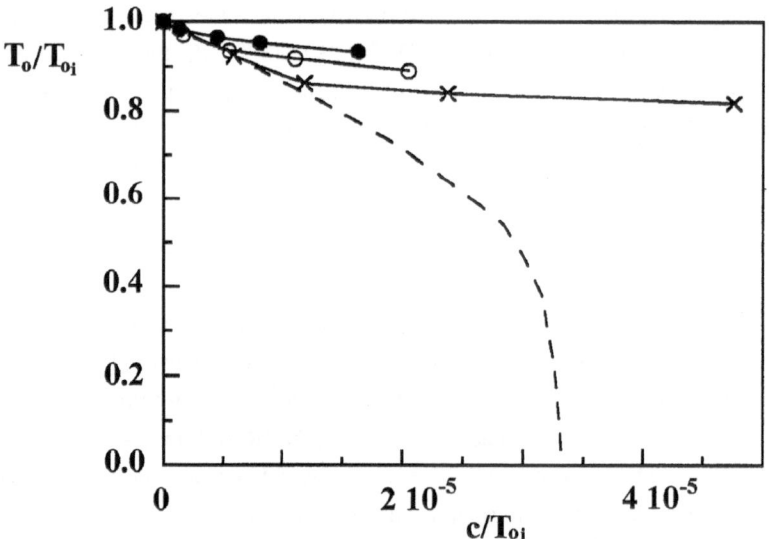

Fig.4 (a) The decrease of the Peierls transition temperature T_0 versus defect concentration for three representative quasi-1D CDW systems. Data are shown for TMTSF-DMTCNQ (x, from [46]) $K_{0.3}MoO_3$ (o, from [24]) and TaS_3.(•, from [67]). As a common feature one can note the levelling off of the decrease, associated with the smearing of the transition (see Fig. 7). (b) Normalized critical temperature does not follow the universal law determined by the scattering rate (dashed line, cf. Fig. 3). The deviation from the initial rate of decrease observed at low concentration is clearly system dependent, stronger for higher initial T_0. Symbols as in (a).

due to the presence of defects. Furthermore, another mean field type treatment in which the defect pinning is introduced as an input parameter suggests a possible positive curvature for the dependence of T_O on c [50].

In fact one expects that the observed variation of the critical temperature is due to a delicate balance between the different contributions: a decrease of the generalized susceptibility, an increase of the defect-induced order parameter and the concomitant phase disorder of the modulation. Random defect positions that hold the phase locally at a fixed value do not permit the establishment of a long-range order except with the cost of an extra deformation energy that reduces the stability of the CDW phase. At small defect concentrations, the static CDW amplitude is small above the critical temperature. Accordingly the CDW-defect interaction must be small as well. Up to a certain level this permits the phase transition to take place with a critical (at least for experiment) divergence of correlation length. However, at high defect concentrations the increase of the defect induced static amplitude increases the global interaction energy, the correlations become defect limited and the transition is lost. An important consequence of this description is that the ordering of the defect distribution with respect to the CDW modulation can influence the onset transition temperature [51].

3.2. PERTURBED STATIC ORDER AND PINNED LOW TEMPERATURE PHASE

The occurrence of the local CDW distortions around defects is the reason for the pinning of the CDW below the onset temperature, a feature that influences all further CDW phenomena at any defect concentration. It splits the CDW in two loosely defined categories: The "clean" CDW resulting from a more or less well defined phase transition, and the "defect dominated" CDW with no practically definable critical temperature and a strongly perturbed long range order. A point to stress is that for any finite defect concentration the CDW is pinned - and it does not necessarily have an infinite correlation length at low temperature. The defects are always there and they can limit the structural perfection. At low defect concentrations these effects can be quite subtle but they are the heart of the sliding CDW problem for example.

The defect perturbed phase coherence in pinned CDW below the critical temperature has been studied in a number of experiments. The early neutron and electron diffraction experiments on quasi-2D compounds showed the limited correlation length observed in doped systems ($c \geq 10^{-2}$) [52]. Later, in-situ irradiation experiments demonstrated how rather small defect concentrations ($c \leq 10^{-3}$) were able to break the structural coherence [53]. However, quantitative characterization of the low defect concentration limit had to await the outcome of the very high resolution synchrotron radiation work. Only quite recently it was shown that in samples whose bulk critical properties are indistinguishable with the usual methods, i.e. critical temperature, the defects can limit the phase coherence of the CDW. Varying defect concentration, both in doped $NbSe_3$ [54] and in irradiated $K_{0.30}MoO_3$ [55] was indeed shown to decrease the phase coherence lengths as observed by the width of CDW diffraction lines. This effect is clearly 3D in the sense that the phase coherence decreases in all directions. The number of defects in coherent domains is an interesting quantity and will be discussed further in section 3.4(b).

The global strain effects on the spatial coherence as observed by broadening of diffraction peaks are of course only one aspect of the influence of defects. One may expect that in addition to the local charge oscillations around a charged defect, there can exist extended defects that break the uniform CDW. Discommensurations, dislocations and stacking faults can occur. These objects will be introduced in more detail in section 5.

3.3. FRIEDEL OSCILLATIONS - PRETRANSITIONAL EFFECTS

In CDW systems the anomalous screening of the charged defect is the origin of the enhancement of the well-known Friedel oscillation found around defects in metals. This local response to defects was evoked early as a possible cause for suppressing the order [56]. More recently, theoretical work has detailed this response with the aim of establishing a more exact form for the long-range potential which has been assumed to be sinusoidal in the more phenomenological treatments of the CDW pinning [57,58]. According to these theories, for strong enough defects the Friedel oscillations in the CDW phase can be seen as locally pinned "inclusions" in the CDW matrix, quite in the way described earlier on more intuitive grounds [59], as depicted in Figure 5. Note, however, that the sharp boundaries are for illustration and do not necessarily exist. The dimensions of the defect associated inclusions are of the order of the CDW amplitude coherence length, that is a few tens of interatomic distances along the 1D chain and much less, perhaps just one chain wide in the transverse direction [60].

The aspect of local oscillations in the metallic phase above the bulk transition temperature, even though quite interesting from the point of view of the effect of defects on the critical properties, has not been investigated in great detail. However, some experimental evidence does exist. For example, the extensively studied Peierls system KCP(Br) seems to show the above mentioned defect limited coherence [56,61,62]. It has been suggested that the reason for this is the disordered Br sublattice but unfortunately this disorder cannot be varied. Similarly X-ray scattering on the blue bronze with about 2×10^{-2} V impurities points out a pretransitional order parameter enhanced by a factor of five with respect to pure material [63].

Further convincing evidence has been obtained by NMR. In the quasi-2D system $2H$-NbSe$_2$ NMR shows pretransitional line broadening above the critical temperature [64], indicating the existence of static (in the NMR time scale) CDW order parameter (Figure 6). Similar NMR observations were made for the quasi-1D blue bronze [65]. Unfortunately there is no study on the defect concentration dependence of these phenomena; therefore the quantitative characterization is still missing. It is noteworthy, however, that the results on the blue

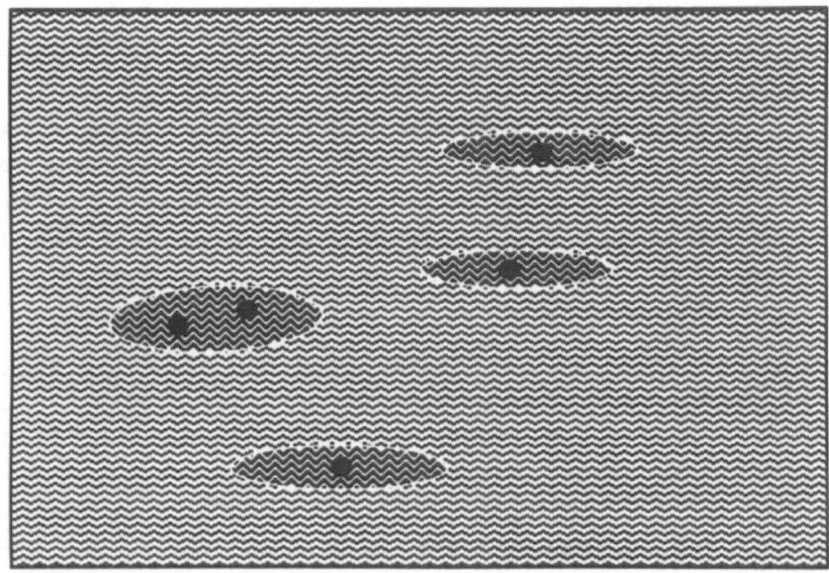

Fig. 5 The landscape of the pinned CDW in the 3D ordered phase consists of a largely unperturbed fraction with the strong elastic deformation of the CDW restricted to a limited range around the defects. At low defect concentrations a large fraction of connected volume of "defect free" CDW exists. This unperturbed fraction might show a long-range order in a scale far superior to the average distance between defects. At high defect concentration the strongly perturbed volumes start to touch and overlap, and lead to a qualitatively different picture where the CDW amplitude is expected to be locally perturbed. Such a defect dominated CDW could correspond to the transitionless regime that usually exists at high defect concentration.

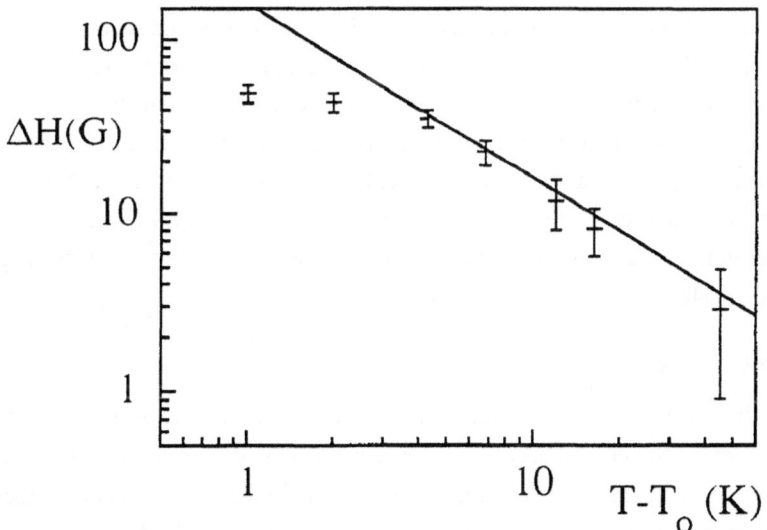

Fig.6 The broadening of the Nuclear Magnetic Resonance line in $NbSe_2$ indicates a static CDW amplitude above the bulk transition temperature T_o, data from [64]. This can be understood as a consequence of strong defect induced Friedel oscillations, amplified by the enhanced electronic susceptibility.

bronze suggest that the whole volume of the sample already has a small CDW amplitude above the bulk transition [65].

Such oscillations around defects could be visible by scanning tunneling microscopy and their characterization by images presents a real challenge. Some observations compatible with this idea do exist [66]. Again it would be useful to establish a connection with varying defect concentrations. Ideally such imaging should identify also the defects of the atomic lattice simultaneously.

3.4. DECREASE OF THE CRITICAL TEMPERATURE: THE EFFECT OF DEFECTS ON THE CDW AMPLITUDE AND COHERENCE

3.4(a) *Critical Temperature and Smearing of the Phase Transition*

The decrease of the critical temperature of the CDW transition has been measured in several compounds as function of the defect concentration. The effects range from the very rapid decrease typical of irradiation defects to the small perturbation of certain cases of alloying. A concomitant effect is the broadening of the transition, usually conveniently observed through some physical property that shows critical behaviour. For example, the magnetic susceptibility or the temperature derivative of the resistivity can be used.

The range of the observed phenomena can be seen in Figure 7 where we have collected data from irradiated [67] and doped [68] samples of TaS_3 and blue bronze [17]. Sometimes apparently very similar effects can be induced by very different defects and at distinctively different concentrations. The initial rate of decrease of the critical temperature varies by orders of magnitude, showing the variation of perturbation associated with specific defects.

A possible reason for the complete disappearance of the phase transition is a gradual coalescence of the strongly perturbed CDW regions (with volume V_d) around the defects. If we suppose this happens when the perturbed volume fraction is $F_p = 1-\exp(-cV_d) \leq 0.5$, then the defect concentration for such an event can be estimated to be in the $10^{-3} < c < 10^{-2}$ range for $100 < V_d < 200$ unit cells.

3.4(b) *Loss of Structural Coherence*

Clearly, small enough defect concentration may have no visible effect either on the phase transition temperature or on the critical broadening, except possibly in the very vicinity of the transition [44]. However, a decrease of the critical temperature associated with considerable smearing of the transition does occur at high enough concentration. This leads to the disappearance of the well-defined transition after the initial decrease has leveled off. The overall behaviour suggests that the CDW amplitude has not vanished in this process. Indeed, what is lost is the phase order of the CDW, leading to a transitionless state in which the CDW has a

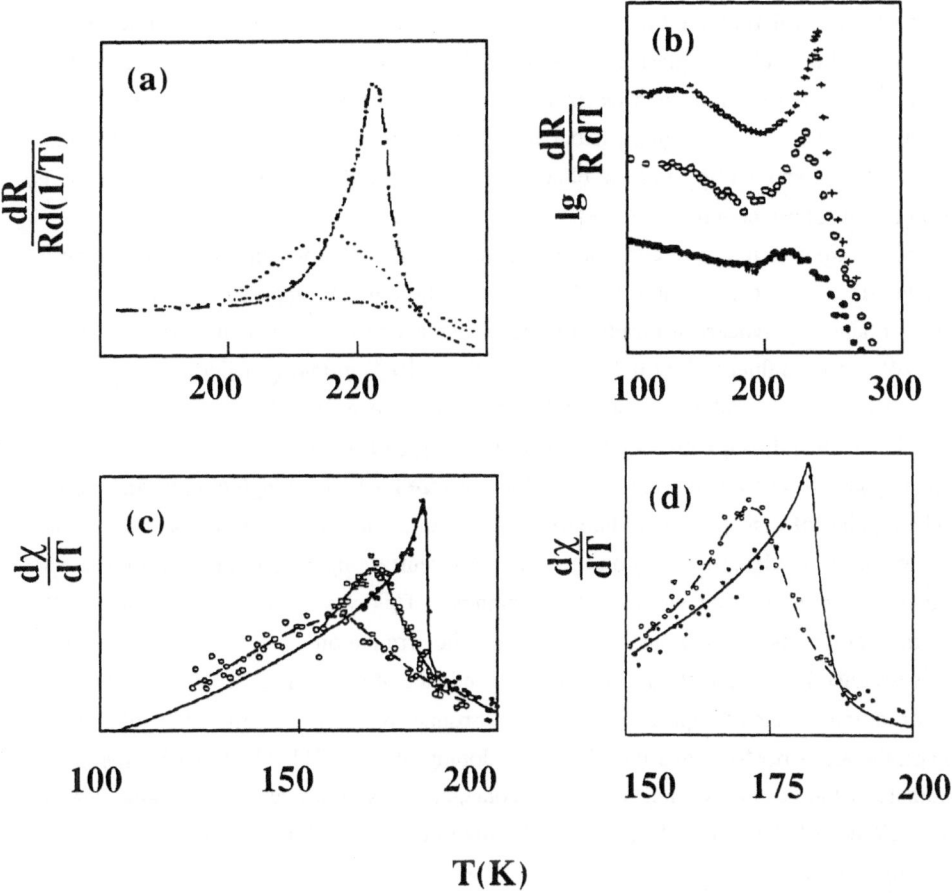

Fig. 7 Examples of the defect induced decrease of the critical temperature and of the smearing of the critical features in several systems. Similar features are observed with quite different kinds of defects, but in concentration ranges differing by orders of magnitude. (a) Temperature derivative of resistivity of electron beam irradiated orthorhombic TaS_3, from [67]. (b) Temperature derivative of resistivity of Nb doped orthorhombic $Ta_{1-x}Nb_xS_3$, from [68].   (c) Temperature derivative of magnetic susceptibility of W-doped blue bronze $K_{0.30}Mo_{1-x}W_xO_3$, for $x=0$, 0.004 and 0.01, from [17]. (d) Temperature derivative of magnetic susceptibility of alkali metal alloyed blue bronze, $Rb_{0.30}MoO_3$ compared with $K_{0.15}Rb_{0.15}MoO_3$, from [17].

very short range order. Typically the quasi-1D systems show a severe loss of transverse order even though the 1D correlations still persist. This has been observed in the case of TaS_3 by electron diffraction, also by *in-situ* measurements. Doped systems of the blue bronze have given evidence on such diffraction line broadening, too. The quasi-2D systems show similarly broadened CDW satellites in the disordered CDW regime.

The first quantitative work on the loss of phase coherence of the quasi-1D CDW was the X-ray study of the organic system TMTSF-DMTCNQ [47]. It confirmed the above picture and the decrease of the correlation lengths was compatible with a model where each defect breaks the phase correlation, i.e. there is essentially one defect in the coherent volume. Earlier it was found in the irradiated quasi-2D system $1T$-TaS_2 that this defects per ordered volume count was about 25 from dark-field electron microscopy observation of the in-plane ordered domains, by consequence many more occur in 3D [69]. In recent observations on the inorganic quasi-1D systems $NbSe_3$ and the blue bronze, from hundreds up to 10^4 - 10^5 defects, were found in the phase coherent volume.

The recent experiments have concentrated on the very small concentration limit where no appreciable effects are seen on the critical features or the critical temperature. The very high resolution obtained by synchrotron radiation was necessary to reveal that the phase coherence of the CDW can be influenced by defects in the 10^{-4} - 10^{-3} atomic concentration range. In addition quantitative estimates for the number of defects in a phase coherent volume were obtained. For $NbSe_3$ doped with Ta this number was huge, of the order of 10^5 [54].

Similar types of experiments on irradiated blue bronze have given highly interesting results [55]. The number of defects, several hundreds in the coherent volume, as well as the dependence of the correlation length ($c^{-0.4}$) on the defect concentration, were found to disagree with the standard theories of either weak or strong pinning. This was even more surprising in the light of the earlier results that had characterized the irradiation defects as strong pinning centers as revealed by the defect concentration dependence of the sliding CDW conductivity.

The structural effects of impurities on the blue bronzes have been examined by alkali metal alloying on the $K_{1-x}Rb_xMoO_3$ and with W and V doping [43, 70, 71]. The loss of coherence is quite notable in the case of W and V doping; concentrations approaching 10^{-2} range lead to coherent volumes of the order of $V_C \approx 10^5$ Å^3, while in the mixed system $K_{0.15}Rb_{0.15}MoO_3$ one finds $V_C \approx 10^8$ Å^3.

4. Structural Variations in the Pinned CDW

4.1. DEFECT PINNING AND LATTICE PINNING

Commonly the CDW modulation whose periodicity is defined by the special nesting Fermi-wave vector is incommensurate with respect to the period of the atomic lattice. However the interaction with the lattice potential favours a commensurate superstructure with a multiplied unit cell. When lowering the temperature the modulation wave vector often tends towards a commensurate value, and a phase transition to a commensurate phase can occur [43, 72, 73]. Among the different CDW phenomena this issue of variation of the CDW modulation with temperature (or another external parameter) is maybe the most dependent on the individual materials. This is not surprising since the original periodicity is determined by the Fermi wave vector which, in turn, depends on the details of the electronic structure and hence on the arrangement of the atomic units and their interactions. Furthermore, the commensurate phase is bound to obey well-defined symmetry relations with the underlying atomic lattice, and therefore the incommensurate to commensurate transition is associated with a rather elaborate 3D rearrangement of the modulated structure.

The pinning by defects has serious consequences on the variation of the incommensurate wave vector and on the stability of the commensurate phase. First, random defect positions are not compatible with a long range ordered commensurate structure. Second, the commensurate modulation is rigid because it has a gap for the long wavelength phase excitations. In other words, intrinsically the commensurate phase is pinned by the lattice, and to each CDW unit corresponds a given set of atomic unit cells. Together these ingredients create a frustrated situation with competing interactions. So it is not surprising to find hysteresis, dependence on thermal history and non-standard relaxation properties [74].

For the description of the physical phenomena these circumstances pose severe problems. The pure system limit with lattice pinning only, is already very intricate because of non-linear effects that can induce discommensurations or the devil's staircase [75, 76]. In the first situation the incommensurate CDW can split into commensurate domains, separated by an ordered array of localized phase slips, discommensurations, that make the global period incommensurate with the lattice. In the second case the temperature variation of the wave vector is discontinuous, jumping from one commensurate value to another by elimination of discommensurations. The inevitable existence of defects makes it difficult to judge whether or not the observed phenomena are intrinsic, defect induced or something in between. In the latter case, discommensurations, even if created by the intrinsic physics can be pinned by defects, and thus their number, mobility and spatial configuration can be governed by defects. In principle the defect-induced part can be investigated by controlling defect concentration, which in turn makes it conceivable to find out what would be left if the defects were absent.

The stability of the commensurate phase in the presence of defects has not been extensively studied, maybe because examples of well-defined incommensurate-commensurate transition are not too easy to find, especially for the quasi-1D case. Where data exists, it shows a high sensitivity of the commensurate phase to defects. More typical is the case where a commensurate phase seems to exist only as an asymptotic limit of the variation of the wave vector with temperature, as revealed by accurate diffraction experiments [43, 77]. On the other hand local probes such as magnetic resonance may show that inhomogeneous situations can persist, with coexistence of commensurate and incommensurate regions. This is exactly the ambiguous case presented above: one has to question whether the final deviation from commensurate is due to defects; would there be a clear transition in the truly pure material? If this is the case, it shows the extreme fragility of the commensurate phase, because by other means it is known that the purity of the materials is rather good; defect concentration must be down at the ppm level. On the other hand deviation from the commensurate in the defected systems can be so small, of the order of the finite width of the broadened diffraction peak [70], that it becomes difficult to define a well ordered, homogeneous commensurate phase.

A vast variety of different phenomena have been associated with the inhomogeneous variations of the CDW modulation wave vector in pinned situations(see e.g. sect. V in [78], [19]). Such features are commonly found in macroscopic properties with well-developed hysteresis and metastability. For example a temperature gradient can break a sample up into different macroscopic states that are metastable and could be characterized by different CDW wave vectors, or different concentrations of discommensurations. Similarly a homogeneous change of temperature can induce metastability that relaxes very slowly and with non-expo-nential laws, without a well-defined energy scale [79, 80]. Different kinds of memory effects have been found, and explained as consequences of the metastable CDW states. Some of them are understood by supposing that only the CDW degrees of freedom, after the application of an external action, have metastable configurations. Other cases have led to a very interesting proposition of metastable defect configuration, and to the concept of defect-density waves (DDW).

4.2 DEFECT DENSITY WAVES: INFLUENCE OF THE CDW ON THE DEFECTS

DDW is the counterpart of the pinning of the CDW by the defects; we might call it the trapping of the defects by the CDW. In the case of appreciable defect mobility, the CDW-defect inter-action can have an effect on the preferred positions of the defects, being minimized for defects separated by roughly integer number of CDW wavelengths. This leads to an ordered arrange-ment of defect positions for which the global interaction is minimized. Such a configuration could be "written" on the defect population by a prolonged treatment at a given temperature (for a given value of modulation periodicity) during which the defect configuration relaxes towards the optimum. Later this particular state could be observed as an anomaly of some CDW property at a subsequent passage across the same temperature. The DDW concept was originally proposed to explain such thermal memory effects in insulating incommensurate systems, and later generalized to the CDW case [81].

Within the DDW concept the random defect distribution approach to the CDW pinning has to be abandoned in favour of a less static picture where adjustments of the defect positions can help to optimize the energy of the pinned CDW. One has to remember that the relaxation of the defect distribution does not need long range diffusion of the defects, which may be rather difficult in the temperature range of typical CDW phenomena. The important spatial scale is the CDW wavelength, typically of the order of a few unit cells of the average crystal structure. Just a few diffusion jumps are enough to explore the available configurations that essentially change the pinning energy, and it has also been suggested that even a local modulation of the defect positions without any diffusion can occur [51]. At high defect concentrations the situation may be different due to interaction between defects that could lead to the necessity of collective behaviour, i.e. the simultaneous movement of many defects for reaching an optimum.

These interesting ideas are in principle verifiable by microscopic experiments on the CDW period with simultaneous characterization of the structural perfection. Nevertheless they have not been examined in detail up to now. For example there is no structural evidence for the DDW, for the microscopic pinning of the CDW period to a preferential value, or for relaxation of the CDW wavelength and diffraction linewidth. Experimentally these are not easy issues. For example during a very high precision diffraction experiment the initial temperature stabilization and sample alignment after temperature change may take enough time that the major part of the relaxation amplitude is missed. The continuing slow relaxation may be within the experimental error. Of course the verification of the DDW concept should also include a demonstration of the defect mobility and ordering which may be even a more difficult task.

As we already mentioned in the end of the previous section the strained CDW may well prefer a localized response instead of a uniform deformation. Accordingly CDW defects may play an important role in the variation of CDW modulation periodicity. The elastic and plastic response of the CDW under external perturbations has been developed in some detail in the theory [82, 83]. There is full analogy between the plastic response of dislocation containing crystals under stress and CDW under an electric field, and one can imagine that a similar dislocation response could happen when internal strains of CDW are present due to temperature variation of the modulation periodicity while in a pinned situation for example.

4.3. INHOMOGENEITY AND METASTABILITY

4.3(a) *Partly Commensurate CDW by ESR*

In the rubidium and potassium blue bronzes, a very interesting study of the defect pinning was carried out using the ESR of defects found in the as-grown material [84]. The simultaneous observation of the defects and their influence on the CDW was realized with this defect sensitive local probe in a very detailed study. It was found that at temperatures below $T= 50$ K the defects were located in two different environments, one with a continuous modulation of the g-factor typical of an incommensurate situation and the other giving a set of absorption lines with discrete g-factors characteristic of a commensurate structure. The commensurate fraction was evaluated quantitatively and its temperature dependence was determined. These observations suggest that the defect pinning hinders the formation of a homogeneous commensurate phase. Unfortunately the defect concentration dependence was not examined. Other ESR work on the blue bronzes has pointed out time-dependent phenomena that corroborate the ideas of mobile defects and defect density waves [19].

4.3(b) *Memory effect in blue bronze - irradiation effect*

The remarkable metastable phenomena observed in the blue bronzes are easily visible in the resistivity measurements. The temperature hysteresis and relaxation amplitude were shown to be sensitive to irradiation-induced disorder in the 10^{-5} to 10^{-3} defect concentration range [85]. This is illustrated in Figure 8 which shows the relevant features for irradiated $K_{0.30}MoO_3$. The relaxation induces a completely modified temperature dependence of resistivity, and a notable memory effect is found after a temperature excursion. As the effects are controlled by defect concentration, the association with the DDW ideas was suggested. A microscopic connection to the structural properties was conjectured but without experimental proof.

5. Local Perturbations of the CDW

5.1. THE LOCAL PICTURE OF THE PINNED CDW

The macroscopically averaged observations of the CDW pinning through bulk properties can be complemented by local spectroscopic probes as shown in the example of ESR presented above. These may help to elucidate the inhomogeneous situations where the local response of

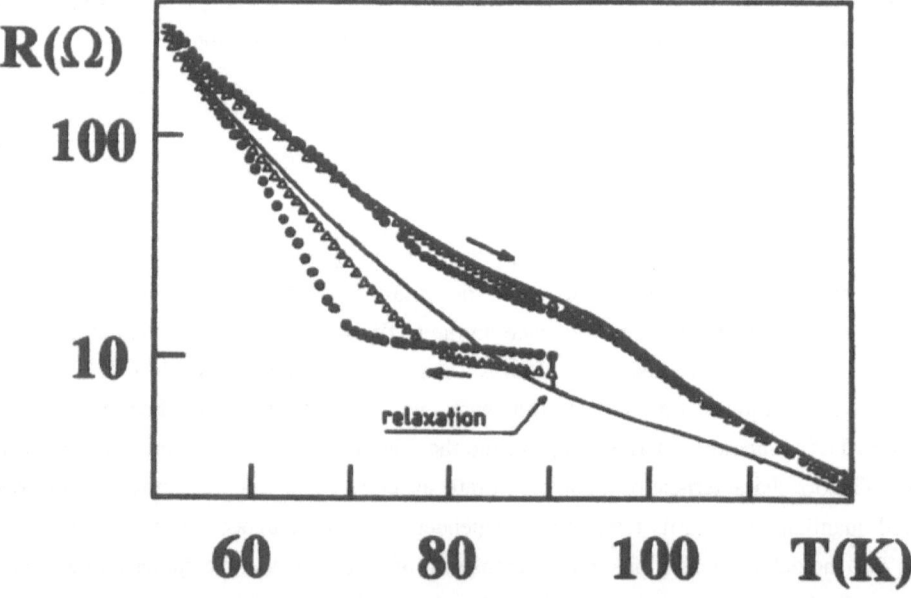

Fig.8 In the irradiated blue bronze $K_{0.30}MoO_3$ the amplification of the relaxation phenomena is accompanied by the enhancement of the memory effects. After a relaxation at a given temperature that proceeds by variation of the resistivity, a modified temperature dependence is found. In spite of the wide temperature excursion the system remembers the relaxation temperature, an anomaly is found on subsequent passage, from [85].

the CDW differs from the average state, for example the coexistence of incommensurate and commensurate phases. Nevertheless, this is not enough for obtaining a complete picture of the pinning. The pinning by defects is expected to lead to internally strained situations in which the CDW has to respond either by long-range elastic deformation or locally, by plastic deformation of its regular structure. In the examination of such phenomena microstructural characterization is invaluable and the various microscopy techniques play an important role since they allow the possibility of examining the order of the CDW at the scale of its primitive periodicity using real space images. One can then observe the defects that concern only the local order of the CDW even in the regions where the underlying atomic lattice is not defected. These defects of the CDW bear a full analogy with the lattice defects that are found in all crystals. Similarly they are essential for the plastic response that is quite analogous with the plasticity of atomic crystals. Extended defects such as dislocations and stacking faults are expected, and more exotic perturbations such as discommensurations can be found. One may also expect that the intrinsic discommensurations, if present, can be pinned by defects in a disordered configuration.

In principle transmission electron microscopy is an ideal tool for the observation of the CDW defects but in practice many difficulties are often encountered. Putting aside the purely technical problems concerning sample preparation and obtaining reasonable intensity for imaging, a real physical problem is the disordered nature of the defects. In fact to be observed and interpreted the defect must have a certain unperturbed 3D-structure over a large enough scale on the order of the sample thickness. This requirement is in fact somewhat in contradiction with the low dimensional properties of the CDW materials, especially at higher defect concentrations where the CDW defects may be poorly defined and often overlapping. In these circumstances one might expect that the recently developed surface sensitive probes, scanning tunneling and atomic force microscopy [86], could do very well because they intrinsically see only 2D or even just the 1D properties. This could offer an advantage especially for the strongly disordered cases where the 3D correlations are strongly perturbed. A highly interesting feature is the combination of the scanning tunneling microscopy with the local spectroscopy of the electronic states. The single particle excitation gap, or the localized states in this gap are visible and the effects of defects can be followed [87].

A good part of the CDW physics has centered on the discommensuration due to its analogy with the soliton, a continuum concept that has stimulated a lot of theoretical work. Periodic arrays of discommensuration defects were evoked as an alternative for the single sinusoidal incommensurate wave model, as being an energetically more favourable configuration at relatively high distortion amplitude [72, 75]. The observation of these objects in the quasi-2D systems by transmission electron microscopy using the satellite reflection dark field method was one of the most remarkable successes in CDW physics. Rather well organized arrangements of discommensurations clearly showed that they are an important part of the intrinsic physics of these systems [88].

Only rare observations of discommensurations in quasi-1D systems have been done in spite of the relative importance that they have in many theoretical explanations of the CDW phenomena. Recent work on $TaTe_4$ and on the $Ta_{1-x}Nb_xTe_4$ alloys [89, 90], reported in detail in this volume, is of interest for the defect aspect. It is the only reported example of defect induced proliferation of discommensurations in CDW systems. As such it constitutes a fundamental case in microstructural aspects of the quasi-1D systems that are rather difficult to reach by transmission electron microscopy. Systems like $NbSe_3$ or TaS_3 gave rather poorly defined "domain" contrasts whose interpretation and association with defects are not clear, even though small defect concentrations do not have any striking effect as shown for irradiated TaS_3 [91]. Other cases of microstructural characterization of defect induced local perturbation phenomena are found in quasi-2D materials. They complement in real space the picture observed by diffraction methods as satellite broadening and in bulk material as smearing of the phase transitions. The full structural variety of incommensurate and commensurate CDW have been thus observed within a wide defect concentration range.

5.2. CDW DEFECTS AND DISORDER IN THE IRRADIATED LAYERED COMPOUNDS TaS_2 AND $TaSe_2$

The layered transition metal dichalcogenides are the materials in which the defect induced CDW disorder has been microscopically characterized in a wide defect concentration range and for different kinds of defects, using various methods ranging from neutron scattering to electron microscopy and recently also scanning tunneling microscopy. Defects from substitutional doping with isoelectronic and non-isoelectronic impurities, electron or fast neutron irradiation can have very similar effects on the bulk properties even though the specific perturbation of the various defects varies quantitatively. For example, practically all kinds of defects can suppress the commensurate low-temperature phase (often called IT_3 phase) but the concentration needed varies from a few 10^{-3} atomic fraction in the case of irradiation defects up to several 10^{-2} atomic fraction for certain dopants. This one order of magnitude difference must reflect a similar variation in the local perturbing potential of the defects, i.e. between irradiation induced vacancies and interstitials with respect to substitutional impurities. In fact the high value of the formation volume of the vacancy type defect, determined for the V-sublattice vacancy in IT-VSe_2 indicates the strength of this type of defect that must be the strongest local perturbation that can created in these compounds.

The effect of a very low concentration of irradiation induced defects on the commensurate phase of IT-TaS_2 and IT-$TaSe_2$ was revealed by electron microscopy [69, 92-94]. Small "CDW defect" contrasts were observed in irradiated IT-TaS_2 and IT-$TaSe_2$ by satellite dark field imaging and high resolution lattice fringes showed a CDW-dislocation type defect associated with these objects. The density of these defects grows in proportion with the defect concentration. Meanwhile, it was also observed that the boundaries of the symmetry related commensurate CDW domains were perturbed by defects.

After suppressing the commensurate phase at higher defect concentrations ($c > 10^{-2}$) a granular contrast remains. The characteristic size of the observed CDW granularity was seen to decrease rapidly with growing defect concentration. Within this regime the granular contrasts showed no particular temperature dependence in spite of the fact that the electron diffraction patterns still had temperature dependence reminiscent of the incommensurate-nearly commensurate transition of the pure material. Invariably, and quite in contrast with the pure system, the two orientation variants of the nearly commensurate phase occurred simultaneously in a rather small spatial scale and could be distinguished by dark field imaging using the split CDW satellite reflection. At very high doses a featureless "glassy" CDW was observed at all temperatures even though some temperature dependent transverse broadening of the CDW satellite was still observable.

More recently the rather similar regime of defect induced disordering was observed by scanning tunneling microscopy [33, 95-97]. Ti- and Nb- doped samples of $1T$-TaS$_2$ and $1T$-TaSe$_2$ were investigated and the effects on the CDW microstructure were in many ways similar to the ones demonstrated earlier in the case of irradiation defects, with the already mentioned difference of concentration range. Here too, a defected CDW structure was evident. It must be noted that in several scanning tunneling microscopy observations a kind of CDW "vacancy" has appeared, in the form of a missing modulation maximum [86, 98]. The relation of this kind of CDW defect with the atomic lattice defects is not yet completely clear.

6. The Microstructural Aspect of the Pinning Strength

6.1 STRONG OR WEAK PINNING?

The defect pinning of the CDW is the central problem in the sliding CDW conductivity and has been the motivation for a major part of the work concerning defects in the CDW. The striking manifestation of the pinned state is the finite threshold field of the CDW conductivity. The full phenomenology, concerning the dc and frequency dependent electric response of the pinned CDW condensate has been the source of ample literature and several reviews [6-8] and these items are not touched in detail here. Considering the easy perturbation of the sliding CDW phenomena with various defects one might expect that the strength of defect CDW interaction is quickly derived from experiments in this domain. Unfortunately the high sensitivity is not of much help in understanding the individual defect properties because the global outcome of the microscopic defect pinning results from a non-trivial average over the defect distribution.

Due to this complicated relation, the work done in characterizing the properties of the pinned CDW has been in many ways detached from the defects that are responsible for the pinning. The understanding of the main ingredient of the problem - defects - has progressed much slower than the assessment of the associated phenomena. The latter can be explained in great detail with highly sophisticated models, where the effect of defects is introduced as a global parameter that has only a remote connection with the defects, their number, their nature, or their thermodynamics. Even the most up to date efforts deal with the problem of knowing whether the pinning is strong or weak [99-105], along the classification proposed by the early theoretical work. The results have often been contradictory and the analysis indecisive, reflecting the complexity of the problem.

The strength of the pinning was originally distinguished by comparing the individual defect interaction with the elastic response of the CDW, resulting in the weak (collectively) pinned state when the latter dominates [106-109]. The argument was that the CDW phase is locally adjusted at each defect site under weak pinning conditions. More recent theoretical work pointed out that such local phase adjustment is possible only for very weak defect potentials [57] and this ingredient appeared in the renewed pinning theories [60, 110]. This has not really clarified the experimental situation, the difference between the strong and weak pinning being even less distinct from the threshold field measurements in which size effects may play a non-negligible role. Experiments have now turned to structural indications of the pinning for characterizing the strong with respect to weak limits [54, 55, 96].

6.2 BASIC PINNING THEORY

Predictions for the perturbation of the structural coherence have been obtained by the Fukuyama-Lee-Rice pinning theory that treats the random defect distribution case at the low temperature limit supposing that the CDW phase deformation is the only degree of freedom [106-109]. The total pinning energy is

$$E_{pin} = \frac{K}{2} \int (\nabla \phi)^2 \, dV - \sum_d U \rho_o \cos(\phi - \phi_d) \tag{3}$$

The local interaction with a defect potential U gives an energy gain $U \rho_o \cos(\phi - \phi_d)$, where ϕ_d is the CDW phase at the defect site. This has to be compared to the global elastic deformation energy of the CDW obtained by integrating the phase deformation energy $(K/2)(\nabla \phi)^2$, where K is the stiffness of the CDW. The elastic energy loss depends sensitively on the range h in which the phase gradient $\nabla \phi$ is contained, as well as on the dimensionality d of the system [111, 112]. At the dilute limit when the phase is optimized on the defect site one finds the per defect pinning energy E_p

$$E_p = \frac{K}{2} h^{d-2} - U \rho_o \tag{4}$$

To minimize the deformation loss in 3D it is advantageous to keep the smallest possible $h = h_o$, determined by the upper limit of the phase excitation spectrum. For $d < 3$ the optimal phase deformation distance is the distance between defects, i.e. $h = 1/c^{1/d}$ where c is the defect concentration. As a consequence, the strong pinning situation persists when each impurity has the optimal phase, for pinning energies

$$U\rho_o > \frac{K}{2}h_o^{d-2} \text{ or } > \frac{K}{2}c^{-1+2/d}$$
(5)

This is depicted in the Figure 9 for the 3D case. The pinned CDW is schematically pictured in the Figure 5, showing the strongly perturbed volumes around defect positions, as well as the anisotropy of h_o.

One has to note that in the 3D case the validity of the treatment breaks down when the distance between defects becomes equal to the cut-off of the phase deformation $h_o = 1/c^{1/d}$. In other words, this can be stated as the limit when the fraction of perturbed volume around the defects is $F_p = 1 - \exp{-cV_d} = 1 - \exp{-c4\pi h_o^3/3} \approx 0.5$. Strong CDW amplitude deformations must then occur and result in a qualitatively different pinned state.

In the weak pinning case there are many impurities in the phase deformation volume that occur on a length scale $L > 1/c^{1/d}$. Part of the defects are offset from their optimal positions and due to these fluctuations the average energy gain at the defect site is reduced by a factor of $(cL^d)^{1/2}$. The pinning energy per defect is

$$E_p = \frac{K}{2cL^2} - \frac{U\rho_o}{(cL^d)^{1/2}}$$
(6)

Minimizing this energy with respect to L gives the length scale of the phase deformation the so-called Fukuyama-Lee-Rice coherence length

$$L_{FLR} = c^{\frac{1}{d-4}}\left(\frac{d}{2}\frac{U\rho_o}{K}\right)^{\frac{2}{d-4}}$$
(7)

and the minimized pinning energy is

$$E_{p,min} \approx -c^{\frac{d-2}{d-4}}\left(\frac{U\rho_o}{K}\right)^{\frac{d}{4-d}}\left(\frac{U\rho_o}{K}\right)$$
(8)

Of course, the weak pinning persists only when the characteristic length scale is greater than the average distance between defects or the cut-off length of the phase deformation, $L_{FLR} > 1/c^{1/d}$, $L_{FLR} > h_o$. Crossing over these limits leads to the regime where amplitude deformations must occur (Figure 9). Note also that the weak pinning energy depends on defect concentration and the stability limit depicted in Figure 9 is a zeroth order approximation.

6.3. RELATION TO EXPERIMENT

A semi-quantitative notion of the pinning conditions may be obtained using the basic theory presented above. Let us start by considering that the transitionless, disordered CDW regime is

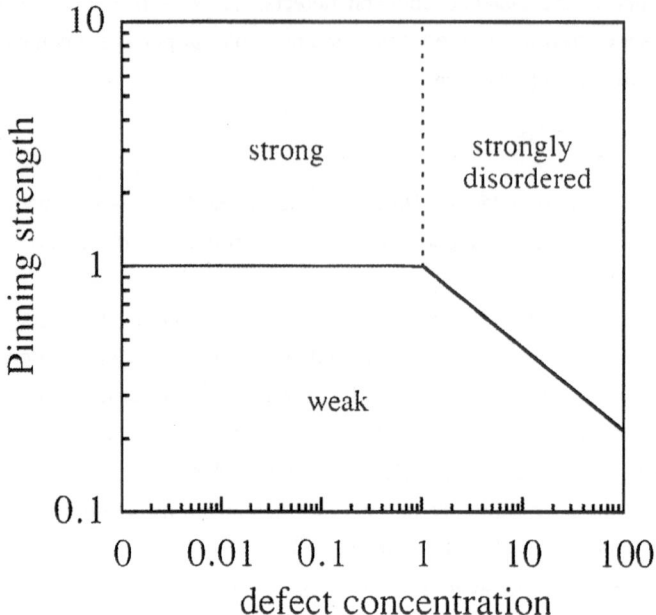

Fig.9 The ratio of the local energy gain to elastic constant of the phase deformation, $U\rho_o/K$ controls the pinning conditions (from [112]). In the 3D case depicted here, the normal strong pinning persists up to a critical concentration, determined by the cut-off length of the phase deformation, above which amplitude deformations must occur. Similar behaviour is observed in the case of weak pinning when the Fukuyama-Lee-Rice coherence length becomes less than the average distance between defects.

indeed due to the coalescence of the perturbed volumes around individual defects, according to the suggestion introduced in section 3.4(a). The perturbed volume of $100<V_d<200$ unit cells gives an indication of the phase deformation cut-off length $h_o \approx 5$ to 6 , and on the critical pinning strength $U\rho_o / K \approx 3$. Equation (7) can be then used to calculate the number of defects in the coherent volume; in 3D

$$N_c = cL_{FLR}^3 = c^{-2}\left(\frac{3}{2}\frac{U\rho_o}{K}\right)^{-6} \tag{9}$$

at the critical limit as well as for weaker pinning conditions. The concentration dependence of this quantity N_c is plotted in the Figure 10 for different pinning strengths. It shows the kind of variation that could be expected if the defect concentration and the perturbing potential could be varied at will.

For the weak pinning case the variation of the number of defects per coherent volume decreases as c^{-2}, and the absolute value of N_c can be high even quite close to the strong pinning limit. Experimentally, for NbSe$_3$ doped with Ta in the 10^{-3} range [54] an elevated value of N_c has been observed: $N_c = 10^5$ at $c=1.8\times10^{-3}$. However, the true concentration dependence of N_c (or L_{FLR}) has not been fully demonstrated, even though the data (only two defect concentrations) is not incompatible with equation (9). In the quasi-2D $1T$-TaS_2 doped with Nb [96, 97], conditions attributed to weak pinning lead to formation of localized defects in form of dislocations which is clearly out of the validity of the long-range phase deformation only approach.

For proper characterization of the coherence a large dynamic range in defect concentration is desirable. An approach in that direction was the study of the irradiated blue bronze [55] even though barely two orders of magnitude were covered from $c = 3 \times 10^{-5}$ to 10^{-3}. This study, concerning the strong pinning case by reference to other measurements [113, 114], pointed out an apparently puzzling behaviour. While, as expected, the number defects in the phase coherent volume had a weak dependence on defect concentration the number itself was high, of the order of 10^3. This is in strong contradiction with the expected value which is of the order of one defect per coherent volume [55]. Earlier, the one defect per coherent volume was found in the irradiated organic compound TMTSF-DMTCNQ [47]. However ,this investigation was done at rather high defect concentrations $c = 10^{-3}$ to 10^{-2}, approaching the limit where the perturbed volume fraction is significant (see Figure 10) and for this reason the results may not be representative of true strong pinning at the dilute limit.

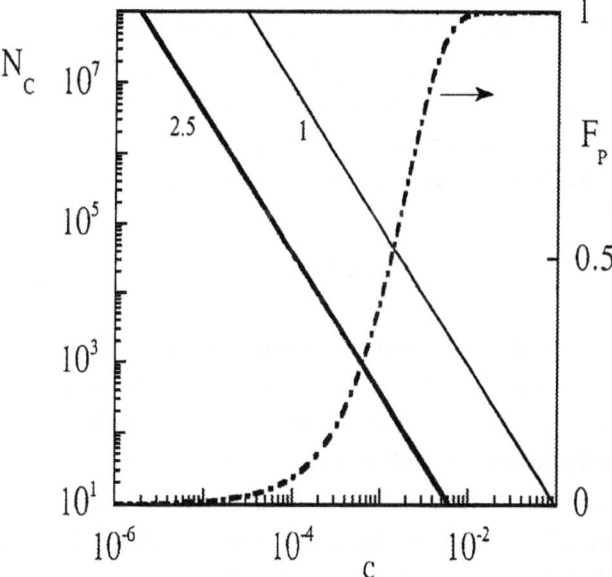

Fig.10 The weak and strong pinning conditions are schematized here as a function of the defect concentration supposing that the phase deformation cut-off length is $h_o = 5$ and the critical pinning strength $U\rho_o/K = 2.5$. Pinning without strong disorder can persist up to the limit where the defect perturbed volume fraction $F_p = 1 - \exp{-4\pi h_o^3/3}$ (right hand scale) fills a considerable proportion of the sample. For pinning weaker than the critical limit the number of defects in the coherent volume N_c is very high (left hand scale, $U\rho_o/K$ as parameter), however with a pronounced decrease with increasing defect concentration.

In fact one should consider the possibility that the number of defects per coherent volume is large at the weak concentration limit since a coherently connected 3D volume can withstand a large number of defects with short range deformations around them ($h_o \ll 1/c^{1/3}$), as illustrated in Figure 5. This point of view is the basis of the modified strong pinning theory [60, 110]. The dependence of the coherent volume on defect concentration has not been worked out in this theory.

Problems in the experimental investigation might be due to the characteristics of the defect distribution. Non-random fluctuations (clustering, ordering) will influence the phase coherence both for weak and strong pinning conditions. Clearly the chemistry/crystallography of the defects plays an important role. The atomic scale defects in the inorganic crystals with simple structures must have a good chance of finding a relaxed configuration. For impurities this could happen at the crystal growth temperatures. Certain irradiation induced damage has been found to heal at cryogenic temperatures, indicating highly mobile defect species reminiscent of those found in simple metals. In such conditions it is not surprising if the defects order to minimize the static interaction with the CDW, with decreased influence on the static coherence and still maintain a strong pinning in dynamic situations.

Clearly the relation between the defect species, DDW and the CDW's elastic and plastic response is a key factor in understanding the microscopic nature of the pinning. Combining the powerful microstructural methods with the more global diffraction techniques should enable light to be shed on the problem of the pinning strength. Knowing that concentrations in the 10^{-6} atomic fraction range can have a significant effect, it is not easy to characterize the defect configuration. In fact, defect phenomena exist in the purest obtainable materials. However, controlled variation of the defect concentration is necessary to highlight these effects.

7. Conclusion

Charge-density wave materials are synthetic compounds that often have complicated structures. They are typically produced under conditions in which the control of stoichiometry, of purity and of structural perfection are not very precise. On the other hand, a good part of the CDW physics is highly sensitive to defects and methods to establish a connection between the microscopic interactions with defects with the global properties are not evident. This is of course a general problem of the physics of disordered systems, with the particularity that in the CDW materials a small proportion of microscopic disorder can control the global properties due to the collective nature of the phenomena.

As a result we have to deal with a problem with many facets which has to be approached carefully to avoid drawing hasty conclusions. At the moment, the characterization of the CDW transport and electrodynamics is far more advanced than the evaluation of the microscopic defect properties. The description of the effects of defects on the CDW thermodynamics, on the structural properties and the understanding of the defects themselves are only partly elucidated. For example there is no adequate quantitative explanation for the suppression of the critical temperature of the CDW onset, except at the limit of very weak CDW in a close to 3D metallic system. Experimental data suggests that the defects are definitely not a small perturbation but are involved with the whole CDW modulation, modifying the properties over the whole range of its existence.

For further progress it is necessary to characterize the defects better. For example the thermodynamic properties of defects that affect the stability, mobility, the distribution statistics and, by consequence, the interactions with the CDW are not well known. Without adequately defining the defect properties it is impossible to validate the theoretical approaches. Progress in this field should help in developing a less static picture of the pinning that is necessary for understanding the interesting problems of defect ordering, to tell when the defects are pinning the CDW and when the CDW is trapping the defects.

Acknowledgements

Several figures were reproduced from previously published works by the kind permission of the American Physical Society (Figures 2, 7b-d), Taylor and Francis Ltd. (Figure 2), Les Editions de Physique (Figures 7a, 8) and the Physical Society of Japan (Figure 9).

References

1. J.A. Wilson, F.J. Di Salvo and S. Mahajan, *Adv. Phys.*, **24**, 117 (1975).
2. D. Jérome and H.J. Schultz, *Adv. Phys.*, **31**, 299 (1982).
3. J. Rouxel, Ed., *Crystal chemistry and properties of materials with quasi-one-dimensional structures*, D. Reidel Publ. Co., Dordrecht, (1986).
4. C. Schlenker, Ed., *Low-dimensional Electronic Properties of Molybdenum Bronzes and Oxides*, Kluwer Academic Publishers, Dordrecht, (1989).
5. P. Monceau, Ed., *Electronic Properties of Inorganic Quasi-One-Dimensional Compounds. Part II: Experimental*, D. Reidel Publ. Co., Dordrecht/Boston/Lancaster, (1985).
6. P. Monceau, in *Electronic Properties of Inorganic Quasi-One-Dimensional Compounds. Part II: Experimental*, Ed. P. Monceau, D. Reidel Publ. Co., Dordrecht, (1985), p. 138.
7. G. Grüner, *Rev. Mod. Phys.*, **60**, 1129 (1988).
8. L. Gorkov and G. Grüner, Ed., *Charge Density Waves in Solids*, Modern Problems in Condensed Matter Science, ed. V.M. Agranovich and A.A. Maradudin. Vol. 25, North Holland, Amsterdam, (1989).
9. G. Grüner and P. Monceau, in *Charge Density Waves in Solids*, Ed. L. Gorkov and G. Grüner, North Holland, Amsterdam, (1989), p. 137.
10. J.H. Crawford and L.M. Slifkin, Ed., *Point defects in solids*, Vol. 1,2 &3, Plenum Press, New York, 1972, 1975, 1978.
11. A.M. Stoneham, *Theory of defects in solids*, Clarendon Press, Oxford, (1975).
12. W. Hayes and A.M. Stoneham, *Defects and defect processes in nonmetallic solids*, J. Wiley & Sons, New York, (1985).
13. F. Agullo–Lopez, C.R.A. Catlow, and P.D. Townsend, *Point defects in materials*, Academic Press, London, (1988).
14. J.W. Brill, N.P. Ong, J.C. Eckert, J.W. Savage, S.K. Khanna, and R.B. Somoano, *Phys. Rev. B*, **23**, 1517 (1981).

15. P.M. Chaikin, W.W. Fuller, R. Lacoe, J.F. Kwak, R.L. Greene, J.C. Eckert and N.P. Ong, *Solid. St. Commun.*, **39**, 553 (1981).
16. M.P. Everson and R.V. Coleman, *Phys. Rev. B*, **28**, 6656 (1983).
17. L.F. Schneemeyer, F.J. DiSalvo, S.E. Spengler and J.V. Waszczak, *Phys. Rev. B*, **30**, 4297 (1984).
18. B.T. Collins, K.V. Ramanujachary, M. Greenblatt and J.V. Waszczak, *J. Sol. St. Chem.*, **77**, 384 (1988).
19. C. Schlenker, J. Dumas, C. Escribe-Filippini and H. Guyot, in *Low-dimensional Electronic Properties of Molybdenum Bronzes and Oxides*, Ed. C. Schlenker, Kluwer Academic Publishers, Dordrecht, (1989), p. 159.
20. R.E. Thorne, *Phys. Rev. B*, **45**, 5804 (1992).
21. M.C. Aronson and M.B. Salamon, *Phys. Rev. B*, **38**, 10476 (1988).
22. J.Y. Veuillen, R. Chevalier, J. Marcus and C. Schlenker, *Solid. St. Commun.*, **63**, 587 (1987).
23. D.J. Lesueur, J. Morillo, H. Mutka, A. Audouard and J.C. Jousset, *Rad. Effects*, 77, 125 (1983).
24. H. Mutka, S. Bouffard, M. Sanquer, J. Dumas and C. Schlenker, *Mol. Cryst. Liq. Cryst.*, **121**, 133 (1985).
25. H. Vichery, F. Rullier-Albenque and S. Bouffard, *J. Phys. France*, **50**, 685 (1989).
26. G. Mihály and L. Zuppiroli, *Phil. Mag.*, **A45**, 549 (1982).
27. L. Zuppiroli, *Rad. Effects*, **62**, 53 (1982).
28. L. Zuppiroli, in *Low Dimensional Conductors and Superconductors*, Ed. D. Jérome and L.G. Caron, Plenum Press, New York, (1987), p. 307.
29. L. Zuppiroli, in *Highly Conducting Quasi-one-dimensional Organic Crystals*, Ed. E.M. Conwell, Academic Press, New York, (1988), p. 437
30. L.F. Schneemeyer, Ph. D. Thesis, Cornell University, 1978.
31. F.J. DiSalvo and J.V. Waszczak, *Phys. Rev. B*, **23**, 457 (1981).
32. H. Mutka and P. Molinié, *J. Phys. C: Solid State Phys.*, 15, 6305 (1982).
33. X.-L. Wu, P. Zhou and C.M. Lieber, *Phys. Rev. Lett.*, **61**, 2604 (1988).
34. Y. Imry and Shang-keng Ma, *Phys. Rev. Lett.*, **35**, 1399 (1975).
35. L.J. Sham and B.R. Patton, *Phys. Rev. B*, **13**, 3151 (1976).
36. H. Schuster, *Solid. St. Commun.*, **14**, 127 (1974).
37. W.A. Roshen, *Phys, Rev. B*, **31**, 7296 (1985).
38. J.A.R. Stiles, D.L. Williams, , and M.J. Zuckermann, *J. Phys. C: Solid State Phys.*, **9**, L489 (1978).
39. H. Mutka, Thèse Docteur-és-Sciences Physiques, Université de Paris-Sud Orsay, 1983.
40. H. Mutka, N Housseau,. J. Pelissier, R. Ayroles, and C. Roucau, *Solid St. Commun.*, **50**, 161 (1984).
41. J.R. Long, S.P. Bowen and N.E. Lewis, *Solid. St. Commun.*, **22**, 363 (1977).
42. W.L. McMillan, *Phys. Rev. B*, **16**, 643 (1977).
43. J.-P. Pouget, in *Low-dimensional Electronic Properties of Molybdenum Bronzes and Oxides*, Ed. C. Schlenker, Kluwer Academic Publishers, Dordrecht, (1989), p. 87.
44. R.S. Kwok, G. Grüner, and S.E. Brown, *Phys. Rev. Lett.*, **65**, 365 (1990).
45. M.D. Núñez-Regueiro, J.M. Lopez-Castillo and C. Ayache, *Phys. Rev. Lett.*, **55**, 1391 (1985).
46. L. Forró, A. Jánossy, L. Zuppiroli, and K. Bechgaard, *J. de Physique*, **43**, 977 (1982).
47. L. Forró, L. Zuppiroli, J.P. Pouget, and K. Bechgaard, *Phys. Rev. B*, **27**, 7600 (1983).
48. B.I. Halperin and C.M. Varma, *Phys. Rev. B*, **14**, 4030 (1976).
49. F. Schwabl and U.C. Täucher, *Phys, Rev. B*, **43**, 11112 (1991).
50. I. Bâldea and M. Badescu, *Phys. Rev. B*, **48**, 8619 (1993).
51. I. Bâldea, *Physica Scripta*, **42**, 749 (1990).
52. D.E. Moncton, F.J. DiSalvo, J.D. Axe, L.J.Sham and B.R. Patton, *Phys. Rev. B*, **14**, 3432 (1976).
53. H. Mutka, *Phase Trans.*, **11**, 221 (1988).
54. E. Sweetland, C.-Y. Tsai, B.A. Wintner, J.D. Brock and R.E. Thorne, *Phys. Rev. Lett.*, **65**, 3165 (1990).
55. S.M. DeLand, G. Mozurkewich, and L.D. Chapman, *Phys. Rev. Lett.*, **66**, 2026 (1991).
56. L.J. Sham and B.R. Patton, *Phys. Rev. Lett.*, **36**, 733 (1976).
57. I. Tüttö and A. Zawadowski, *Phys. Rev. B*, **32**, 2449 (1985).
58. L. Jian-cheng, *J. Phys. C: Solid State Physics*, **20**, 4917 (1987).
59. L. Zuppiroli, H. Mutka, and S. Bouffard, *Mol. Cryst Liq. Cryst.*, **81**, 1 (1982).
60. J.R. Tucker, W.G. Lyons, and G. Gammie, *Phys. Rev. B*, **38**, 1148 (1988).
61. B. Renker, L. Pintschovius, W. Gläser, H. Rietschel, R. Comès, L. Liebert and W. Drexel, *Phys. Rev. Lett.*, **32**, 836 (1974).
62. K. Carneiro, in *Electronic Properties of Inorganic Quasi-One-Dimensional Compounds. Part II: Experimental*, Ed. P. Monceau, D. Reidel Publ. Co., Dordrecht, 1985, p.
63. S. Girault, A.H. Moudden, J.P. Pouget, and J.M. Godard, *Phys. Rev. B*, **38**, 7980 (1988).

64. C. Berthier, D. Jérome, and P. Molinié, *J. Phys. C: Solid State Physics*, **11**, 797 (1978).
65. P. Butaud, P. Ségransan, C. Berthier, J. Dumas, and C. Schlenker, *Phys. Rev. Lett.*, **55**, 253 (1985).
66. G.P.E.M. van Bakel and J.T.D. Hosson, *Phys. Rev. B*, **46**, 2001 (1992).
67. H. Mutka, S. Bouffard, G. Mihály, and L. Mihály, *J. Phys. (Paris) Lettres*, **45**, L113 (1984).
68. D. Reagor and G. Grüner, *Phys. Rev. B*, **39**, 7626 (1989).
69. H. Mutka and N. Housseau, *Phil. Mag.*, 797 (1983).
70. T. Tamegai, K. Tsutsumi and S. Kagoshima, *Synt. Met.*, **19**, 923 (1987).
71. S. Girault, A.H. Moudden and J.P. Pouget, *Phys. Rev. B*, **39**, 4430 (1989).
72. W.L. McMillan, *Phys. Rev. B*, **12**, 1187 (1975).
73. J.P. Pouget and R. Comes, in *Charge Density Waves in Solids*, Ed. L. Gorkov and G. Grüner, North-Holland, Amsterdam, (1989), p. 85.
74. P. Prelovsek, *Phase. Trans.*, **11**, 203 (1988).
75. P. Bak, *Rep. Prog. Phys.*, **45**, 587 (1982).
76. S. Aubry and P. Quemerais, in *Low-dimensional Electronic Properties of Molybdenum Bronzes and Oxides*, Ed. C. Schlenker, Kluwer Academic Publishers, Dordrecht, (1989), p. 295.
77. A.H. Moudden, J.D. Axe, P. Monceau and F. Lévy, *Phys. Rev. Lett.*, **65**, 223 (1990).
78. G. Hutiray and J. Sólyom, Ed.,*Charge Density Waves in Solids*, Lecture Notes in Physics, Vol. 217, Springer-Verlag, Berlin Heidelberg New York Tokyo, (1985).
79. G. Mihály and L. Mihály, *Phys. Rev. Lett.*, **52**, 149 (1984).
80. R.J. Cava, R.M. Fleming, E.A. Rietman, R.G. Gunn, and L.F. Schneemeyer, *Phys. Rev. Lett.*, **53**, 1677 (1984).
81. J.P. Jamet, *Phase Trans.*, **11**, 335 (1988).
82. D. Feinberg and J. Friedel, *J. Phys. France*, **49**, 485 (1988).
83. D. Feinberg and J. Friedel, in *Low-dimensional Electronic Properties of Molybdenum Bronzes and Oxides*, Ed. C. Schlenker, Kluwer Academic Publishers, Dordrecht, (1989), p. 407.
84. A. Janossy, G.L. Dunifer and J.S. Payson, *Phys. Rev. B*, **38**, 1577 (1988).
85. H. Mutka, F. Rullier-Albenque, and S. Bouffard, *J. Phys. (Paris)*, **48**, 425 (1987).
86. R.V. Coleman, Z. Dai, W.W. McNairy, C.G. Slough, and C. Wang, in *Surface properties of layered structures*, Ed. G. Benedek, Kluwer Academic Publishers, Dordrecht, (1992), p. 27.
87. Z. Dai, C.G. Slough, and R.V. Coleman, *Phys. Rev. B*, **45**, 9469 (1992).
88. R.L. Withers and J.A. Wilson, *J. Phys. C: Solid State Phys*, **19**, 4809 (1986).
89. J.C. Bennett, F.W. Boswell, A. Prodan, J.M. Corbett, and S. Ritchie, *J. Phys.: Condens. Matter*, **3**, 6959 (1991).
90. J.C. Bennett, S. Ritchie, A. Prodan, F.W. Boswell, and J.M. Corbett, *J. Phys.: Condens. Matter*, **4**, 2155 (1992).
91. H. Mutka, S. Bouffard and L. Zuppiroli, *Lecture Notes in Physics*, **217**, 55 (1985).
92. G. Salvetti, C. Roucau, R. Ayroles, H. Mutka and P. Molinié, *C.R. Acad. Sc. Paris*, **299**, 843 (1985).
93. G. Salvetti, R. Ayroles, C. Roucau, H. Mutka and P. Molinié, *Lecture Notes in Physics*, **217**, 92 (1985).
94. G. Salvetti, C. Roucau, R. Ayroles, H. Mutka and P. Molinié, *J. Phys. (Paris) Lettres*, **46**, L507 (1985).
95. X.L. Wu and C.M. Lieber, *Phys. Rev. B*, **41**, 1239 (1990).
96. H. Dai and C.M. Lieber, *Phys. Rev. Lett.*, **66**, 3183 (1991).
97. H. Dai and C.M. Lieber, *Phys. Rev. Lett.*, **69**, 1576 (1992).
98. B. Giambattista, C.G. Slough, W.W. McNairy and R.V. Coleman, , *Phys. Rev. B*, **41**, 10082 (1990).
99. J. McCarten, M. Maher, T.L. Adelman, and R.E. Thorne, *Phys. Rev. Lett.*, **63**, 2841 (1989).
100. J.R. Tucker, *Phys. Rev. Lett.*, **65**, 270 (1990).
101. J.C. Gill, *Phys. Rev. Lett.*, **65**, 271 (1990).
102. R.E. Thorne and J. McCarten, *Phys. Rev. Lett.*, **65**, 272 (1990).
103. D.A. DiCarlo, J. McMarten, T.L. Adelman, M. Maher and R.E. Thorne, *Phys. Rev. B*, **42**, 7643 (1990).
104. J.R. Tucker, *Phys. Rev. B*, **47**, 7614 (1993).
105. D.A. DiCarlo, J. McMarten and R.E. Thorne, *Phys. Rev. B*, **47**, 7614 (1993).
106. H. Fukyama and P.A. Lee, *Phys. Rev. B*, **17**, 535 (1978).
107. P.A. Lee and H. Fukuyama, *Phys. Rev. B*, **17**, 542 (1978).
108. P.A. Lee and T.M. Rice, *Phys. Rev. B*, **19**, 3970 (1979).
109. Y. Fukuyama and H. Takayama, in *Electronic Properties of Inorganic Quasi-One-Dimensional Compounds. Part I: Theoretical*, Ed. P. Monceau, D. Reidel Publ. Co., Dordrecht, (1985), p. 40.
110. J.R. Tucker, *Phys. Rev. B*, **40**, 5447 (1989).
111. S. Abe, *J. Phys. Soc. Japan*, **54**, 3494 (1985).

112. S. Abe, *J. Phys. Soc. Japan*, **55**, 1987 (1986).
113. H. Mutka, S. Bouffard, J. Dumas and C. Schlenker, *J. Phys. (Paris) Lettres*, **45**, L729 (1984).
114. T. Chen and J.R. Tucker, *Phys. Rev. B*, **41**, 7402 (1990).

ANALYSIS OF SCANNING TUNNELING AND ATOMIC FORCE MICROSCOPY IMAGES

MYUNG-HWAN WHANGBO, JINGQING REN

Department of Chemistry, North Carolina State University, Raleigh, North Carolina 27695-8204, USA

SERGEI N. MAGONOV and HARDY BENGEL

Materials Research Center, Albert-Ludwigs University, Stefan-Meier-Str. 21, D-79104 Freiburg, Germany

1. Introduction

The quantum mechanical phenomenon of electron tunneling was used for the imaging of a conducting surface in atomic scale in 1982 when the first scanning tunneling microscope was built [1]. The most important feature of scanning tunneling microscopy (STM) is the real-space visualization of the surfaces of metals and semiconductors in atomic scale. The tunneling current decreases exponentially with increasing the tip-sample distance. This makes it possible to visualize surface structures with sub-angstrom resolution and detect various atomic-scale defects inaccessible by diffraction and spectroscopic techniques [2-4]. An atomic force microscope was invented to enable the detection of atomic-scale features of insulating surfaces [5]. In atomic force microscopy (AFM) the repulsive force between the tip (located at the end of a cantilever) and sample is commonly measured on the basis of the cantilever deflection. Because the repulsive force is universal, AFM is applicable to conducting as well as insulating materials.

To characterize sample surfaces of chemical interests by STM and AFM, it is necessary to distinguish genuine features from experimental artifacts in observed images and interpret the observed images. Atomic scale STM and AFM images are routinely recorded for many crystalline surfaces, but their interpretation is by no means straightforward. It is tempting to assign the atomic-size spots of STM and AFM images to the surface atomic or molecular structures. This interpretation can be misleading especially for STM because the electron tunneling involves only the energy levels of the sample lying in the vicinity of the Fermi level e_f. When the tip-sample interactions are neglected, the STM image is described by the partial electron density plot $\rho(r_0, e_f)$ of the sample surface [6]. In contact-mode AFM measurements, all the electrons of the surface atoms are involved in the repulsive interactions with the tip so

F. W. Boswell and J.C. Bennett (eds.),
Advances in the Crystallographic and Microstructural Analysis of Charge Density Wave Modulated Crystals, 185–224.
© 1999 *Kluwer Academic Publishers.*

that the AFM image is described by the total electron density plot $\rho(r_0)$ of the surface. In this review, we discuss the theoretical basis of interpreting STM and AFM images and then survey the STM and AFM images of several layered inorganic compounds.

2.Simulations of STM and AFM images

To a first approximation, the STM and AFM images of many crystalline compounds are described by their partial and total electron density plots. The theoretical and computational aspects of the electron density plot calculations are discussed in this section.

2.1 ELECTRONIC STRUCTURES OF SOLIDS

The electronic structures of crystalline solids are described by their energy bands. The electronic band structures of solids can be calculated by using the tight-binding electronic band structure method [7], in which the crystal orbitals of a solid are represented as linear combinations of its atomic orbitals. As an example, consider a two-dimensional (2D) rectangular lattice with repeat vectors $\mathbf{a}$ and $\mathbf{b}$ (and lengths a and b, respectively) shown in Figure 1a. The positions of the unit cells in the lattice are described by the lattice vectors $\mathbf{R} = m\mathbf{a} + n\mathbf{b}$, where m and n are integers. Suppose that each unit cell has a set of M atomic orbitals $\{\chi_1, \chi_2, \chi_3, \cdots, \chi_M\}$. An atomic orbital χ_μ ($\mu = 1, 2, 3, \cdots, M$) located at the unit cell at $\mathbf{R}$ is written as $\chi_\mu(\mathbf{r}-\mathbf{R})$. The electronic structure of the solid can be constructed in terms of the Bloch orbitals $\{\phi_1(\mathbf{r},\mathbf{k}), \phi_2(\mathbf{r},\mathbf{k}), \phi_3(\mathbf{r},\mathbf{k}), \cdots, \phi_M(\mathbf{r},\mathbf{k})\}$, where $\phi_\mu(\mathbf{r},\mathbf{k})$ ($\mu = 1, 2, 3, \cdots, M$) is defined by linearly combining the atomic orbitals χ_μ located at all different unit cells,

$$\phi_\mu(\mathbf{r},\mathbf{k}) = \frac{1}{\sqrt{N}} \sum_{\mathbf{R}} \exp(i\mathbf{k}\cdot\mathbf{R})\, \chi_\mu(\mathbf{r}-\mathbf{R}). \tag{1}$$

Here N is the total number of unit cells in the lattice ($N \to \infty$), and $\mathbf{k}$ is the wave vector. A 2D lattice with repeat vectors $\mathbf{a}$ and $\mathbf{b}$ has the corresponding reciprocal vectors $\mathbf{a}^*$ and $\mathbf{b}^*$ [7]. For the rectangular lattice (Figure 1a), the directions of $\mathbf{a}^*$ and $\mathbf{b}^*$ are identical with those of $\mathbf{a}$ and $\mathbf{b}$, respectively, and their lengths are $2\pi/a$ and $2\pi/b$, respectively. In general, the wave vector $\mathbf{k}$ can be expressed as $\mathbf{k} = x\mathbf{a}^* + y\mathbf{b}^*$, where x and y are dimensionless numbers, so the term $\exp(i\mathbf{k}\cdot\mathbf{R})$ is expressed as

$$\exp(i\mathbf{k}\cdot\mathbf{R}) = \exp(ik_a ma)\exp(ik_b nb) \tag{2}$$

where $k_a = x(2\pi/a)$ and $k_b = y(2\pi/b)$. Since the exponential term of Eq. 2 is a periodic function of $\mathbf{k}$, the Bloch orbitals can be constructed by using only those wave vectors $\mathbf{k} = (k_a, k_b)$ belonging to the region defined by $-\pi/a \le k_a \le \pi/a$ and $-\pi/b \le k_b \le \pi/b$. This region of $\mathbf{k}$ values is known as the first primitive zone (FPZ) (Figure 1b) [7].

The crystal orbitals $\psi_i(\mathbf{r},\mathbf{k})$ ($i = 1, 2, 3, \cdots, M$) of the lattice are expressed as linear

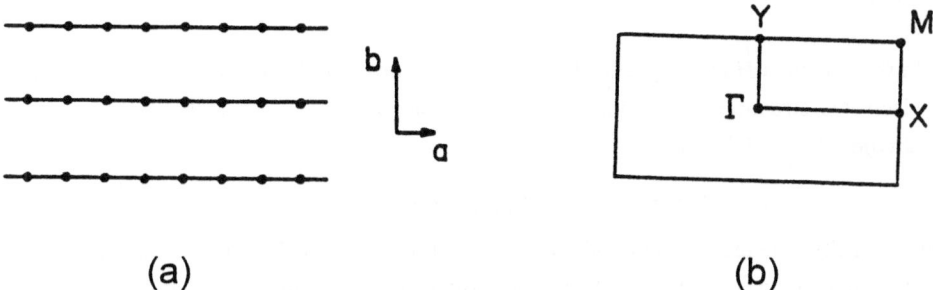

Fig. 1. (a) Schematic representation of a rectangular 2D lattice. (b) FPZ associated with the rectangular 2D lattice in (a). $\Gamma = (0, 0)$, $X = (a^*/2, 0)$, $Y = (0, b^*/2)$, and $M = (a^*/2, b^*/2)$.

combinations of the Bloch orbitals

$$\psi_i(\mathbf{r},\mathbf{k}) = \sum_\mu C_{\mu i}(\mathbf{k})\phi_\mu(\mathbf{r},\mathbf{k}). \tag{3}$$

The coefficients $C_{\mu i}(\mathbf{k})$ and the energies $e_i(\mathbf{k})$ of the crystal orbitals $\psi_i(\mathbf{r},\mathbf{k})$ ($i = 1, 2, 3, ..., M$) are determined by solving the eigenvalue problem associated with the effective Hamiltonian H^{eff}.

$$H^{eff} \psi_i(\mathbf{r},\mathbf{k}) = e_i(\mathbf{k}) \psi_i(\mathbf{r},\mathbf{k}). \tag{4}$$

For a given $\mathbf{k}$, this eigenvalue problem is solved by using the variation principle to obtain the secular determinant

$$\begin{vmatrix} H_{11}(\mathbf{k}) - e_i(\mathbf{k})S_{11}(\mathbf{k}) & \cdots & H_{1M}(\mathbf{k}) - e_i(\mathbf{k})S_{1M}(\mathbf{k}) \\ & \cdot & \\ & \cdot & \\ & \cdot & \\ H_{M1}(\mathbf{k}) - e_i(\mathbf{k})S_{M1}(\mathbf{k}) & \cdots & H_{MM}(\mathbf{k}) - e_i(\mathbf{k})S_{MM}(\mathbf{k}) \end{vmatrix} = 0 \tag{5}$$

where the matrix elements $H_{\mu v}(\mathbf{k})$ and $S_{\mu v}(\mathbf{k})$ (μ, $v = 1, 2, 3, \cdots, M$) are expressed as

$$H_{\mu v}(\mathbf{k}) = \langle\phi_\mu(\mathbf{k})|H^{eff}|\phi_v(\mathbf{k})\rangle =$$
$$\langle\chi_\mu(\mathbf{r})|H^{eff}|\chi_v(\mathbf{r})\rangle + \sum_\mathbf{R}\{ \exp(-i\mathbf{k}\cdot\mathbf{R}) \langle\chi_\mu(\mathbf{r}-\mathbf{R})|H^{eff}|\chi_v(\mathbf{r})\rangle + \exp(i\mathbf{k}\cdot\mathbf{R}) \langle\chi_\mu(\mathbf{r})|H^{eff}|\chi_v(\mathbf{r}-\mathbf{R})\rangle \}$$

$$\tag{6}$$

and

$$S_{\mu v}(\mathbf{k}) = \langle\phi_\mu(\mathbf{k})|\phi_v(\mathbf{k})\rangle = \langle\chi_\mu(\mathbf{r})|\chi_v(\mathbf{r})\rangle + \sum_\mathbf{R}\{ \exp(-i\mathbf{k}\cdot\mathbf{R}) \langle\chi_\mu(\mathbf{r}-\mathbf{R})|\chi_v(\mathbf{r})\rangle$$

$$+ \exp(i\mathbf{k}\cdot\mathbf{R}) \langle\chi_\mu(\mathbf{r})|\chi_v(\mathbf{r}-\mathbf{R})\rangle \} \tag{7}$$

The lattice sum in Eqs. 6 and 7 (i.e., the sum over **R**) can be limited to the terms involving the reference and n-th nearest neighbors (n = 1, 2, 3, etc.) when the remaining terms are negligibly small.

In the extended Hückel tight binding method [8], only the valence atomic orbitals are used for electronic structure calculations, and the atomic orbitals $\chi_\mu(r,\theta,\phi)$ are approximated by Slater type orbitals (STO's),

$$\chi_\mu(r,\theta,\phi) \propto r^{n-1} \exp(-\zeta r) \, \Phi(\theta,\phi) \tag{8}$$

where n is the principal quantum number of the atomic orbital, ζ is the orbital exponent, and $\Phi(\theta,\phi)$ is the spherical harmonics determining the angular variation of the atomic orbital. The STO's are normalized so that $\langle\chi_\mu|\chi_\mu\rangle = 1$, and the integrals between two orbitals $\langle\chi_\mu|\chi_\nu\rangle$ ($\mu \neq \nu$) define the overlap integrals $S_{\mu\nu}$. The energy matrix elements $H_{\mu\nu}$ between two different orbitals χ_μ and χ_ν are approximated by the Wolfsberg-Helmholz formula

$$H_{\mu\nu} \propto (H_{\mu\mu} + H_{\nu\nu})S_{\mu\nu} \tag{9}$$

where $H_{\mu\mu}$ and $H_{\nu\nu}$ are the valence-shell ionization potentials of the atomic orbitals χ_μ and χ_ν, respectively.

The energy levels belonging to M different energy bands by solving the eigenvalue problem Eq. 4 for a fine mesh of **k**-points covering the FPZ. The electronic band structure of the solid thus obtained is presented by plotting $e_i(\mathbf{k})$ vs **k** along certain directions of the FPZ. These plots are known as dispersion relations. Alternatively, the electronic band structure can be presented by plotting n(e) vs e, where n(e) is the electronic density of states (DOS) of the lattice at a given energy e [7]. The orbital compositions making up n(e) can be found by plotting the projected density of states (PDOS).

Any given band can accommodate up to two electrons per unit cell. Whether a given solid is a metal or not depends upon the number of electrons per unit cell and the nature of the energy bands. The highest occupied band of a metal is not completely filled, so that there is no energy gap between the highest occupied and lowest unoccupied band levels (Figure 2a). A normal semiconductor and a normal insulator has no partially filled bands, so there is a band gap between the highest occupied (i.e., valence) band and lowest unoccupied (i.e., conduction) band levels (Figure 2b). A magnetic semiconductor and a magnetic insulator have unpaired spin orbitals [9]. In the electronic band picture, they are represented by energy bands whose levels are all singly filled (Figure 2c) [9c,d].

2.2 FERMI SURFACE NESTING AND CHARGE DENSITY WAVE

A metal has at least one partially filled band. All the energy levels of a band are represented by the wave vectors of the FBZ. Thus, for a partially filled band, some wave vectors of the FBZ are associated with the occupied band levels, and the remaining wave vectors with the

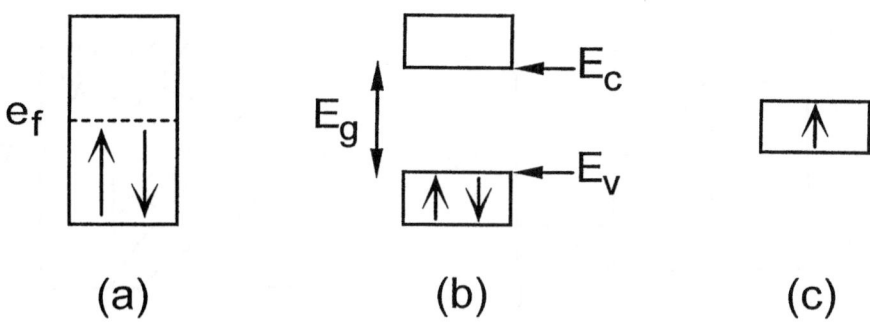

Fig. 2. Band filling patterns: (a) Half filled band leading to a metal.   (b) Completely filled and completely empty bands leading to a normal semiconductor or insulator. (c) Half filled band leading to a magnetic semiconductor or insulator.

unoccupied band levels. The Fermi surface of a partially filled band is the boundary surface separating the occupied wave vectors from the unoccupied wave vectors. As an example, Figure 3a shows the dispersion relation of a half filled band dispersive only along the $\Gamma \rightarrow X$ direction (that is, metallic along the a-direction). Figure 3b illustrates the occupied and unoccupied wave vector regions (shaded and unshaded, respectively) of the FPZ for this one dimensional (1D) band. The Fermi surface of this band consists of two parallel lines perpendicular to the $\Gamma \rightarrow X$ direction at $\pm \pi/2a$ (i.e., $\pm k_f$). When translated by a vector $\mathbf{q}$, the left-hand-side piece of the Fermi surface is superposed to the right-hand-side piece of the Fermi surface. When a piece of a given Fermi surface is superposed to another piece by a translational vector $\mathbf{q}$, the Fermi surface is said to be nested by the vector $\mathbf{q}$. For an ideal 1D surface such as the one in Figure 3b, many different nesting vectors are possible. These vectors differ in their components along $\Gamma \rightarrow Y$, but their components along $\Gamma \rightarrow X$ are identical (that is, $2k_f$).

The normal metallic state of a 1D metal is susceptible to a metal-to-insulator transition such as a charge density wave (CDW) formation [10]. To illustrate this point, one may consider a simple case in which the 2D lattice of Figure 1a has one atomic orbital χ per unit cell. Then, this lattice has only one band, and its crystal orbitals are identical with the Bloch orbitals. Suppose this band is half filled, so that the orbitals are grouped into occupied and unoccupied ones $\phi(\mathbf{r},\mathbf{k})$ and $\phi(\mathbf{r},\mathbf{k}')$, respectively. Under a perturbation H', an occupied band level $\phi(\mathbf{r},\mathbf{k})$ of a normal metallic state interacts with an unoccupied band level $\phi(\mathbf{r},\mathbf{k}')$ leading to new orbitals [7,9,11],

$$\varphi(\mathbf{r},\mathbf{k}) \propto \phi(\mathbf{r},\mathbf{k}) + \lambda\,\phi(\mathbf{r},\mathbf{k}')$$

$$\varphi(\mathbf{r},\mathbf{k}') \propto -\lambda\,\phi(\mathbf{r},\mathbf{k}) + \phi(\mathbf{r},\mathbf{k}'), \tag{10}$$

where λ is a mixing coefficient. It is equal to 1 when the orbitals $\phi(\mathbf{r},\mathbf{k})$ and $\phi(\mathbf{r},\mathbf{k}')$ are

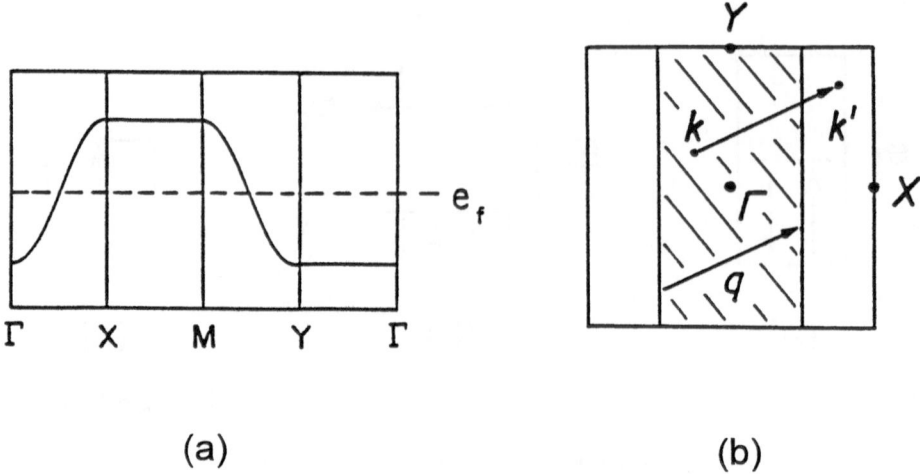

(a)   (b)

Fig. 3. (a) Dispersion relation expected when the 2D lattice of Figure 1 has no intersite interaction
along the b-direction. The dashed line represents the Fermi level. (b) Occupied wave vectors
(shaded) and unoccupied wave vectors (unshaded) of the FPZ associated with the half-filled
band in (a). The Fermi surface consists of two parallel lines, perpendicular to the chain direction, separated
by the distance of $q = a^*/2$.

degenerate, and smaller than 1 otherwise. The electron density distributions associated with
$\varphi(\mathbf{r},\mathbf{k})$ and $\varphi(\mathbf{r},\mathbf{k}')$ are given by

$$\varphi(\mathbf{r},\mathbf{k})\,\varphi^*(\mathbf{r},\mathbf{k}) \propto \phi(\mathbf{r},\mathbf{k})\,\phi^*(\mathbf{r},\mathbf{k}) + \gamma^2\,\phi(\mathbf{r},\mathbf{k}')\,\phi^*(\mathbf{r},\mathbf{k}') + \Delta\rho$$

$$\varphi(\mathbf{r},\mathbf{k}')\,\varphi^*(\mathbf{r},\mathbf{k}') \propto \gamma^2\,\phi(\mathbf{r},\mathbf{k})\,\phi^*(\mathbf{r},\mathbf{k}) + \phi(\mathbf{r},\mathbf{k}')\,\phi^*(\mathbf{r},\mathbf{k}') - \Delta\rho \qquad (11)$$

where

$$\Delta\rho = \gamma\,[\phi^*(\mathbf{r},\mathbf{k})\,\phi(\mathbf{r},\mathbf{k}') + \phi(\mathbf{r},\mathbf{k})\phi^*(\mathbf{r},\mathbf{k}')] \qquad (12a)$$

If we expand the $\phi^*(\mathbf{r},\mathbf{k})\phi(\mathbf{r},\mathbf{k}')$ and $\phi(\mathbf{r},\mathbf{k})\phi^*(\mathbf{r},\mathbf{k}')$ terms using Eq. 1 and keep only the diagonal
terms $\chi(\mathbf{r}-\mathbf{R})\,\chi^*(\mathbf{r}-\mathbf{R})$, it can be shown that

$$\Delta\rho \propto \sum_{\mathbf{R}} \cos(\mathbf{q}\cdot\mathbf{R})\,\chi(\mathbf{r}-\mathbf{R})\,\chi^*(\mathbf{r}-\mathbf{R}) \qquad (12b)$$

With respect to $\phi(\mathbf{r},\mathbf{k})$ or $\phi(\mathbf{r},\mathbf{k}')$, therefore, $\varphi(\mathbf{r},\mathbf{k})$ and $\varphi(\mathbf{r},\mathbf{k}')$ each have density wave character whose periodicity in real space is governed by the term $\cos(\mathbf{q}\cdot\mathbf{R})$. $\varphi(\mathbf{r},\mathbf{k})$ leads to density
accumulation where $\varphi(\mathbf{r},\mathbf{k}')$ has density depletion, and vice versa. If the density distribution
arising from $\phi(\mathbf{r},\mathbf{k})$ or $\phi(\mathbf{r},\mathbf{k}')$ is represented by a straight line, then the density accumulation and
depletion associated with $\varphi(\mathbf{r},\mathbf{k})$ and $\varphi(\mathbf{r},\mathbf{k}')$ occur as a wave form as shown in Figure 4, where
shaded and unshaded half waves represent density accumulation and depletion, respectively.

A CDW state is obtained when the orbitals $\varphi(\mathbf{r},\mathbf{k})$ are each doubly occupied. For a nested
Fermi surface with vector $\mathbf{q}$, the orbitals $\phi(\mathbf{r},\mathbf{k})$ and $\phi(\mathbf{r},\mathbf{k}')$ related by $\mathbf{q} = \mathbf{k}' - \mathbf{k}$ are degenerate
when $\mathbf{k}$ lies on the Fermi surface ($\mathbf{k}$ with the $\Gamma \to X$ component of $\pm\,k_f$, that is, k_f), but

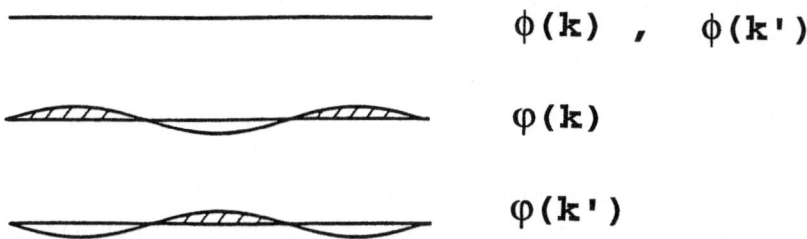

$$\phi(k) \quad , \quad \phi(k')$$

$$\varphi(k)$$

$$\varphi(k')$$

Fig. 4. Electron density distributions of the orbitals $\varphi(r,k)$ and $\varphi(r,k')$ with respect to those of the orbitals $\phi(r,k)$ and $\phi(r,k')$.

nondegenerate otherwise (Figure 3b). In general, an interaction between two orbitals becomes stronger as the energy difference between the two decreases [12]. If a Fermi surface is nested, the occupied and unoccupied band levels related by the nesting vector $q = k' - k$ have no energy difference for all the vectors k on the Fermi surface, and very small energy difference for the vectors k in the vicinity of the Fermi surface. For a 1D metal with nesting vector q, the interactions between the occupied and unoccupied introduces an energy gap 2Δ at the Fermi level. Here $\Delta = <\phi(r,k_f)|H'|\phi(r,k'_f)>$, $k'_f = q + k_f$, and the perturbation H' causing the orbital mixing is lattice vibration (Figure 5). Due to the energy lowering associated with the energy gap opening, a low-dimensional metal with Fermi surface nesting is susceptible to a CDW formation.

2.3 THEORETICAL ASPECTS OF STM

Suppose that the tip and sample are both metals and form a metal-insulator-metal junction in the STM configuration. The gap between the tip and sample provides an insulating barrier (e.g., air, vacuum, etc.). When V_{bias} between the electrodes is zero, their Fermi levels become

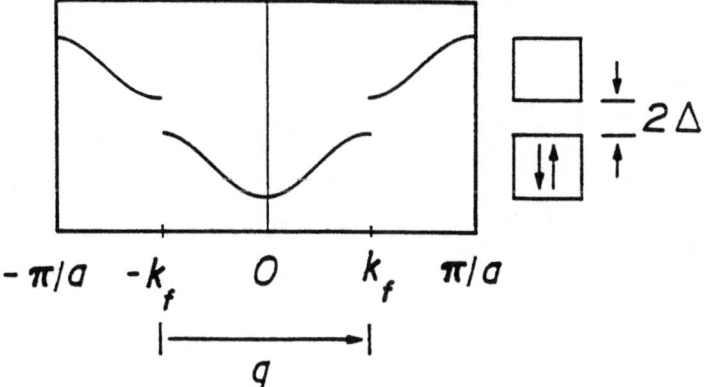

Fig. 5. Band gap opening caused by a CDW associated with the nesting vector $q = 2k_f$ for a 1D metallic system.

equal and there is no tunneling current between them (Figure 6a). When V_{bias} is positive (with the tip grounded), the energy levels of the sample are lowered by eV_{bias} so that the electrons in the occupied levels of the tip (between e_f and $e_f - eV_{bias}$) tunnel into the unoccupied levels of the sample (Figure 6b). When V_{bias} is negative (with the sample grounded), the energy levels of the sample are raised by $e\,|V_{bias}|$ so that the electrons in the occupied levels of the sample (between $e_f + e\,|V_{bias}|$ and e_f) tunnel into the unoccupied levels of the tip (Figure 6c).

The transmission probability of the tunneling is the largest for the electron at the Fermi level of the negatively biased electrode and steadily decreases as the energy is lowered from that level (shown by the lengths of the arrows in Figures 6b and 6c). In general, most of tunneling electrons comes from within 0.3 eV of the Fermi level of the negatively charged electrode [13].

Suppose that a pristine semiconductor sample forms a semiconductor-insulator-metal (SIM) junction with the metallic tip in the STM configuration. The bottom of the conduction band and the top of the valence band of the semiconductor may be referred to as E_c and E_v,

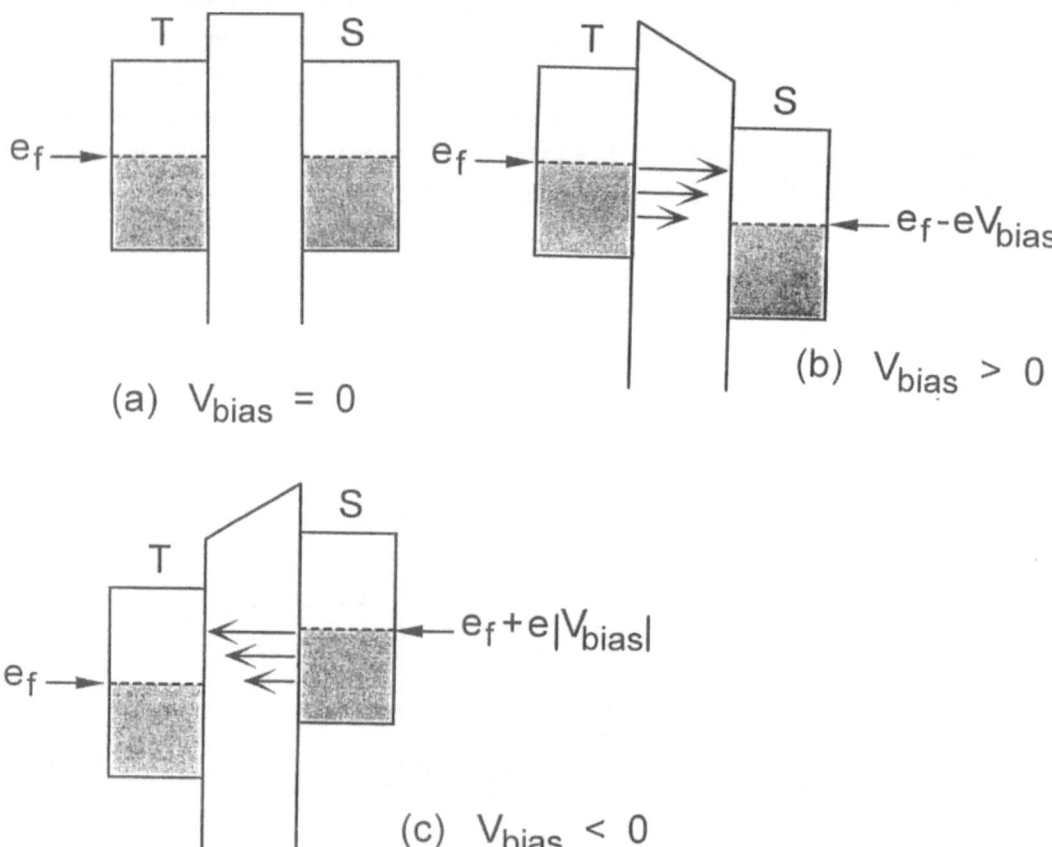

Fig. 6. Energy bands associated with the metal-insulator-metal junction between a metallic tip and a metallic sample: (a) $V_{bias} = 0$. (b) $V_{bias} > 0$. (c) $V_{bias} < 0$.

respectively, so that the band gap is $E_g = E_c - E_v$. When V_{bias} is zero, the Fermi level of the tip lies at the midpoint of the semiconductor band gap [i.e., $e_f = (E_c + E_v)/2$] (Figure 7a) and there is no tunneling current between them. When the positive bias greater than $E_g/2$ so that $E_c - eV_{bias} < e_f$, the electrons of the tip tunnel into the empty levels of the conduction band (between e_f and $E_c - eV_{bias}$) (Figure 7b). When the magnitude of the negative bias is greater than $E_g/2$ so that $E_v + e|V_{bias}| > e_f$, the electrons in the valence band of the sample (between $E_v + e|V_{bias}|$ and e_f) tunnel into the tip (Figure 7c).

A semiconductor with n-type dopants have donor levels near the bottom of the conduction band, and the thermal excitation of electrons from the donor levels populates the bottom of the conduction band. For an SIM junction with such an n-type semiconductor, the Fermi level is close to the donor level E_D of the semiconductor when $V_{bias} = 0$ (Figure 8a). A semiconductor with p-type dopants have acceptor levels near the top of the valence band. The thermal excitation of electrons from the valence band populates the acceptor level. For an SIM

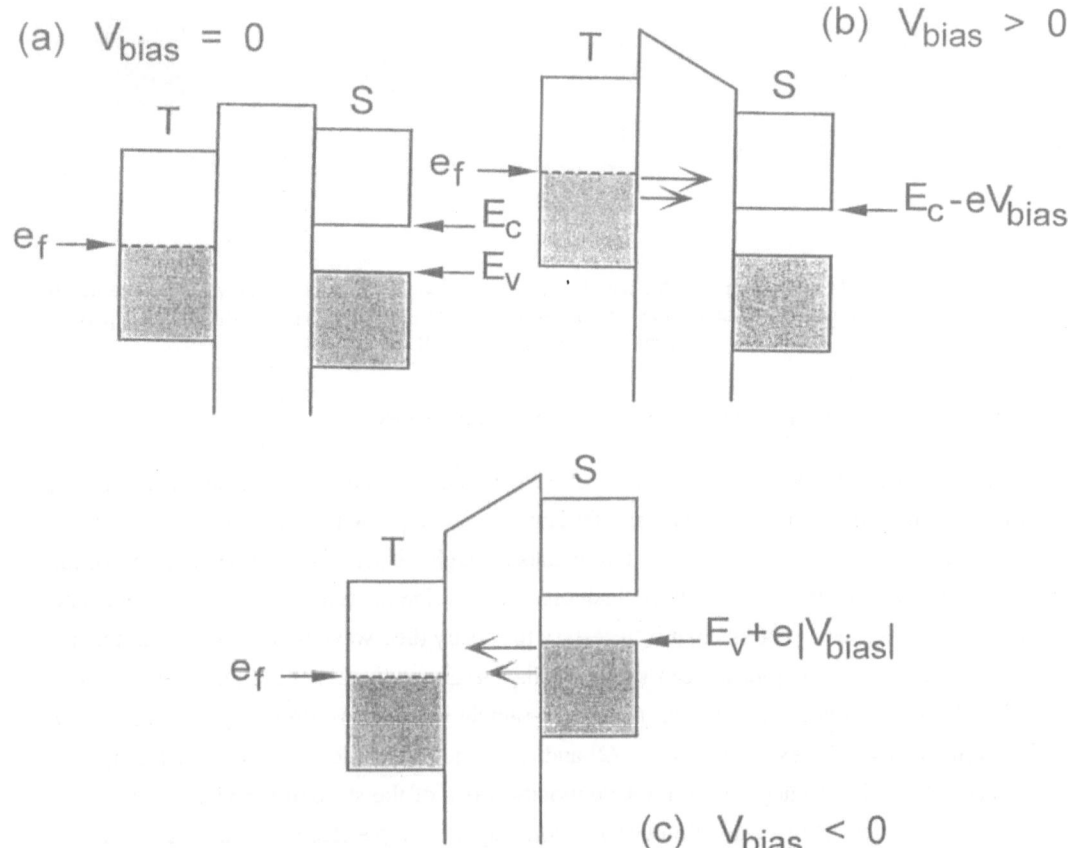

Fig. 7. Energy bands associated with the metal-insulator-semiconductor junction between a metallic tip and a semiconductor sample: (a) $V_{bias} = 0$. (b) $V_{bias} > 0$. (c) $V_{bias} < 0$.

junction with such a p-type semiconductor, the Fermi level is close to the acceptor level E_A of the semiconductor when $V_{bias} = 0$ (Figure 8b). Depending upon the sign and magnitude of the bias voltage, the STM current of doped semiconductors may involve not only the bottom of the conduction band and the top of the valence band but also the donor and acceptor levels lying in the band gap. In addition, the wave functions of the donor and acceptor levels are spatially localized in the regions around the donor and acceptor dopants [14], unlike those of the valence and conduction band levels which are delocalized. Therefore, STM images of doped semiconductors or semiconductors with point defects can exhibit local imperfections [15].

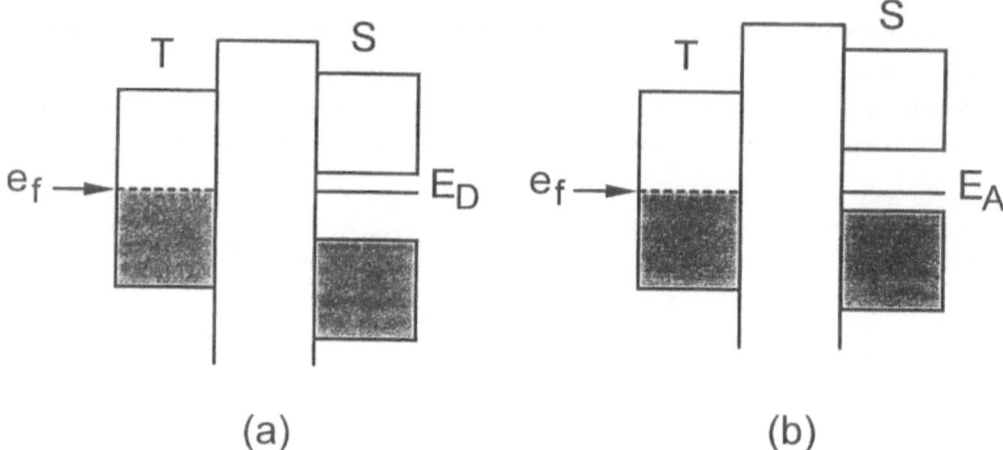

(a)   (b)

Fig. 8. Energy bands associated with the metal-insulator-semiconductor junction between a metallic tip and a semiconductor sample: (a) For an n-type semiconductor at $V_{bias} = 0$. (b) For a p-type semiconductor at $V_{bias} = 0$.

2.4 TERSOFF-HAMMAN THEORY AND ITS EXTENSION

The simplest and most practical theory of STM was developed by Tersoff and Hamman,[6] who based their theory on Bardeen's transfer Hamiltonian approach [16]. In Bardeen's perturbation treatment, the tunneling between two metals through a barrier is described in terms of the overlap between the tails of their wave functions. To describe the tunneling between the tip and sample using Bardeen's approach, it is necessary to specify their wave functions. In their theory, Tersoff and Hamman [6] introduced three simplifying assumptions: *(1)* the tip can be approximated by an atom with an s-orbital, *(2)* the tip-sample interactions are negligible, and *(3)* the bias voltage is small. *Assumptions (1), (2) and (3)* respectively allow one to treat the tip as a mathematical point, to neglect the possible modifications of the structures and wave functions of both electrodes, and to take into consideration only the sample electronic states at the Fermi level e_f for the tunneling. Under these assumptions, Tersoff and Hamman showed that the STM current I_{tun} is proportional to the local density of states (LDOS) of the sample at the position of the tip r_t,

$$I_{tun} \propto \sum \int \, dk \, |\psi_i^s(r_t,k)|^2 \, \delta(e_s\text{-}e_f) \tag{13}$$

where $\psi_i^s(r,k)$ is the i-th electronic band orbital of the sample. For simplicity of notation, the band orbital energy $e_i^s(k)$ of the sample is written as e_s. The delta function insures that only the band orbitals of the sample at the Fermi level contribute to the LDOS. The equi-current contours observed in constant-height STM measurements correspond to the constant contours of the partial electron density $\rho(r_0,e_f)$ of the sample surface at the Fermi level (evaluated at the position of the tip-to-sample distance r_0). Therefore, for a given sample surface, the brightness pattern of the STM image is related to the high electron density (HED) pattern of the $\rho(r_0,e_f)$ plot. Certainly, the s-orbital tip approximation oversimplifies the electronic structure of the tip, but Tersoff-Hamman theory has been successful in interpreting the STM images of a large variety of organic and inorganic materials [17].

Ou-Yang et al. developed a more general but still practical theory of STM [18], in which they kept *assumptions (2)* and *(3)* of Tersoff-Hamman theory. Using the time-dependent first-order perturbation theory, Ou-Yang et al. deduced an expression for the STM current as

$$I_{tun} \propto \sum_m \sum_i \int \, dk \, |<\psi_m^t|H|\psi_i^s(r_t,k)>|^2 \, \delta(e_s\text{-}e_f) \tag{14}$$

when the tip is treated as a semi-infinite linear chain of transition metal atoms (perpendicular to the surface). The ψ_m^t denotes the m-th d orbital of the first tip atom (i.e., the atom closest to the sample), and H is the Hamiltonian appropriate for an electron moving in a system consisting of the tip and sample in STM. Eq. 14 is reduced to Eq. 13 by introducing the following approximations [18a]:

$$<\psi_m^t|H|\psi_i^s(r_t,k)> \propto <\psi_m^t|\psi_i^s(r_t,k)> \propto \psi_i^s(r_t,k) \tag{15}$$

Namely, the energy matrix element $<\psi_m^t|H|\psi_i^s(r_t,k)>$ is proportional to the overlap integral $<\psi_m^t|\psi_i^s(r_t,k)>$, which is in turn proportional to the value of the orbital $\psi_i^s(r_t,k)$. These approximations are generally valid and form the basis of semi-empirical quantum mechanical computational methods [8a].

For the STM measurements of insulating molecular species adsorbed on a metallic substrate, the $\psi_i^s(r_t,k)$ refer to the orbitals of the adsorbate/substrate system [18a]. From the viewpoint of Tersoff-Hamman theory, this means that the STM images of an adsorbate/substrate system can be simulated in terms of the partial electron density plot $\rho(r_0,e_f)$ calculated for the adsorbate/substrate system with the tip positioned above the adsorbate layer [18,19].

In principle, the perturbation treatments of the electron transfer in STM are not valid when the overlap between the wave functions of the tip and sample surface becomes significant. The theoretical analysis of an idealized tip-sample system by Ciraci [20] suggests that this electronic contact occurs before the mechanical contact as the tip approaches the sample surface. In practice, however, the tip-sample force interaction occurs through the contamination layer and hence leads to macroscopic and microscopic deformations of the surface even before the tip and sample experience the electronic contact [21,22]. Therefore, the perturbation approaches are still useful, and the observed STM image is related to the partial electron density plot of the deformed surface.

In a general theory of STM, it is attempted to reliably treat the quantum properties of electrons of a few eV kinetic energy, because such electrons may interact with structures of dimensions of the order of their wavelength. For this purpose, one should not only use realistic potentials but also solve the associated three-dimensional scattering problem. However, it is a difficult problem to treat both parts rigorously [23].

2.5 THEORETICAL ASPECTS OF AFM

The forces between closed shell atoms and molecules are primarily determined by the outer regions of the atoms, in which the atomic electron densities overlap [24]. Therefore, if the AFM tip is approximated by an atom (by analogy with Tersoff-Hamman theory) and if the surface topography is negligibly affected by the tip-sample interactions, it is expected that the repulsive force F_{rep} felt by the tip in the contact mode AFM is proportional to the LDOS of the sample at the position of the tip r_t,

$$F_{rep} \propto \sum_{e_i(k) \leq e_f} \int dk \ |\psi_i{}^s(r_t, k)|^2 \qquad (16)$$

where the sum is over all the occupied band levels. Thus the equi-force contours in the constant-height AFM measurements correspond to constant contours of the total electron density $\rho(r_0)$ of the sample surface at the position of the tip-to-sample distance r_0. Therefore, for a given sample surface, the brightness pattern of the AFM image is related to the HED pattern of the $\rho(r_0)$ plot. It should be recalled that this approach is valid when the tip-force induced surface corrugation is negligible.

2.6 IMAGE SIMULATION BY DENSITY PLOT CALCULATIONS

The partial density $\rho(r, e_f)$ necessary for the simulation of the STM image of a solid surface is obtained by summing the density contributions of the band orbitals $\psi_i(k)$ of the sample

$$\rho(r, e_f) = \sum_{|e_i(k) - e_f(k)| \leq \Delta} \psi_i(k)^* \psi_i(k) \qquad (17)$$

where the sum includes only those orbitals whose energies $e_i(k)$ belongs to the energy window, $|e_i(k) - e_f| \leq \Delta$, appropriate for the problem. Here Δ is a small positive number. When the energy bands associated with tunneling are several eV wide, the energy window Δ can be chosen to be about 0.25 eV. (In most cases, for metallic compounds, the variation of Δ in the 0.05 - 0.30 eV range does not qualitatively affect the results.) For metals, for which the highest occupied and lowest occupied levels are degenerate, the sum includes all the occupied band orbitals whose energies lie between e_f and $e_f - \Delta$ for the sample-to-tip tunneling (Figure 9a), and all the unoccupied band orbitals whose energies lie between $e_f + \Delta$ and e_f for the tip-to-sample tunneling (Figure 9b). Finally, the $\rho(r_0, e_f)$ plot for STM image simulation is obtained by calculating the values of the partial density $\rho(r, e_f)$ at the tip-to-surface distance r_0 and presenting the results in a 2D contour plot or in a three-dimensional (3D) surface plot. Thus, three important steps of $\rho(r_0, e_f)$ plot calculations are (a) the electronic band structure calculations to obtain $\psi_i(r, k)$ and $e_i(k)$, (b) the selection of the band levels $\psi_i(r, k)$ appropriate for the partial density $\rho(r, e_f)$ with a certain energy window Δ, and (c) the evaluation of $\rho(r, e_f)$ at the tip-to-sample r_0.

For the calculations of $\rho(r_0, e_f)$ plots, the tip-to-surface distance r_0 value may be taken to be small (e.g., 0.5 Å from the atoms closest to the tip), because the orbital amplitudes of the wave functions of a surface decrease exponentially with distance from the surface so that the $\rho(r_0, e_f)$ values become too small for meaningful comparisons if r_0 is large (e.g., 4 Å). The $\rho(r_0, e_f)$ plots calculated for $r_0 = 0.5$ Å have been found to reflect the essential patterns of the observed STM images rather well [17]. Furthermore, a simultaneous STM/AFM study of 1T-TaS$_2$ at ambient conditions shows [25] that the STM image obtained with the tip in contact with the surface is essentially identical with that obtained in a traditional STM study. Therefore, it is justified to use the small r_0 value.

For the sample-to-tip tunneling of an intrinsic semiconductor, the occupied band orbitals lying between E_v and $E_v - \Delta$ should be selected (Figure 10a). For the tip-to-sample tunneling of an intrinsic semiconductor, the unoccupied band orbitals lying between $E_c + \Delta$ and E_c should be selected (Figure 10b). A magnetic semiconductor has unpaired spin orbitals responsible for its magnetic properties. These levels can either donate electrons to the tip or receive electrons from the tip, and the band representing the localized electron in each unit cell is narrow (Figure 2c) [9]. Thus, to calculate the partial density plot of a magnetic semiconductor, e_f should be set equal to the top of the partially filled band(s) and all the levels of the partially filled band(s) may be selected for both the sample-to-tip and tip-to-sample tunneling.

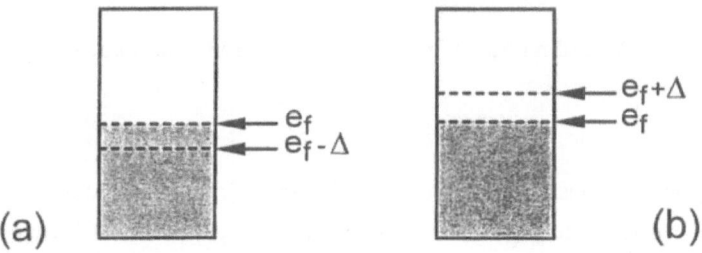

Fig. 9. Energy windows appropriate for the partial density plot calculations of a metal: (a) For $V_{bias} < 0$. (b) For $V_{bias} > 0$.

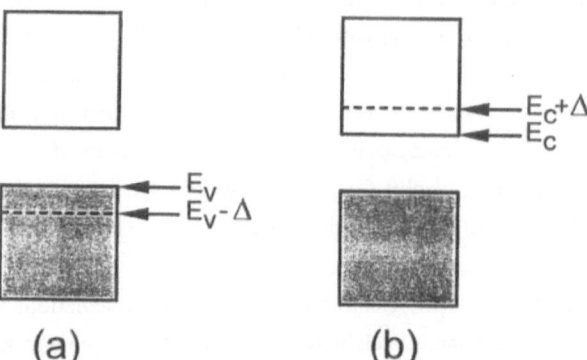

Fig. 10. Energy windows appropriate for the partial density plot calculations of a normal semiconductor:
(a) For $V_{bias} < 0$. (b) For $V_{bias} > 0$.

The wave functions of the donor and acceptor states of a normal semiconductor are spatially localized around the defect sites [14] and hence are not described by electronic band structure calculations. Depending upon the polarity and magnitude of the applied bias voltage, the electron tunneling process may involve the defect states near the top of the valence band and the bottom of the conduction band. This complicates the STM image interpretation [15].

2.7 AFM IMAGE SIMULATION

The total density $\rho(\mathbf{r})$ necessary for the simulation of the AFM image of a solid surface is obtained by summing the density contributions of all the occupied band orbitals $\psi_i(\mathbf{r}, \mathbf{k})$,

$$\rho(\mathbf{r}) = \sum_{e_i(k) \leq e_f} \psi_i(\mathbf{r,k})^* \psi_i(\mathbf{r,k}). \tag{18}$$

The $\rho(r_0)$ plot for AFM image simulation is obtained by calculating the values of the total density $\rho(\mathbf{r})$ at the tip-to-sample distance r_0, which may be taken to be short (e.g., 0.5 Å) as in the case of the $\rho(r_0, e_f)$ plot calculations, and plotting the results in a 2D contour plot or in a 3D surface plot.

3. STM and AFM images of several layered compounds

3.1 MoOCl$_2$

The MoOCl$_2$ layer is constructed from MoCl$_4$O$_2$ octahedra (with the oxygen atoms at the trans corners) by sharing the O corners in one direction and the Cl-Cl edges along the perpendicular direction [26], as shown in the perspective view in Figure 11. In each MoOCl$_2$ layer, the Mo-Mo bonds are formed across the shared Cl-Cl edge to give rise to an Mo-Mo⋯Mo

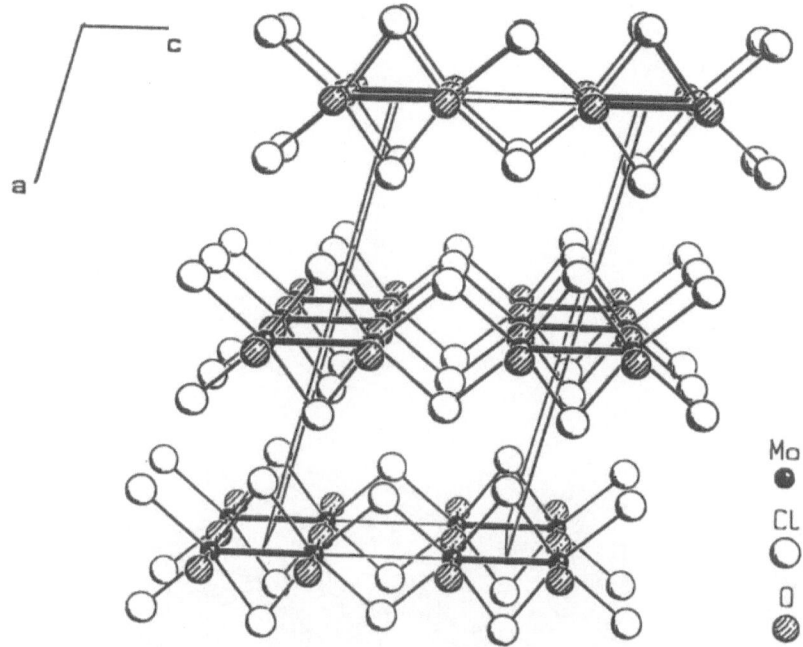

Fig. 11. Perspective view of the crystal structure of $MoOCl_2$.

bond alternation along the crystallographic c-direction. However, there is no bond alternation along the Mo-O-Mo direction (i.e., the b-direction). Due to the Mo-Mo$\cdots$Mo bond alternation, there are two kinds of Cl atoms on the $MoOCl_2$ layer (contained in the bc-plane). The Cl atoms above the Mo-Mo bonds lie 0.3 Å higher than those above the Mo$\cdots$Mo linkage. The electrical resistivity measurements on single crystal samples of $MoOCl_2$ show a metallic conductivity down to 4.2 K [27].

Figure 12a presents the $\rho(r_0)$ plot calculated for the bc-plane surface of an $MoOCl_2$ layer. The high electron density (HED) spots are centered on the surface Cl atoms, and the higher-lying Cl atoms have the higher density. Figure 12b shows the AFM image measured for the bc-plane surface, which consists of rows of bright and less bright spots as in the case of the $\rho(r_0)$ plot. The $\rho(r_0,e_f)$ plot of the bc-plane surface (Figure 13a) is strikingly different from the $\rho(r_0)$ plot. The HED spot on each Cl atom is not spherical but represents in-plane 3p-orbital density whose axis is perpendicular to the Mo-Mo$\cdots$Mo direction. The higher-lying Cl atom has a stronger contribution to the $\rho(r_0,e_f)$ plot. These aspects are indeed found in the STM image of the bc-plane surface (Figure 13b).

3.2 β-Nb₃I₈

Nb_3X_8 (X = Cl, Br, I) is made up of identical Nb_3X_8 layers (Figure 14) [28]. In the Nb

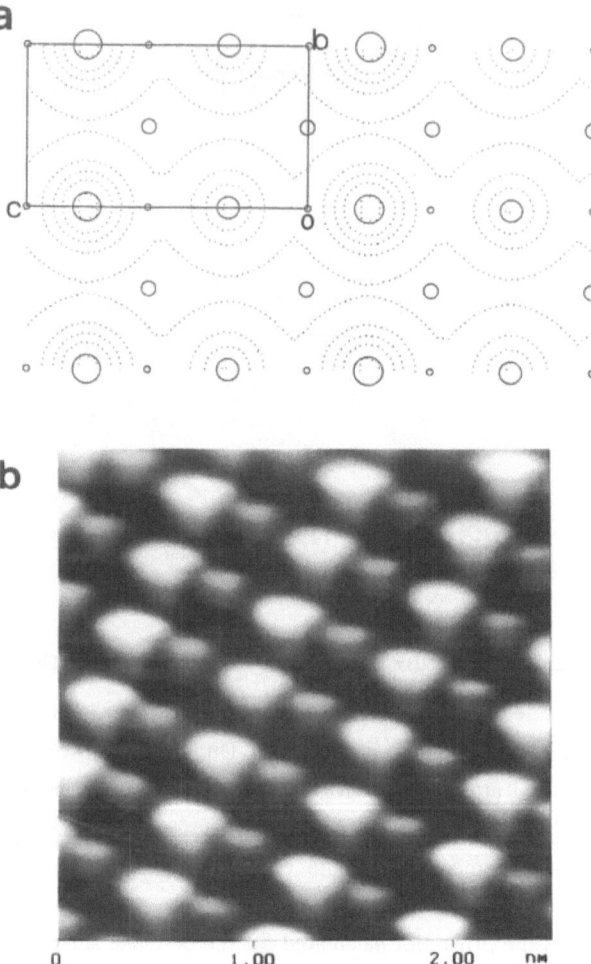

Fig. 12. (a) $\rho(r_0)$ plot calculated for a single $MoOCl_2$ layer, where the contour values used are 15, 10, 5, and 0.5 x 10^{-2} electrons/au^3. (b) AFM height image of $MoOCl_2$. The contrast covers height variations in the 0 - 0.5 nm range.

atom sheet of each Nb_3X_8 layer, the Nb atoms form triangular Nb_3 clusters, so that each Nb atom is under a distorted octahedral environment. In each Nb_3X_8 layer, the top and bottom sheets of X atoms are not equivalent. On the bottom halogen sheet containing the X(2) and X(4) atoms in Figure 14 (surface A), the X(4) atoms lie farther away from the Nb atom sheet than the X(2) atoms (by 0.60, 0.55 and 0.50 Å for X = I, Br and Cl, respectively). On the top halogen sheet containing the X(1) and X(3) atoms in Figure 14 (surface B), the X(1) atoms lie farther away from the Nb atom sheet than the X(3) atoms (by 0.44, 0.33 and 0.35 Å for X = I, Br and Cl, respectively). In principle, the surfaces of Nb_3X_8 (X = Cl, Br, I) samples can be either surface A or B.

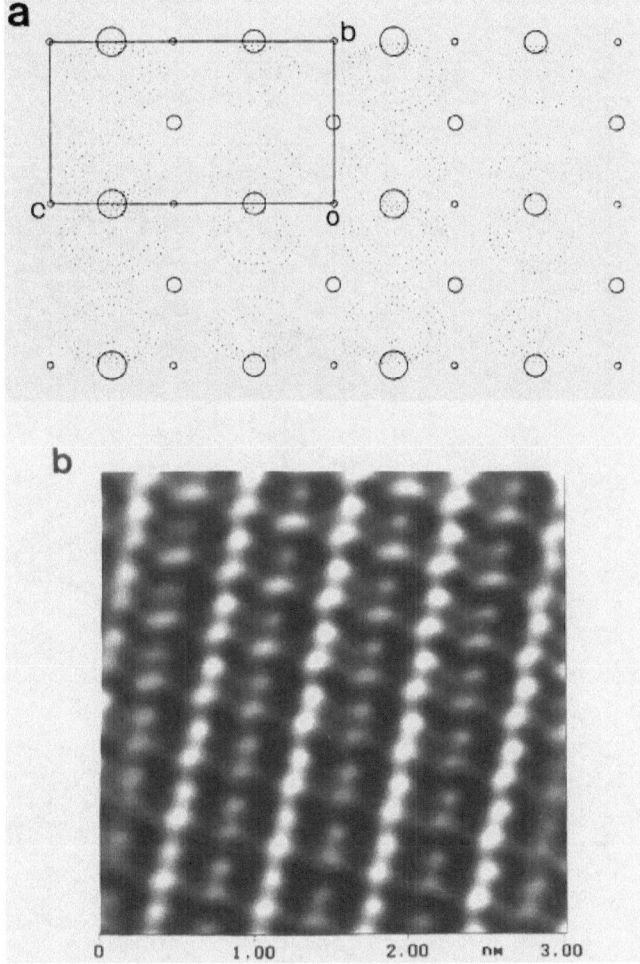

Fig. 13. (a) $\rho(r_0, e_f)$ plot calculated for a single MoOCl$_2$ layer, where the contour values used are 10, 4, 2 and 0.5 x 10^{-4} electrons/au^3. (b) STM current image of MoOCl$_2$ (I$_{set}$ = 5 nA, V$_{bias}$ = 25 mV). The contrast covers current variations in the relative units.

Figure 15 shows atomic resolution AFM image obtained for β-Nb$_3$I$_8$ [29]. This image exhibits the unit cell pattern consisting of one bright spot and three less bright ones. This pattern resembles the atomic arrangement of surface A. The most representative STM images, recorded on the same crystal surface of β-Nb$_3$I$_8$, are shown in Figures 16. The STM images with one big bright spot per unit cell (Figure 16a) are obtained frequently. By changing the tunneling parameters to a smaller gap resistance, a more resolved STM image with four spots per unit cell is obtained (Figure 16b). A zoomed and filtered part of such an image (Figure 16c) reveals three brighter triangular spots and a less pronounced one per unit cell [29].

Figure 17 shows the dispersion relations of the energy bands calculated for a single Nb$_3$I$_8$

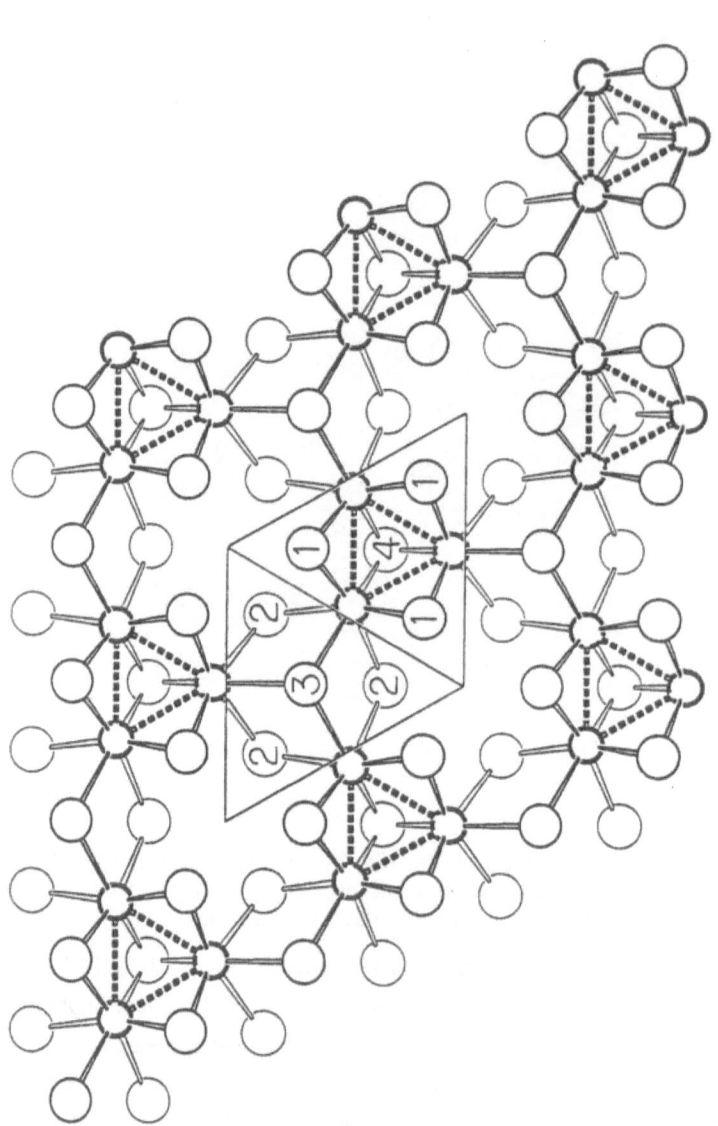

Fig. 14. Schematic projection view, along the crystallographic c-axis direction, of a single Nb_3X_8 layer. The Nb and X atoms are represented by small and large circles, respectively. The Nb_3 clusters in the Nb atom sheet were shown by connecting the Nb atoms with dashed lines.

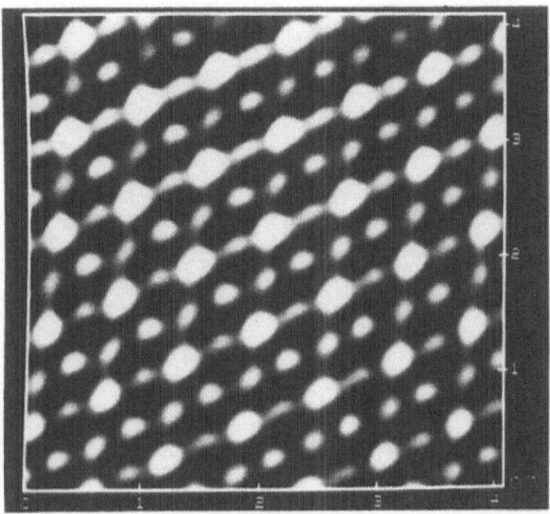

Fig. 15. AFM height image of β-Nb$_3$I$_8$ crystal. The contrast covers height corrugation in the 0 - 0.5 nm range.

layer (nine bands lying in the vicinity of the Fermi level) [29]. With the oxidation state I⁻, there are seven electrons to fill the bottom four bands of Figure 17. The highest-occupied band, which is somewhat narrow, is half filled. The $\rho(r_0)$ and $\rho(r_0,e_f)$ plots of a single Nb$_3$I$_8$ layer calculated for surface A are shown in Figure 18a and 18b, respectively. Within the unit cell of the $\rho(r_0)$ plot, one atom has a higher density than do three other atoms in agreement with the AFM image of Figure 15. The $\rho(r_0,e_f)$ plot, obtained by sampling all levels of the half-filled band of Figure 17, is consistent with the STM pattern of Figure 16c. In each unit cell of this plot, the three I(2) atoms have a greater density than do the I(4) atom, although the I(2) atoms lie farther away from the tip by 0.60 Å. As shown by the DOS analysis of Figure 18c, this reflects the fact that the highest-occupied band has a stronger contribution from the I(2) atoms than from the I(4) atoms. In addition, in the $\rho(r_0, e_f)$ plot, the density peaks of the three I(2) atoms within a unit cell are closer to one another than expected on the basis of the crystal structure. This effect is caused by the hybridization of the I(2) atom p-orbitals in the half-filled band, and it explains the observation of triangular bright spots in the STM image in Figure 16b. When these spots are not resolved, they appear as a big bright spot as shown in Figure 16a.

3.3 1T-TaX$_2$ (X = S, Se)

The layered chalcogenides 2H-MX$_2$ consist of MX$_2$ layers made up of edge-sharing MX$_6$ trigonal prisms (Figure 19a), where M is a transition metal and X is a chalcogen atom. Likewise, the layered chalcogenides 1T-MX$_2$ phases consist of MX$_2$ layers made up of edge-sharing MX$_6$

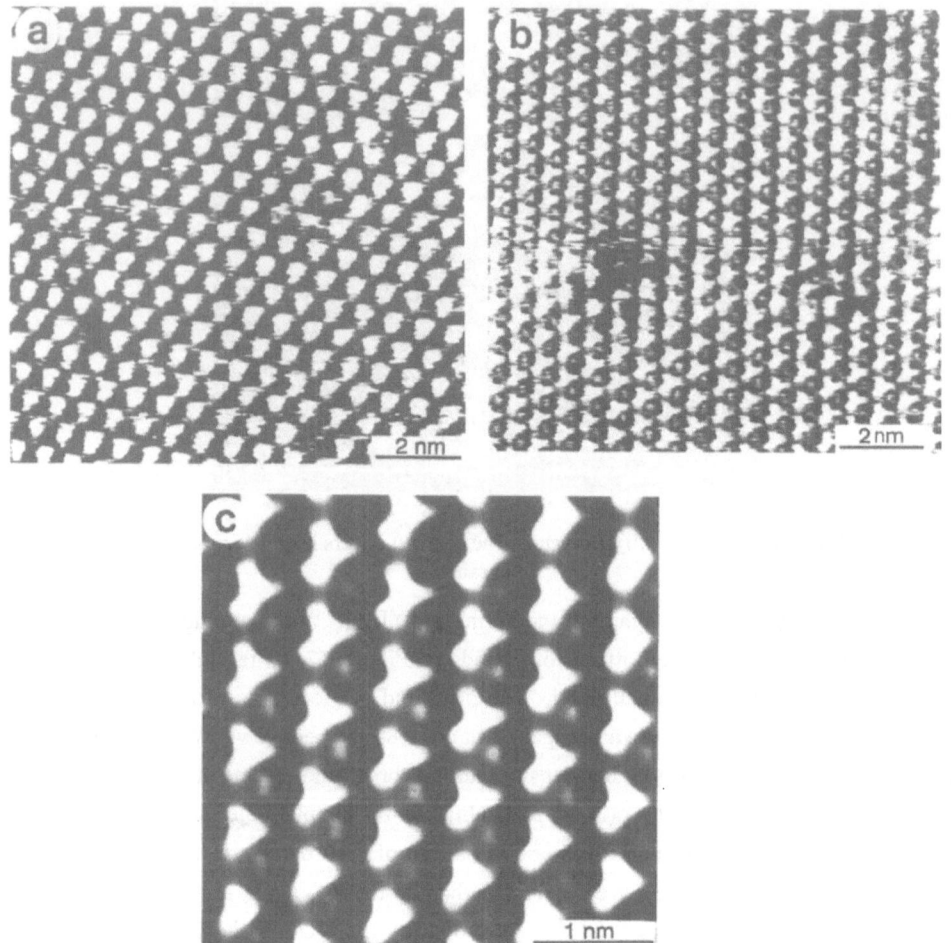

Fig. 16. STM current images of β-Nb$_3$I$_8$: (a) (V_{bias} = 1000 mV, I_{set} = 10 nA). (b) (V_{bias} = 180 mV, I_{set} = 0.75 nA). (c) Zoomed part of the image in (b) after FFT filtering. The contrast covers current variations in the relative units.

octahedra (Figure 19b). The individual MX$_2$ layers of 1T-MX$_2$ and 2H-MX$_2$ phases may be referred to as 1T-MX$_2$ and 2H-MX$_2$ layers, respectively. The metal atoms of undistorted 1T-MX$_2$ and 2H-MX$_2$ layers form a hexagonal lattice (Figure 20a). 1T-MX$_2$ phases with d^1 to d^3 exhibit various patterns of metal atom clustering, which originates from the formation of metal-metal bonding through the shared octahedral edges [30]. For example, the 1T-MX$_2$ compounds with d^1 ions have $\sqrt{13} \times \sqrt{13}$ clusters (Figure 20b). Such a metal clustering and other examples in the 1T-MX$_2$ and 2H-MX$_2$ systems are commonly referred to as CDW's, but not all of them are caused by the electronic instability associated with Fermi surface nesting [30].

In a $\sqrt{13} \times \sqrt{13}$ cluster of 1T-TaX$_2$ (X = S, Se) (Figure 21) [31], the Ta(1) and Ta(2)

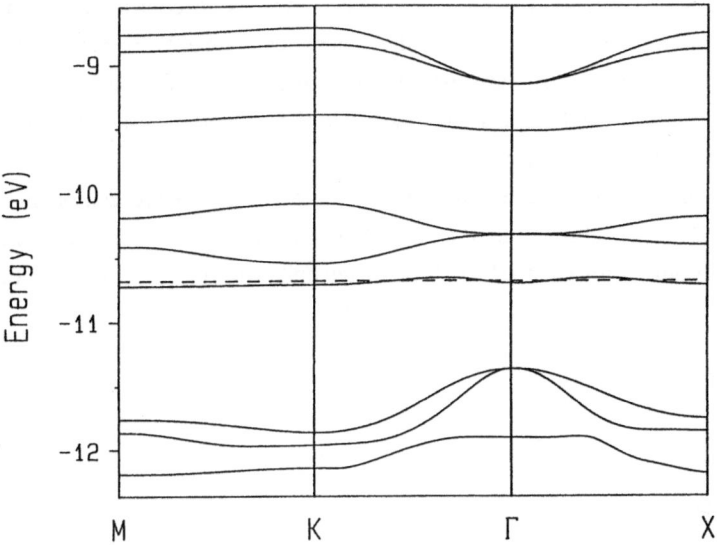

Fig. 17. Dispersion relations of the nine bands, in the vicinity of the Fermi level (shown by the dashed line), calculated for a single Nb_3I_8 layer of β-Nb_3I_8.

atoms form a centered hexagon (i.e., seven-atom cluster) and the Ta(3) atoms cap the edges of the hexagon. The clustering in the metal-atom sheets induces a corrugation in the chalcogen-atom sheets. In the crystal structure of 1T-$TaSe_2$, the Se(1) atoms are farthest away from the metal-atom sheet (at the distance of 1.78 Å). With respect to the Se(1) atoms, the Se(2), Se(3), Se(4) and Se(5) atoms are closer to the metal-atom sheet by 0.05, 0.28, 0.26 and 0.25 Å, respectively. Thus, the height corrugation in the Se atom surface is less than 0.3 Å.

Figure 22 shows the dispersion relations of the bottom portion of the t_{2g}-block bands calculated for a single 1T-$TaSe_2$ layer with the $\sqrt{13} \times \sqrt{13}$ modulation. With 13 d electrons per unit cell, the highest occupied band is half filled and is very narrow [30]. (The electrons in such a narrow band are susceptible to electron localization [9]. Indeed, 1T-TaS_2 is not a metal but a magnetic semiconductor.[32]) This band is largely responsible for the $\rho(r_0, e_f)$ plot and hence the STM image [33]. The orbital compositions of this band are analyzed in the PDOS plots of Figures 23a - 23d [33]. Important *observations* to note from these plots are: *(a)* the highest-occupied band has a larger contribution from the Ta atoms than from the Se atoms, *(b)* the per-atom Ta contribution to the highest-occupied band decreases in the order Ta(1) > Ta(2) > Ta(3), *(c)* the per-atom Se contribution to the highest-occupied band decreases in the order Se(1) > Se(2) >> Se(4), Se(3), Se(5), and *(d)* the contributions of the Se(1) and Se(2) atoms to the highest-occupied band are given almost exclusively by their $4p_z$ orbitals.

The $\rho(r_0, e_f)$ plot of a single 1T-$TaSe_2$ layer, presented in Figure 24a, is obtained by sampling all levels of the highest-occupied band of Figure 22 [33]. The $\rho(r_0, e_f)$ plot shows that the electron density distribution has no contribution from the Ta atoms and arises exclusively from the surface Se(1) and Se(2) atoms. Practically, there is no contribution from the Se(5) atoms, and the electron density decreases in the order, Se(1) > Se(2) >> Se(3), Se(4). This is consistent with *observation (c)* and also with the fact that the Ta atoms are much farther away from the tip than the surface Se atoms. The peaks of the $\rho(r_0, e_f)$ plot for the Se(1) and Se(2) atoms (Figure 24a) are centered at the atomic positions, which arises from *observation (d)*. It is clear from Figure 24a that the bright spots of the STM image are caused mainly by the surface Se(1) and Se(2) atoms, and that the STM image should exhibit a pattern of six-chalcogen-atom triangles for each $\sqrt{13} \times \sqrt{13}$ cluster unit, with the inner-triangle of the Se(1)

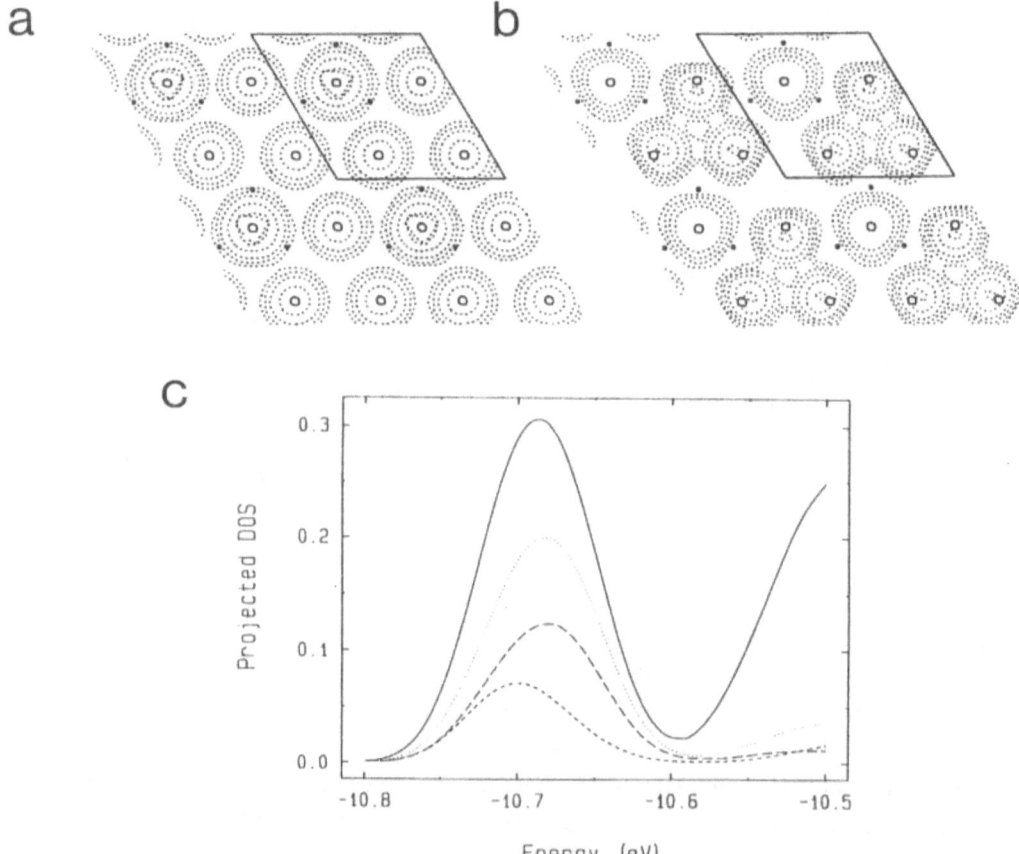

Fig. 18. (a) $\rho(r_0)$ and (b) $\rho(r_0, e_f)$ plots calculated for surface A of a single Nb_3I_8 layer of β-Nb_3I_8. The plot area consists of four unit cells, and a unit cell is indicated by a rhombus. For clarity, the I atoms on the opposite surface are not shown. The Nb and I atoms are shown by small and large circles, respectively. The contour values used are 85, 50, 20, 10 and 5 x 10^{-3} electrons/au^3 in (a), and 100, 50, 20, 10, 5 and 2 x 10^{-5} electrons/au^3 in (b). (c) PDOS values of the iodine atoms of surface A, in the energy region of the half-filled band, calculated for a single Nb_3I_8 layer of β-Nb_3I_8. The p_z orbital of I(2) [dotted line], the p_x and p_y orbitals of I(2) [solid line], the p_z orbital of I(4) [short-dash line], and the p_x and p_y orbitals of I(4) [long-dash line].

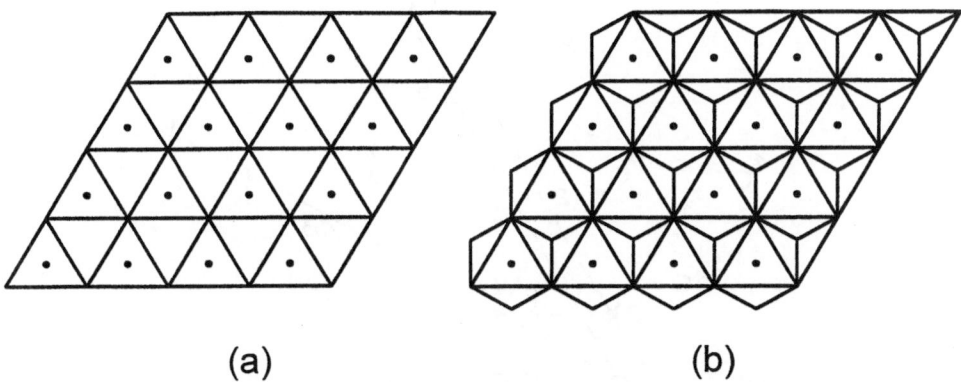

(a)   (b)

Fig. 19. Top projection views of several layers derived from MX_6 trigonal prisms and octahedra: (a) 2H-MX_2 layer. (b) 1T-MX_2 layer.

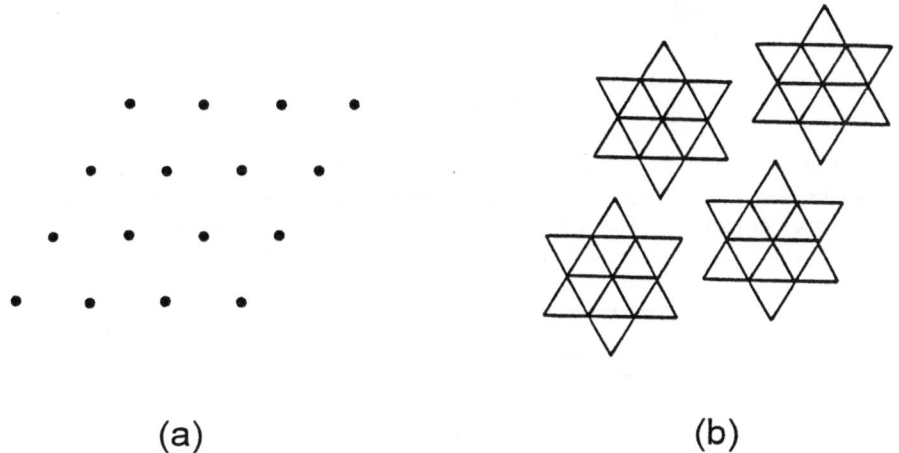

(a)   (b)

Fig. 20. Metal atom arrangements in 1T-MX_2 layers: (a) Hexagonal lattice of d^0 1T-MX_2 layer. (b) $\sqrt{13} \times \sqrt{13}$ clusters of d^1 1T-MX_2 layer.

atoms much brighter than the outer-triangle of the Se(2) atoms. Such a pattern is found in high-resolution STM images of 1T-TaSe$_2$ (Figure 25).

Figure 24b shows the cross-sectional view of the partial electron density $\rho(\mathbf{r}, e_f)$ in the plane B perpendicular to the layer (Figure 21). This cross-section contains the Se(1), Se(2), Se(3), and Se(5) atoms of a $\sqrt{13} \times \sqrt{13}$ cluster. The STM height image is simulated by the contour plot of the constant electron density in Figure 24b. Compared with the Se(1) and Se(2) positions, the electron densities at the Se(3), Se(4) and Se(5) positions are very small. The z-height difference in the contour lines can easily reach 2 Å, which is consistent with the anomalously large surface corrugation deduced from constant-current STM measurements.

Certainly, the $\sqrt{13} \times \sqrt{13}$ clustering of 1T-TaX$_2$ (X = S, Se) are induced by the d electrons of Ta, and their highest occupied bands are dominated by the Ta atoms [30,33]. However, it is

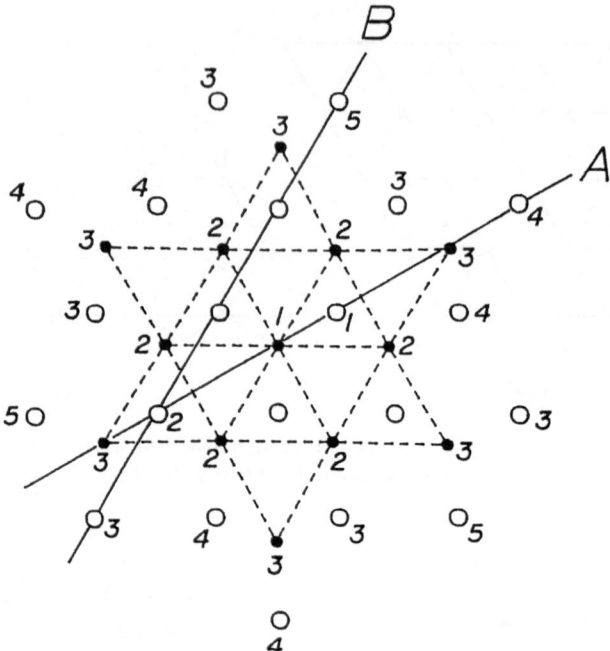

Fig. 21. Schematic representation of the three different Ta atoms and five different Se atoms of a $\sqrt{13} \times \sqrt{13}$ cluster in 1T-TaSe$_2$. The Ta and Se atoms are presented by small filled and large empty circles, respectively. The Se atoms of the bottom Se atom sheet are not shown for clarity. The lines A and B refer to the planes perpendicular to the layer of 1T-TaSe$_2$.

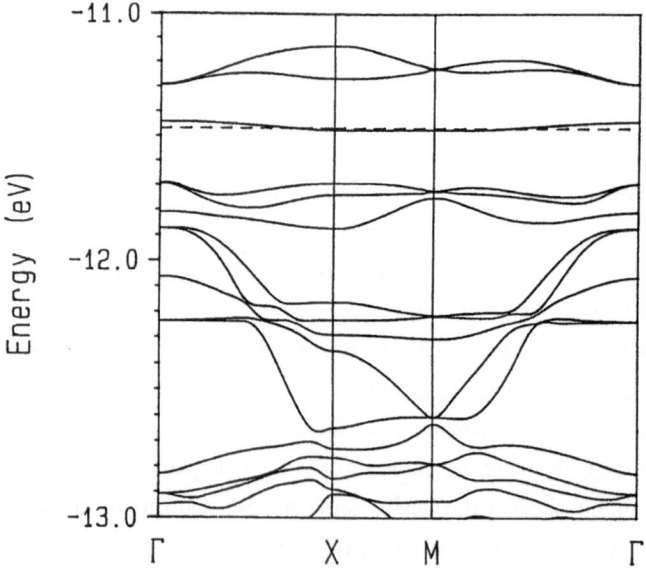

Fig. 22. Dispersion relations of the bottom portion of the t$_{2g}$-block bands calculated for a single 1T-TaSe$_2$ layer with the $\sqrt{13} \times \sqrt{13}$ modulation. The dashed line refers to the Fermi level.

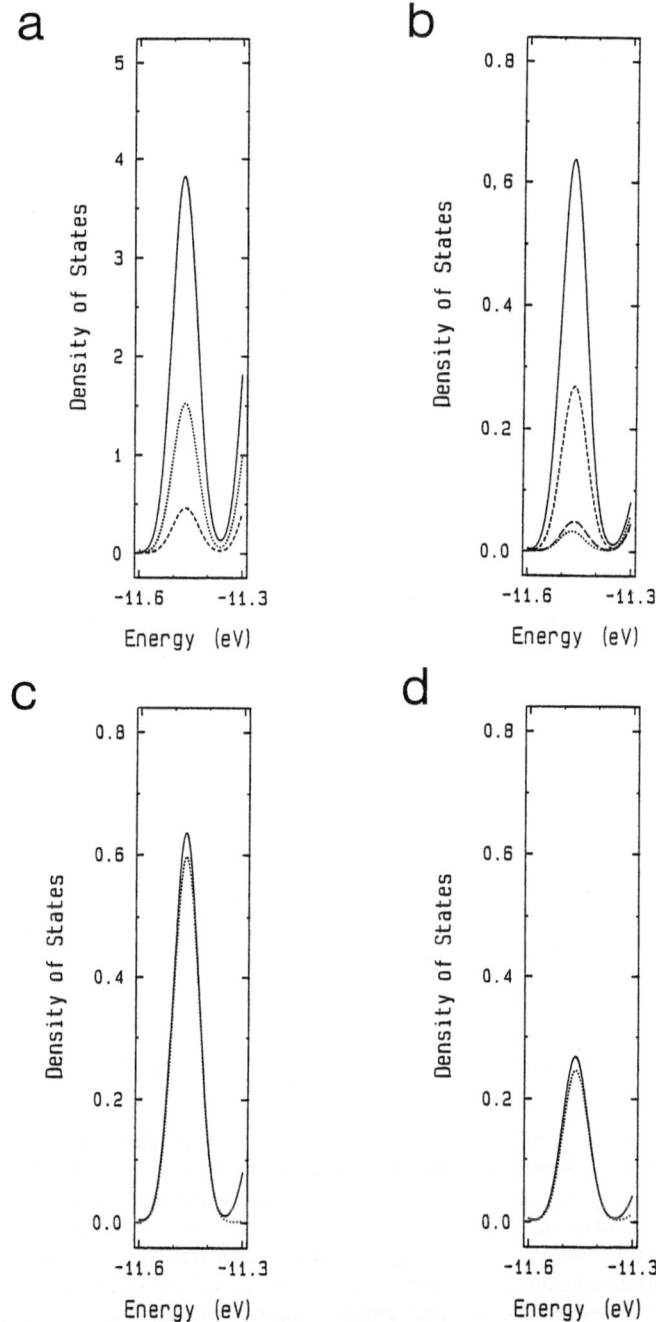

Fig. 23. (a) PDOS values per atom calculated for the Ta(1) (solid line), Ta(2) (dotted line), and Ta(3) (dashed line) atoms of 1T-TaSe$_2$. (b) PDOS values per atom calculated for the Se(1) (solid line), Se(2) (short-dash line), Se(4) (long-dash line), and Se(5) (dotted line) atoms of 1T-TaSe$_2$. The PDOS values per atom for the Se(3) are practically identical with those for the Se(5) atom, and hence are not shown. (c) PDOS values per atom of the Se(1) atom (solid line) and its 4p$_z$ orbital contribution (dotted line). (d) PDOS values per atom of the Se(2) atom (solid line) and its 4p$_z$ orbital contribution (dotted line).

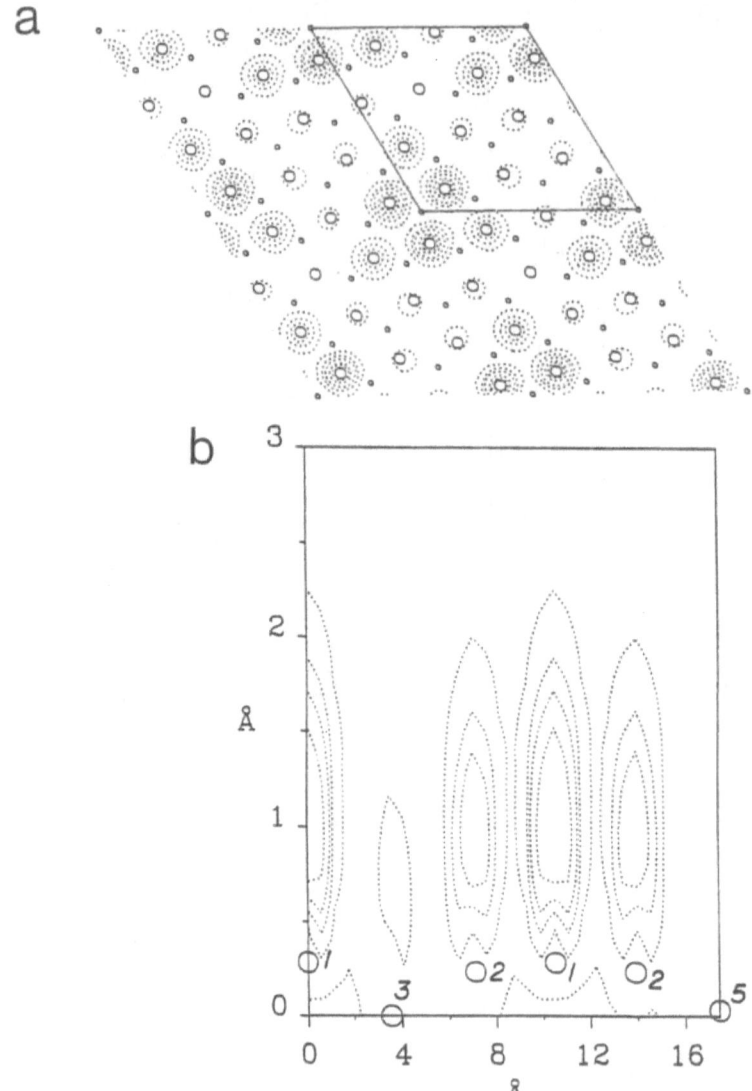

Fig. 24. (a) $\rho(r_0, e_f)$ plots calculated for a single layer of 1T-TaSe$_2$. The plot area consists of four unit cells, and a unit cell is indicated by a rhombus. The contour values used are 25, 20, 10, 5 and 1 x 10^{-4} electrons/au^3. For clarity, the Se atoms on the bottom surface are not shown. The Ta and Se atoms are shown by small and large circles, respectively. (b) Cross-sectional view of the partial electron density distribution of a single 1T-TaSe$_2$ layer (resulting from the band levels at the Fermi level) in the plane B, perpendicular to the layer, defined in Figure 2. The Se atoms are represented by empty circles. This cross-section contains the Se(1), Se(2), Se(3), and Se(5) atoms of a $\sqrt{13} \times \sqrt{13}$ cluster. The contour values used are 20, 10, 5 and 1 x 10^{-4} electrons/au^3.

the surface chalcogen atoms that determine the STM pattern because of their proximity to the scanning tip. In the earlier studies of 1T-TaX$_2$ (S = S, Se), the metal d-orbitals were thought to dominate the STM pattern [34]. The large height amplitudes of the STM images simply

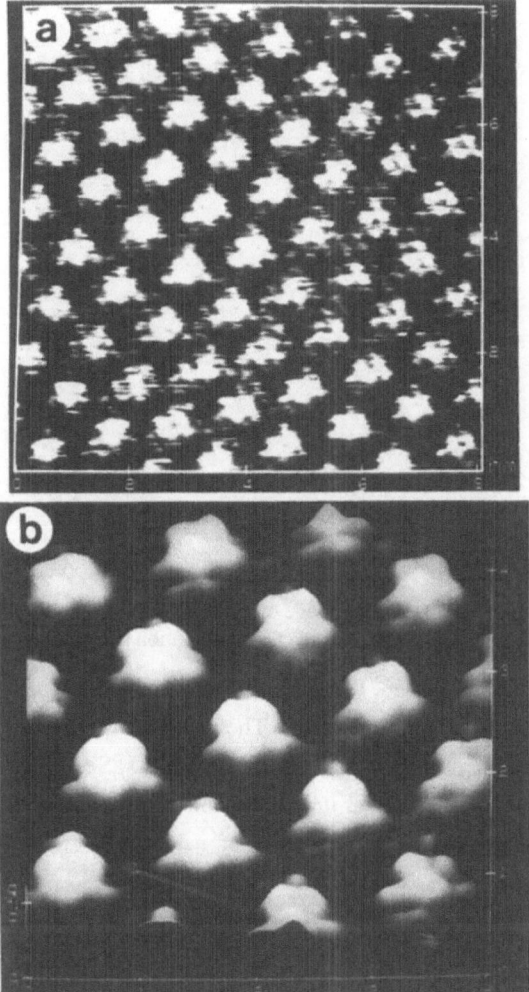

Fig. 25. STM current images of 1T-TaS$_2$ (I_{set} = 6 nA, V_{bias} = -10 mV). The contrast covers current
variations in the relative units. (b) A zoomed part of the image in (a) after FFT filtering.

reflects the density distributions of the $\rho(r_0, e_f)$ plot. *Observations (c)* and *(d)*, together with
the fact that the surface Se(1) and Se(2) atoms are closer to the scanning tip than any other
atoms by about 0.25 Å, explain why the amplitude of the CDW modulation is so large in the
STM image.

The STM images of 1T-TaSe$_2$ show large defects, which appear as if whole $\sqrt{13} \times \sqrt{13}$
cluster units are missing (Figure 26). The STM images of 1T-TaSe$_2$ are primarily determined
by the partial electron density plot $\rho(r_0, e_f)$ associated with the highest occupied band of its
surface layer. To examine a probable cause for such STM defects, it is necessary to analyze the
nature of this band. It should be noted that a 1T-TaSe$_2$ layer with $\sqrt{13} \times \sqrt{13}$ modulation is
made up of the $\sqrt{13} \times \sqrt{13}$ clusters, i.e., the Ta$_{13}$Se$_{42}{}^{32-}$ clusters (Figure 27), which contain 13

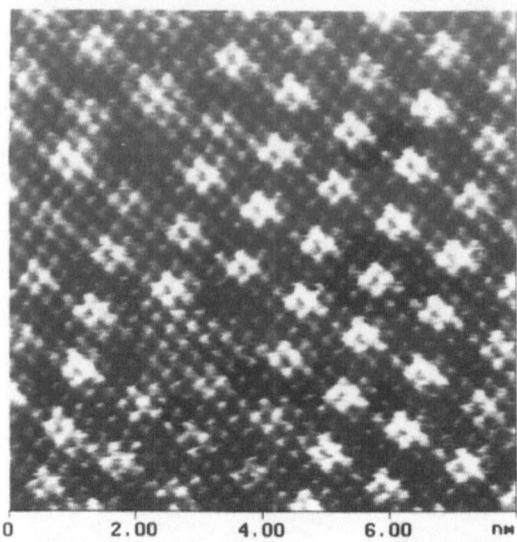

Fig. 26. Large imperfection in the STM current image of 1T-TaSe$_2$ (V_{bias} = -10 mV, I_{set} = 6 nA). The contrast is proportional to the current variation in relative units.

d-electrons. The energy levels of a $\sqrt{13} \times \sqrt{13}$ cluster calculated by the extended Hückel molecular orbital method are shown in Figure 28. The HOMO of this cluster is singly filled. It is important to note that the highest occupied band of the 1T-TaSe$_2$ layers is largely made up of the singly filled HOMO's from its cluster units. This means that the bright spots in the STM images of 1T-TaSe$_2$, obtained with either positive or negative V_{bias}, represent the electron densities associated with the HOMO's of its cluster units.

The tunneling current is sensitive to the partial electron density associated with the Fermi level. Thus, for 1T-TaSe$_2$, when one cluster unit is modified to have a lower-lying HOMO and a higher-lying LUMO than do other neighboring cluster units, it cannot contribute to a tunneling current and appears as dark in the STM image as if the whole cluster is missing.

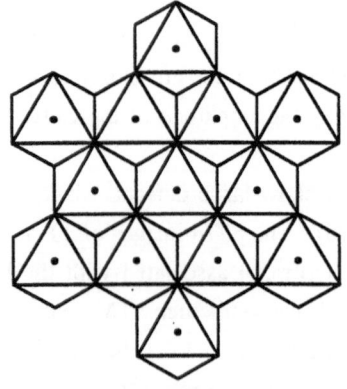

Fig. 27. Schematic representation of a Ta$_{13}$Se$_{42}$$^{32-}$ cluster.

Such a situation is easily created when the metal cluster units have an atom vacancy. For example, Figure 28 compares the energy levels of a $\sqrt{13} \times \sqrt{13}$ cluster with those of the $Ta_{12}Se_{42}^{36-}$ cluster derived from it by deleting the central Ta atom (i.e., the Ta(1) atom in Figure 21). The HOMO and LUMO levels of the $Ta_{12}Se_{42}^{36-}$ cluster are quite different from those of the defect-free cluster. Therefore, the clusters with metal atom vacancies appear as dark holes in the STM images.

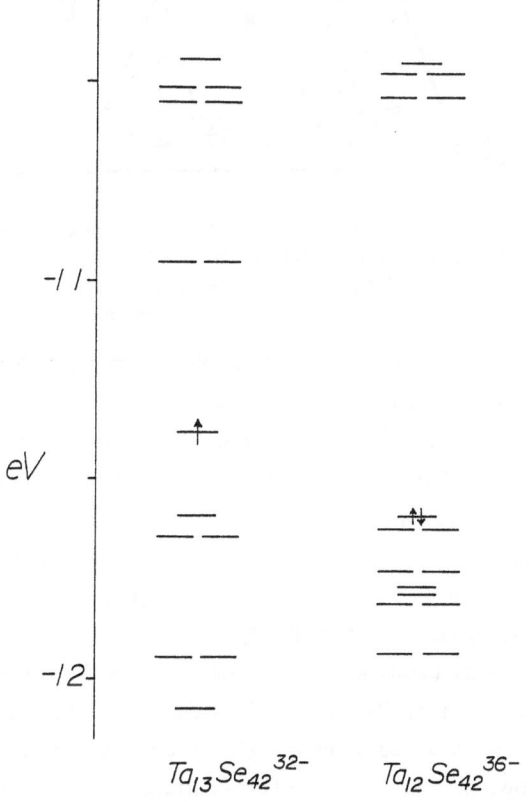

Fig. 28. Low-lying energy levels of a $\sqrt{13} \times \sqrt{13}$ cluster $Ta_{13}Se_{42}^{32-}$ and the $Ta_{12}Se_{42}^{36-}$ cluster derived from $Ta_{13}Se_{42}^{32-}$ by deleting the Ta(1) atom. For the simplicity of presentation, only the highest occupied levels are shown to have electrons.

3.4 NbSe3

Niobium triselenide NbSe3 is a layered compound made up of trigonal prismatic chains. A unit cell of NbSe3 contains three pairs of prismatic chains of types I, II and III (Figure 29). It is a quasi 1D metal and exhibits two CDW transitions with onset temperatures 144 and 59K [35]. ^{93}Nb NMR measurements of NbSe3 show that the 144K CDW primarily affects type III chains, while the 59K CDW primarily affects type I chains.[36] Indeed, the modulated structure of NbSe3 determined by synchrotron-radiation single-crystal X-ray diffraction shows [37] that the

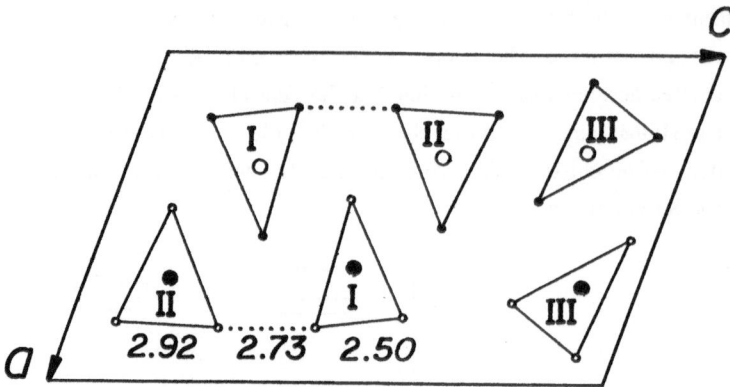

Fig. 29. Projection view, along the b-axis direction, of a unit cell of a single NbSe$_3$ layer. The large and small circles represent the Nb and Se atoms, respectively. The shaded and unshaded atoms differ in their b-axis height by b/2. The dashed lines show the short Se-Se contacts between type I and II chains. Three short Se-Se distances are shown in Å unit.

144 and 59 K CDW's mainly involve the displacements of Nb atoms of type III and I chains, respectively. In contrast, the STM studies of NbSe$_3$ at 4.2K suggest that all three types of chains carry a strong CDW modulation [38]. To understand the apparently conflicting conclusion of the STM studies, it is necessary to calculate the $\rho(r_0, e_f)$ plots, not for the normal metallic state, but for the CDW states of NbSe$_3$.

The dispersion relations of the bottom six d-block bands calculated for a single layer of NbSe$_3$ are shown in Figure 30 [35c,39], which shows four partially filled bands. In terms of the Nb d-orbital character, these four bands are primarily derived from type I and III chains, and the two empty bands from type II chains. As the wave vector moves along Γ to Y in Figure 30, the four partially filled bands at the Fermi level are mainly described by the orbitals of type I, III, III and I chains, respectively. The Fermi surfaces associated with the partially filled bands are shown in Figure 31. The Fermi surfaces associated with the two type III chains are practically flat, while those related to the two type I chains show a slight curvature. The vector q_1 (= 0.248b*) nests the Fermi surfaces of the two type III chains, and the vector q_2 (= 0.262b*) nests those of the two type I chains [40].

In order to simulate the STM images of the two CDW's in NbSe$_3$, one needs to construct the CDW-state band orbitals φ_1 and φ_2 associated with the nesting vectors q_1 and q_2, respectively, and then calculate the $\rho(r_0, e_f)$ plots using these band orbitals. A CDW state expected for the case of one partially filled 1D band was discussed in Section 2.2. This treatment should be extended to construct the two CDW states of NbSe$_3$. For convenience, consider a 1D metal with two partially filled bands $\phi_1(k)$ and $\phi_2(k)$. Without loss of generality, it can be assumed that the band $\phi_1(k)$ is occupied in the wave vector region $-k_{f1} < k \leq k_{f1}$, and the band $\phi_2(k)$ in the region $-k_{f2} < k \leq k_{f2}$. Then, for a small non-negative δ, the occupied orbitals $\phi_1(k_{f1}-\delta)$, $\phi_1(-k_{f1}+\delta)$, $\phi_2(k_{f2}-\delta)$, and $\phi_2(-k_{f2}+\delta)$ are close in energy to the Fermi level, and so are the

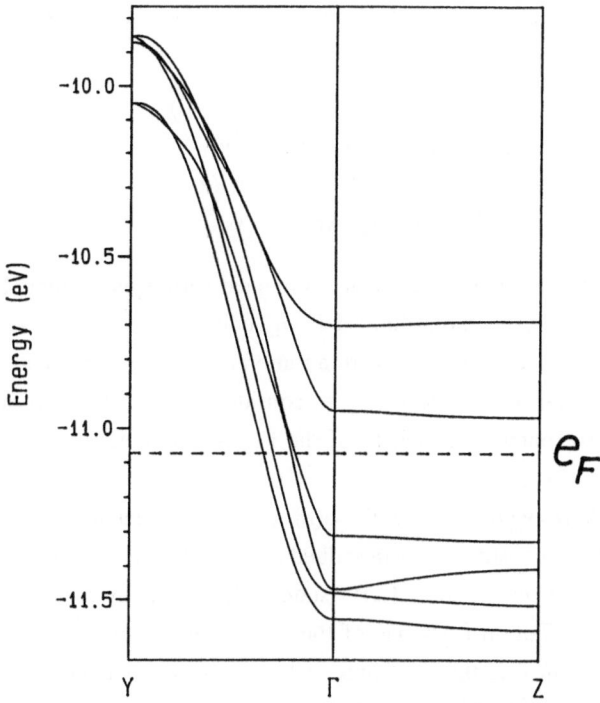

Fig. 30. Dispersion relations of the bottom six d-block bands of a single NbSe$_3$ layer, where $\Gamma = (0, 0)$, $Y = (b^*/2, 0)$, and $Z = (0, c^*/2)$.

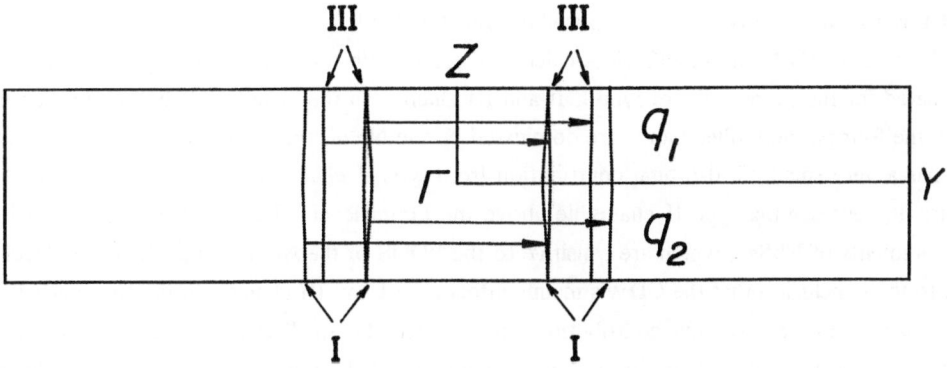

Fig. 31. Fermi surfaces associated with the four partially filled bands of a single NbSe$_3$ layer. On going from Γ to Y, the Fermi surfaces are mainly described by type I, III, III and I chains, respectively. $\Gamma = (0, 0)$, $Y = (b^*/2, 0)$, and $Z = (0, c^*/2)$.

unoccupied orbitals $\phi_1(-k_{f1}-\delta)$, $\phi_1(k_{f1}+\delta)$, $\phi_2(-k_{f2}-\delta)$, and $\phi_2(k_{f2}+\delta)$. The Fermi surfaces of the two bands are nested by the single vector $q = k_{f1} + k_{f2}$, which leads to a single CDW with modulation vector q.   The new orbitals $\varphi(k_{f1}-\delta)$, $\varphi(-k_{f1}+\delta)$, $\varphi(k_{f2}-\delta)$, and $\varphi(-k_{f2}+\delta)$

representing the CDW modulation with the vector q are constructed in terms of the interband orbital mixing [11],

$$\varphi(k_{f1}-\delta) \propto \phi_1(k_{f1}-\delta) + \phi_2(-k_{f2}-\delta)$$
$$\varphi(k_{f2}-\delta) \propto \phi_2(k_{f2}-\delta) + \phi_1(-k_{f1}-\delta)$$
$$\varphi(-k_{f1}+\delta) \propto \phi_1(-k_{f1}+\delta) + \phi_2(k_{f2}+\delta)$$
$$\varphi(-k_{f2}+\delta) \propto \phi_2(-k_{f2}+\delta) + \phi_1(k_{f1}+\delta)$$

where the occupied and unoccupied orbitals are combined with equal weights because they are nearly degenerate. The $\rho(r_0,e_f)$ plots derived from such band orbitals φ simulate the STM images expected for the CDW associated with an interband nesting vector q. In constructing the CDW-state band orbitals φ, it is necessary to consider only small values of δ because only the electrons lying in the immediate vicinity of the Fermi level contribute significantly to the tunneling current.

Figures 32a and 32b respectively show the $\rho(r_0,e_f)$ plots calculated for the q_1 and q_2 CDW states of $NbSe_3$, which are constructed as described above [41]. In both plots, only the surface Se atoms are seen as expected. The q_1 CDW state is dominated by the Se atoms of type III chains, with a small contribution from one of the two Se atoms of type I chains. In the q_2 CDW, the largest contributor is the Se atoms of type I chains, but there is a non-negligible contribution from the Se atoms of type II chains. The $\rho(r_0,e_f)$ plots of the q_1 and q_2 CDW states are combined in Figure 32c, which shows that the brightness of the chain images in the STM picture of $NbSe_3$ should increase in the order: type II < type I < type III. Certainly, from the $\rho(r_0,e_f)$ plots of Figure 32 alone, it is possible to suggest that type II chains are also involved in the CDW formation of $NbSe_3$, as proposed by Dai et al. [38].

Figure 33a shows the PDOS plots calculated for the Nb d-orbitals, and Figure 33b those calculated for the Se orbitals, of type I, II and III chains. In the vicinity just below the Fermi level, the four partially filled bands are dominated by the Nb d-orbitals of type I and III chains and has a very small Nb d-orbital contribution from type II chains because the two bands of Figure 30 representing type II chains lie above the Fermi level. Therefore, the [93]Nb NMR measurements of $NbSe_3$, which are sensitive to the PDOS of the Nb atoms at the Fermi level, lead to the conclusion that the CDW's mainly affect type I and III chains. In the four partially filled bands, the Se orbital contribution is considerably smaller than the Nb d-orbital contribution (Figure 33). Nevertheless, in the vicinity just below the Fermi level, the Se orbital contribution from type II chains is only slightly smaller than that from type I or III chains (Figure 33b). Since the Nb atoms lie considerably below the surface Se atoms, only the surface Se atoms of a single NbSe3 layer can be seen from the $\rho(r_0,e_f)$ plots and hence in the STM images. Consequently, the calculated $\rho(r_0,e_f)$ plots show the contributions of the Se atoms of type II chains, though weaker in intensity than the Se atom images of type I and III chains.

Concerning the question of which chains are responsible for the formation of the two

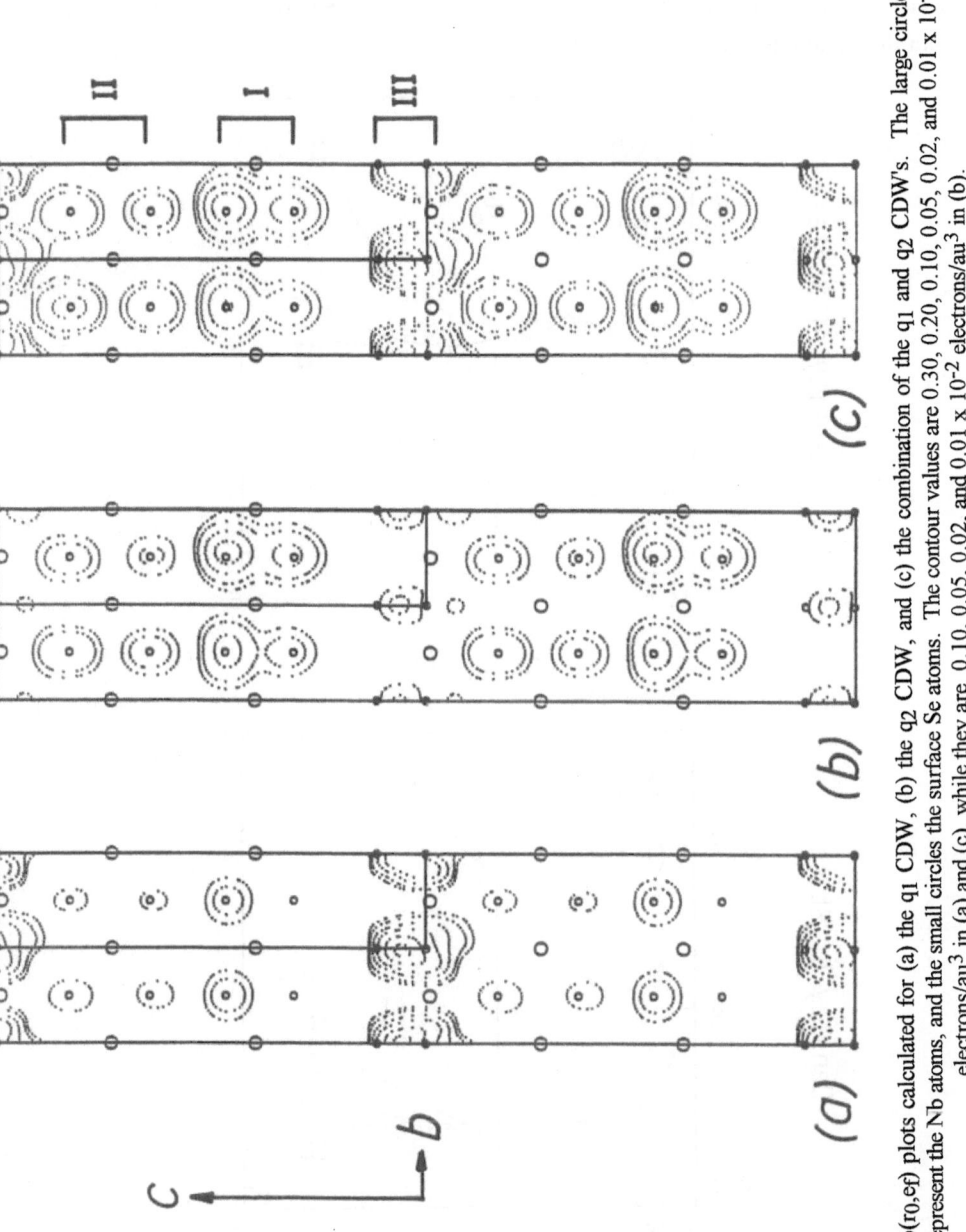

Fig. 32. $\rho(r_0, \epsilon_f)$ plots calculated for (a) the q1 CDW, (b) the q2 CDW, and (c) the combination of the q1 and q2 CDW's. The large circles represent the Nb atoms, and the small circles the surface Se atoms. The contour values are 0.30, 0.20, 0.10, 0.05, 0.02, and 0.01 x 10^{-2} electrons/au^3 in (a) and (c), while they are 0.10, 0.05, 0.02, and 0.01 x 10^{-2} electrons/au^3 in (b).

CDW's in NbSe₃, however, the conclusion of the ^{93}Nb NMR [36] and single crystal X-ray diffraction [37] studies is appropriate because the major character of the electrons at the Fermi level, which are responsible for the CDW formation, is not the Se atoms of all three types of chains but the Nb atoms of type I and III chains.

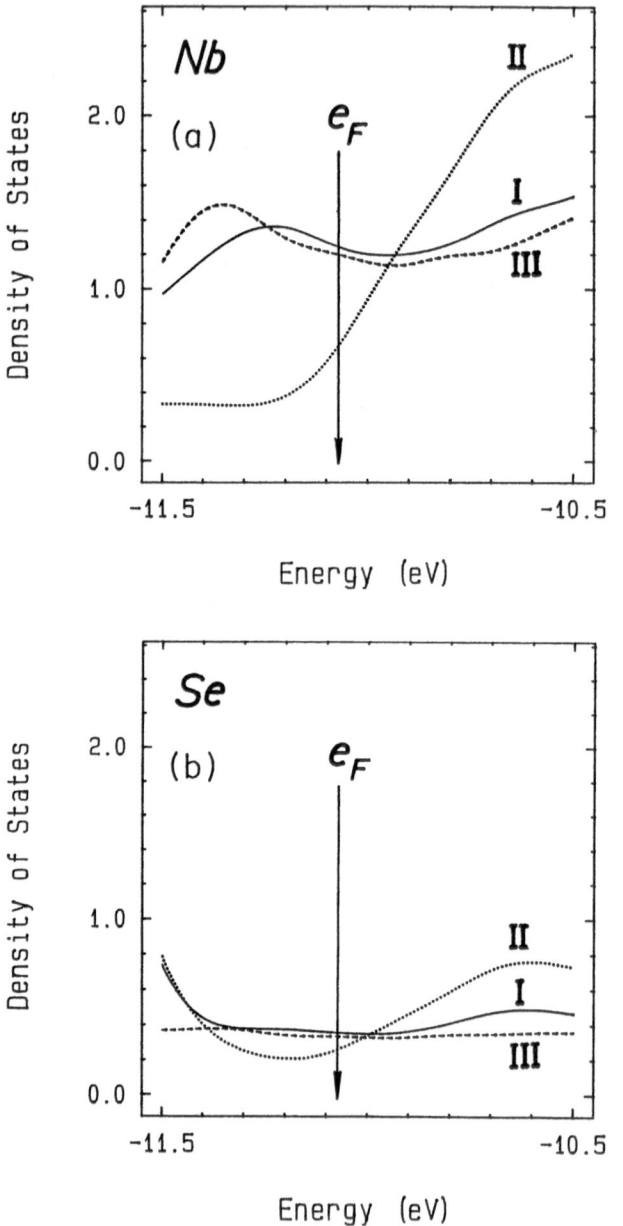

Fig. 33. PDOS plots calculated for (a) the Nb d-orbitals and (b) the Se orbitals of type I chains (solid line), type II chains (dotted line) and type III chains (dashed line) in the vicinity of the Fermi level.

3.5 2H-MoS$_2$ and 2H-WSe$_2$

The layered compounds 2H-MoS$_2$ and 2H-WSe$_2$ consist of 2H-MX$_2$ layers (Figure 19a). Both the metal and chalcogen atom sheets have a hexagonal arrangement, and all the atoms of the surface chalcogen sheet have the same z-height. With d^2 (M^{4+}) ions, each 2H-MX$_2$ layer has two electrons per unit cell to fill its d-block bands. Since the bottom d-block band is separated from the other d-block bands (Figure 34a), 2H-MoS$_2$ and 2H-WSe$_2$ are semiconductors. The AFM image of 2H-MoS$_2$ is shown in Figure 35c, which exhibits a hexagonal pattern. The STM image of 2H-MoS$_2$ is essentially identical in nature with the AFM image (hence not shown). The $\rho(r_0)$ plot calculated for a single 2H-MoS$_2$ layer (Figure 35a) shows that the bright spots of the AFM image represent the surface sulfur atoms. In the earlier interpretation of the STM image of 2H-MoS$_2$, it was thought that the bright spots of the STM image are associated with the d$_{z^2}$-orbitals of Mo [42]. However, as shown in Figure 35b, the $\rho(r_0,e_f)$ plot calculated for a single 2H-MoS$_2$ layer is dominated by the contribution of the surface sulfur atoms. Although the d-block bands of a single 2H-MoS$_2$ layer have small contributions from the sulfur orbitals, the Mo atom sheet is far removed from the surface sulfur atom sheet (by 1.59 Å). Thus, the surface S atoms are responsible for the bright spots of both the STM and the AFM images 2H-MoS$_2$ [17].

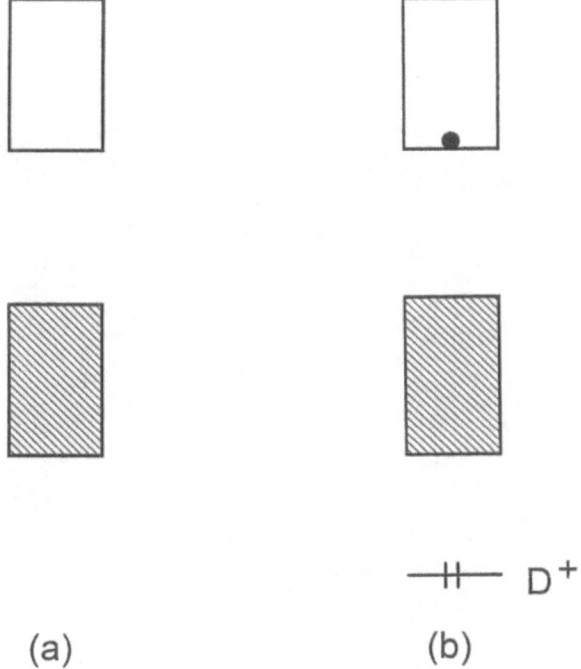

(a)   (b)

Fig. 34. Energy level that a donor dopant can bring to the energy region of the valence and conduction bands of the defect-free 2H-WSe$_2$ layer: (a) free of defects. (b) a donor ligand producing a D$^+$ cation at the Se-site. Shaded and unshaded square boxes represent the valence and conduction bands, respectively.

In the synthesis of 2H-MoS$_2$ and 2H-WSe$_2$, Cl$_2$ and I$_2$ are commonly used as transporting agents, respectively [43]. Thus, it is probable that the chalcogen atom surfaces of 2H-MoS$_2$ and 2H-WSe$_2$ have substitution defects (i.e., Cl at an S site of 2H-MoS$_2$ and I at an Se site of 2H-WSe$_2$). Within a rigid band scheme a donor atom such as I at the Se site (I$_{Se}$) of 2H-WSe$_2$ creates an electron at the bottom of the conduction band (Figure 34b). However, it is energetically more favorable for this electron to be trapped around the donor cation (e.g., I$^+$) by relaxing the geometry around it (Figure 36a) [44]. The valence and conduction bands of 2H-WSe$_2$ are d-block bands, so they are antibonding between the metal W and its ligand atoms L (i.e., L = Se, I) [12]. In the local area around the donor defect where an electron is trapped (i.e., the electron-trapped area), there will be a geometry relaxation which slightly lengthens the W-L bonds to decrease the extent of the antibonding. This creates the singly-filled donor level

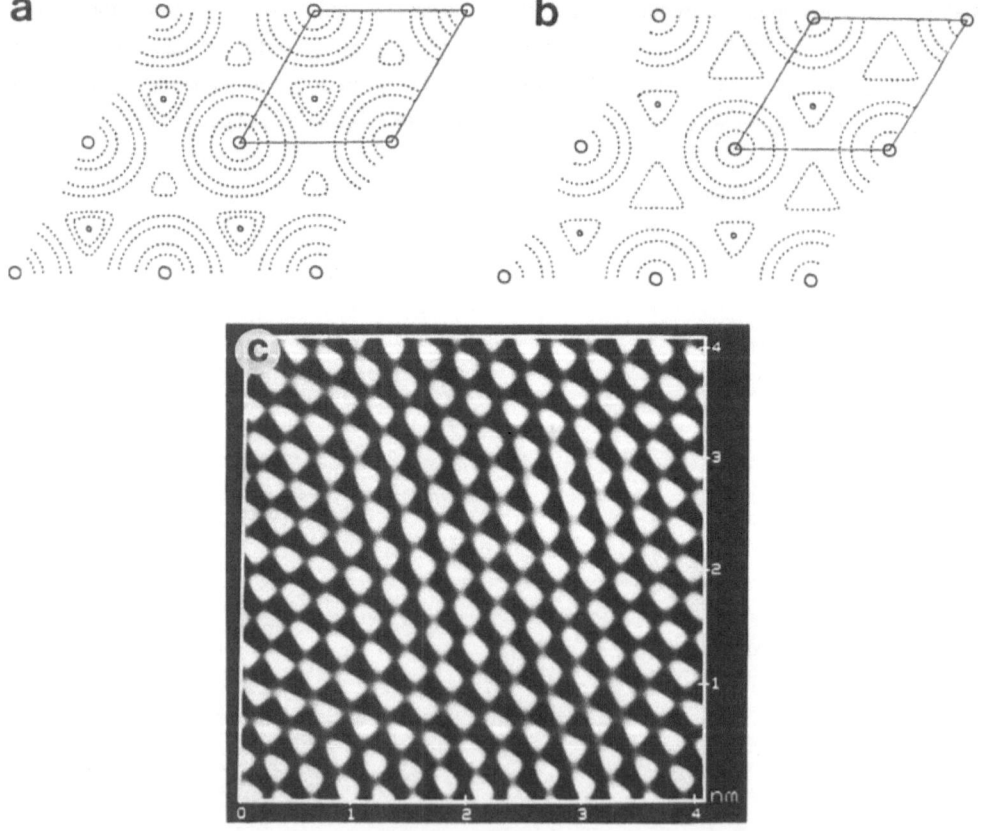

Fig. 35. (a) AFM height image of 2H-MoS$_2$. The contrast variations cover the 0 - 3 Å range. (b) $\rho(r_0)$ plot calculated for a single 2H-MoS$_2$ layer, where the contour values used are 20, 10, 5, 1 and 0.5 x 10^{-2} electrons/au^3. (c) $\rho(r_0, e_f)$ plot calculated for the conduction band bottom of a single 2H-MoS$_2$ layer, where the contour values used are 20, 10, 5, 1 and 0.5 x 10^{-4} electrons/au^3. In (b) and (c), the surface S and the Mo atoms are represented by large and small circles, respectively.

(i.e., one composed mainly of the orbitals of the electron-trapped area) below the bottom of the conduction band (Figure 36b) [44]. The radial distribution of the trapped electron state being analogous to that of a hydrogenic 1s orbital [14], the extent of the geometry relaxation should decrease with the distance from the defect site. By analogy with the donor defects in the elemental semiconductors Ge and Si [14], the electron-trapped area is expected to have a nanometer-scale size. As shown in Figure 37, the STM image of 2H-WSe$_2$ recorded with negative bias voltage exhibit nanometer scale bright spots. The latter are assigned to the electron trapped areas around the sites of the donor defects such as I [44].

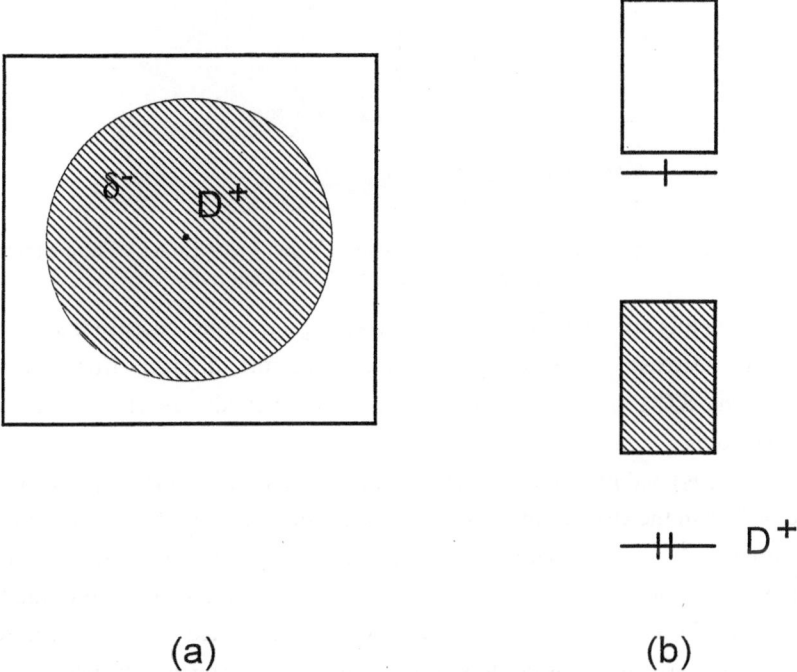

(a)   (b)

Fig. 36. (a) Schematic representation of the electron-trapped area around a donor cation D^+, where the shading indicates that the trapped area has higher electron density than the surrounding. (b) Schematic representation of the donor level associated with the trapped-electron state.

4. Concluding Remarks

To a first approximation, the atomic scale patterns of the AFM and STM images of a sample surface are described by the HED patterns of the $\rho(r_0)$ and $\rho(r_0,e_f)$ plots, respectively. For a variety of organic and inorganic layered materials, the atomic- and molecular-scale features of their STM images have been successfully interpreted on the basis of the $\rho(r_0,e_f)$ plots calculated with the extended Hückel tight binding electronic band structure method [17]. For all the layered transition metal chalcogenides and halides described in this review, the

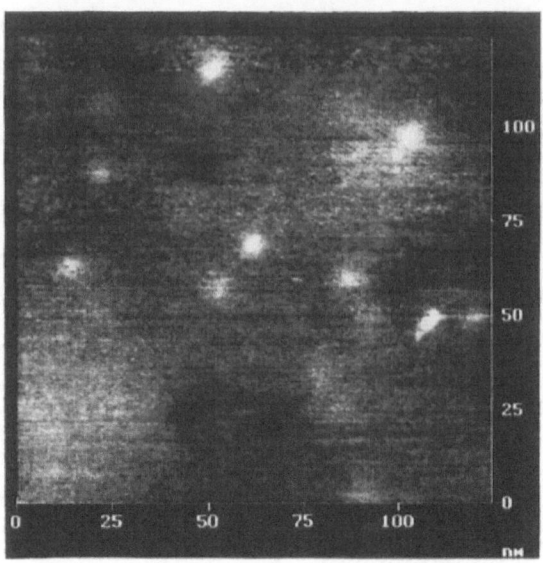

Fig. 37. STM image of WSe$_2$ obtained with V_{bias} = -600 mV and I_{set} = 1.0 nA. The contrast covers the height variation in the 0.0-1.0 nm range.

$\rho(r_0,e_f)$ and $\rho(r_0)$ plots are dominated by their surface ligand atoms, because the amplitudes of atomic orbitals decrease exponentially with increasing the distance from their centers. Consequently, the STM and AFM images of these compounds are all represented by the surface ligand atoms.

What the $\rho(r_0,\ e_f)$ and $\rho(r_0)$ plots simulate are ideal STM and AFM images expected for a defect-free surface in the absence of tip-sample interactions. During the scanning in STM and AFM measurements, a sample surface may undergo a geometry relaxation due to the tip-sample force interactions [22,45]. In such cases, it is not sufficient to interpret the STM and AFM images only in terms of the electron density plots calculated for the unrelaxed surface structure. Nevertheless, the density plots of the unrelaxed surface are invaluable because the discrepancy between the patterns of these plots and observed images provide clues to the surface local hardness variation and hence the nature of the tip-force induced surface relaxation [22,45].

Acknowledgments

The work was supported by the Office of Basic Energy Sciences, Division of Materials Sciences, U.S. Department of Energy, under Grant DE-FG05-86ER45259 and by European Community under the Human Capital and Mobility Project (ERBCHRXCT940675). M.-H. W. thanks Alexander von Humboldt Foundation for a Humboldt Research Award for Senior U.S. Scientists, which made possible his extended stay at Materials Research Center, Albert-Ludwigs University.

References

1. G. Binnig, H. Rohrer, Ch. Gerber, E. Weibel, *Phys. Rev. Lett.* **49**, 57 (1982).
2. R. Wiesendanger, H.-J. Güntherodt (Eds.), *Scanning Tunneling Microscopy I, II and III*, Springer Verlag, Heidelberg, 1992 and 1993.
3. C.J. Chen, *Introduction to Scanning Tunneling Microscopy*, Princeton University Press, Princeton, 1993.
4. D.A. Bonnell (Ed.), *Scanning Tunneling Microscopy and Spectroscopy*, VCH, New York, 1993.
5. G. Binnig, C. Quate, Ch. Gerber, *Phys. Rev. Lett.* **56**, 930 (1986).
6. J. Tersoff, D.R. Hamman, *Phys. Rev. B* **31**, 805 (1985).
7. For a review, see: E. Canadell, M.-H. Whangbo, *Chem. Rev.* **91**, 965 (1991)
8. (a) R. Hoffmann, *J. Chem. Phys.* **39**, 1397 (1963).
    (b) M.-H. Whangbo and R. Hoffmann, *J. Am. Chem. Soc.* **100**, 6093 (1978).
9. (a) N.F. Mott, *Metal-Insulator Transitions*, Barnes and Noble, New York, 1977.
    (b) B.H. Brandow, *Adv. Phys.* **26**, 651 (1977).
    (c) M.-H. Whangbo, *J. Chem. Phys.* **70**, 4963 (1979).
    (d) M.-H. Whangbo, *Acc. Chem. Res.* **16**, 95 (1983).
10. R. Moret and J.-P. Pouget, in *Crystal Chemistry and Properties of Materials with Quasi-One-Dimensional Structures*, J. Rouxel, Ed., Reidel, Dordrecht, The Netherlands, 1986.
11. (a) M.-H. Whangbo, *J. Chem. Phys.* **73**, 3854 (1980).
    (b) M.-H. Whangbo and E. Canadell, *Acc. Chem. Res.* **22**, 375 (1989).
    (c) M.-H. Whangbo, *Adv. Chem. Ser.* **226**, 269 (1990).
12. T.A. Albright, J.K. Burdett and M. -H. Whangbo, *Orbital Interactions in Chemistry*, Wiley, New York, 1985.
13. R.J. Hamers, in *Scanning Tunneling Microscopy and Spectroscopy: Theory, Techniques and Applications*, D. A. Bonnell (Ed.), VCH Publishers, New York, 1993.
14. C. Kittel, *Introduction to Solid State Physics, 2nd ed.*, Wiley, New York, pp 353-358 (1956).
15. (a) Z.F. Zheng, M.B. Salmeron, E.R. Weber, *Appl. Phys. Lett.* **64**, 1836 (1994).
    (b) J.F. Zheng, X. Liu, N. Newman, E.R. Weber, D.F. Ogletree, M. Salmeron, *Phys. Rev. Lett.* **72**, 1490 (1994).
    (c) M. Maboudian, K. Pond, V. Bressler-Hill, M. Wassermeier, P.M. Petroff, G. A.D. Briggs, W.H. Weinberg, *Surf. Sci.* **275**, L662 (1992).
    (d) R.M. Feenstra, J.A. Stroscio, *J. Vac. Tecnnol. B* **5**, 923 (1987).
    (e) M.-H. Whangbo, J. Ren, S.N. Magonov, H. Bengel, B.A. Parkinson, A. Suna, *Surf. Sci.* **326**, 311 (1995).
16. J. Bardeen, *Phys. Rev. Lett.* **6**, 57 (1961).
17. S.N. Magonov, M.-H. Whangbo, *Adv. Mater.* **6**, 355 (1994).
18. (a) H. Ou-Yang, R. Källebring, R.A. Marcus, *J. Chem. Phys.* **98**, 7565 (1993).
    (b) H. Ou-Yang, R.A. Marcus, B. Källebring, *J. Chem. Phys.*, **100**, 7814 (1994).
19. (a) W. Liang, M.-H. Whangbo, A. Wawkuschewski, H.-J. Cantow, S. Magonov, *Adv. Mater.* **5**, 817 (1993).
    (b) A. Wawkuschewski, H.-J. Cantow, S.N. Magonov, M. Möller, W. Liang, M.-H. Whangbo, *Adv. Mater.*, **5**, 821 (1993).
20. C. Ciraci, in *Scanning Tunneling Microscopy III*, R. Wiesendanger, H.-J. Güntherodt (Eds.), Springer-Verlag, Heidelberg, 1993.
21. (a) C.M. Mate, R. Erlandsson, G.M. McClelland, S. Chiang, *Surf. Sci.* **208**, 473 (1989).
    (b) M. Salmeron, D.F. Ogletree, C. Ocal, H.-C. Wang, G. Neubauer, W. Kolbe, G. Meyers, *J. Vac. Sci. Technol. B* **9**, 1347 (1991).
22. (a) M.-H. Whangbo, W. Liang, J. Ren, S.N. Magonov, A. Wawkuschewski, *J. Phys. Chem.* **98**, 7602 (1994).
    (b) H. Bengel, H.-J. Cantow, S.N. Magonov, L. Monconduit, M. Evain, M.-H. Whangbo, *Surf. Sci. Lett.* **321**, L170 (1994).
    (c) H. Bengel, H.-J. Cantow, S. N. Magonov, L. Monconduit, M. Evain, W. Liang, M.-H. Whangbo, *Adv. Mater.* **6**, 649 (1994).
    (d) H. Bengel, H,-J. Cantow, S. N. Magonov, H. Hillebrecht, G. Thiele, W. Liang, M.-H. Whangbo, *Surf. Sci.*, **343**, 95 (1995).
23. For a review, see: G. Doyen, in *Scanning Tunneling Microscopy III*, R. Wiesendanger, H.-J. Güntherodt (Eds.), Springer-Verlag, Berlin (1993).
24. R.G. Gordon, Y.S. Kim, *J. Chem. Phys.* **56**, 3122 (1972).
25. R.C. Barrett, J. Nogami, C. F. Quate, *Appl. Phys. Lett.* **57**, 992 (1990).

26. (a) H. Schäfer, J. Tillack, *J. Less-Common Metals* **6**, 152 (1964).
    (b) H. G. von Schnering, H. Wöhrle, *Angew. Chem.* **75**, 684 (1963).
    (c) H. Schäfer, H.G. von Schnering, *Angew. Chem.* **76**, 833 (1964).
27. P. Zönnchen, G. Thiele, H. Bengel, S.N. Magonov, C. Hess, C. Schlenker, D.-K. Seo, M.-H. Whangbo, submitted for publication.
28. (a) H. Schäfer, H. G. von Schnering, *Angew. Chem.* **76**, 833 (1964).
    (b) A. Simon, H. G. von Schnering, *J. Less Common Met.* **11**, 31 (1966).
    (c) F. Hulliger, *Structural Chemistry of Layer-Type Phases*, F. Lévy, (ed.) Reidel, Dordrecht, The Netherlands (1976).
29. S.N. Magonov, P. Zönnchen, H. Rotter, H.-J. Cantow, G. Thiele, J. Ren, M.-H. Whangbo, *J. Am. Chem. Soc.* **115**, 2495 (1993).
30. M.-H. Whangbo, E. Canadell, *J. Am. Chem. Soc.* **114**, 9587 (1992).
31. R. Brouwer, F. Jellinek, *Physica B* **99**, 51 (1980).
32. (a) E. Tosati, P. Fazekas, P. *J. Physique.* **37**, C4-165, (1976).
    (b) B. Dardel, M. Grioni, D. Malterre, P. Weibel, Y. Baer, F. Lévy, *Phys. Rev. B*, **45**, 1462 (1992).
33. M.-H. Whangbo, J. Ren, E. Canadell, D. Louder, B.A. Parkinson, H. Bengel, S.N. Magonov, *J. Am. Chem. Soc.* **115**, 3760 (1993).
34. (a) R.V. Coleman, B. Giambattista, P. K. Hansma, W.W. McNairy, C.G. Slough, *Adv. Phys.* **37**, 559 (1988).
    (b) R.V. Coleman, Z. Dai, W.W. McNairy, C.G. Slough, C. Wang, in *Scanning Tunneling Microscopy*, J. A. Stroscio, W. J. Kaiser (Eds.), Academic Press, New York, 1993.
35. (a) A. Meerchaut, J. Rouxel, in *Crystal Chemistry and Properties of Materials with Quasi-One-Dimensional Structures*, J. Rouxel, Ed., Reidel, Dordrecht, The Netherlands, 1986.
    (b) P. Monceau, in *Electronic Properties of Inorganic Quasi-One-Dimensional Compounds, Part II*, P. Monceau, Ed., Reidel, Dordrecht, The Netherlands, 1985.
    (c) M.-H. Whangbo, in *Crystal Chemistry and Properties of Materials with Quasi-One-Dimensional Structures*, J. Rouxel, Ed., Reidel, Dordrecht, The Netherlands, 1986.
36. (a) F. Devreux, *J. Phys. (Les Ulis, Fr.)* **43**, 1489 (1982).
    (b) J.H. Ross, Jr., Z. Wang, C. P. Schlichter, *Phys. Rev. Lett.* **56**, 633 (1986).
    (c) J. H. Ross, Jr., Z. Wang, C. P. Schlichter, *Phys. Rev. B* **41**, 2722 (1990).
37. (a) S. van Smaalen, J.L. de Boer, A. Meetsma, H. Graafsma, H.-S. Sheu, A. Darovskikh, P. Coppens, *Phys. Rev. B* **45**, 3103 (1992).
    (b) S. van Smaalen, J. L. de Boer, P. Coppens, H. Graafsma, *Phys. Rev. Lett.* **67**, 1471 (1991).
38. (a) Z. Dai, C.G. Slough, R.V. Coleman, *Phys. Rev. Lett.* **66**, 1318 (1991).
    (b) Z. Dai, C.G. Slough, R.V. Coleman, *Phys. Rev. Lett.* **67**, 1472 (1991).
39. (a) M.-H. Whangbo, P. Gressier, *Inorg. Chem.* **23**, 1305 (1984).
    (b) E. Canadell, I.-E.I. Rachidi, J. P. Pouget, P. Gressier, A. Meerschaut, J. Rouxel, D. Jung, M. Evain, M.-H. Whangbo, *Inorg. Chem.* **29**, 1401 (1990).
40. The calculated CDW vectors $q_1 = 0.248b^*$ and $q_2 = 0.262b^*$ for a single NbSe$_3$ layer are in good agreement with the b^*-components of the experimental CDW vectors $q_1 = (0, 0.241b^*, 0)$ and $q_2 = (0.5a^*, 0.259b^*, 0.5c^*)$, respectively. The fact that the b^*-components are all close to $0.25b^*$ is best explained by the picture in which two electrons are shared by four bands.
41. J. Ren, M.-H. Whangbo, *Phys. Rev. B* **46**, 4917 (1992).
42. (a) G.W. Stupian, M.S. Leung, *Appl. Phys. Lett.* **51**, 1560 (1987).
    (b) M. Weimer, J. Kramar, C. Bai, J.D. Baldeschwieler, *Phys. Rev. B* **37**, 4292 (1988).
    (c) S. Akari, M. Stachel, H. Birk, S. Schreck, M. Lux, K. Dransfeld, *J. Microsc.* **152**, 521 (1988).
43. F. Lévy, *Nuovo Cimento B* **38**, 359 (1977).
44. (a) S.N. Magonov, H.-J. Cantow, M.-H. Whangbo, *Surf. Sci. Lett.* **318**, L1175 (1994).
    (b) M.-H. Whangbo, J. Ren, S.N. Magonov, H. Bengel, B.A. Parkinson, A. Suna, *Surf. Sci.* **326**, 311 (1995).
45. S.N. Magonov, M.-H. Whangbo, *Surface analysis with scanning tunneling and atomic force microscopy. Experimental and theoretical aspects of image interpretation*, J. Wiley &. Sons, New York, 1996.

ELUCIDATING COMPLEX CHARGE DENSITY WAVE STRUCTURES IN LOW-DIMENSIONAL MATERIALS BY SCANNING TUNNELING MICROSCOPY

HONGJIE DAI, JIE LIU AND CHARLES M. LIEBER

Department of Chemistry and Division of Applied Sciences, Harvard University, Cambridge, MA 02138, USA.

1. Applications Of Scanning Probe Microscopy

An important goal of condensed matter research is to understand how microscopic or atomic level structural and electronic characteristics of a solid determine observable properties like superconductivity and magnetism. This goal is motivated by the recognition that such an understanding will enable scientists ultimately to design rationally bulk solids and nanostructures having predictable properties. Instrumental methodologies that provide a real-space picture of the connectivity of atoms in a solid and/or the local electronic structure are perhaps most appealing since they can probe materials directly in the space (real vs. reciprocal) that we often think, and can directly characterize defects and disorder that play significant roles in determining the properties of solids. Furthermore, real-space probes are required to assess the intrinsic structural and electronic properties of very small material structures that are a focus of the burgeoning area of nanotechnology.

In the past decade, scanning probe microscopies (SPMs), such as scanning tunneling microscopy (STM) and atomic force microscopy (AFM), have rapidly grown into uniquely powerful tools for probing microscopic properties of materials [1-23]. STM can be used to probe the surface structure and electronic states of conducting and semiconducting materials directly on the atomic scale [1-15], while AFM can be used to assess the structure and elastic properties of insulators, semiconductors and conductors at the nanometer scale [16-26]. SPMs are also effective tools for manipulating matter on the atomic to nanometer scales, constructing individual nanostructures, and probing the properties of individual nanostructures [11,26-30].

SPMs are by their very nature highly surface sensitive techniques. The ability to probe the uppermost layers of a material is an important advantage in studies of very small supported structures, such as nanocrystals on surfaces, and in fundamental investigations of growth and catalysis on metal and semiconductor surfaces [31-34]. Surface sensitivity can, however, be a severe limitation in studies designed to probe the bulk properties of a material since the coordinatively

F. W. Boswell and J.C. Bennett (eds.),
Advances in the Crystallographic and Microstructural Analysis of Charge Density Wave Modulated Crystals, 225–257.
© 1999 *Kluwer Academic Publishers.* .

unsaturated surface atoms of three-dimensional (3D) solids typically adopt a different geometrical arrangement (i.e., reconstruct) from those in the bulk. However, highly anisotropic materials--- for example, layered solids possessing strong covalent bonding in two-dimensional (2D) sheets with weak noncovalent bonds holding these layers together--- cleave preferentially along planes defined by the weak noncovalent bonds. The coordination of atoms at the surface cleavage plane in such a low-dimensional solid is similar to that in the bulk (i.e., the covalent bonding is unchanged), and thus these surfaces do not reconstruct. SPM studies of such low-dimensional materials offer the opportunity to probe atomic level structural and electronic properties that are representative of the bulk of a material.

Low-dimensional solids also offer a great richness and complexity of physical phenomena that make them generally an important focal point of condensed matter research. For example, many two-dimensional and one-dimensional metal chalcogenide materials are known to exhibit complex charge density wave (CDW) and spin density wave phases [35], and furthermore it is well-known that the layered structure of the copper oxide superconductors is central to the high-temperature superconductivity that these materials exhibit [36]. STM has been extremely effective in elucidating, not only the atomic structures of these fascinating materials, but also collective phenomena such as the pinning of CDW and magnetic vortex lattices [10-15,37-39]. In the case of charge density wave systems, STM can provide detailed structural information about the atomic lattice and the electron density modulation with a resolution of ~1 Å. Detailed real-space STM pictures also provide an unambiguous approach to understanding complex CDW structures that have been long-standing and controversial problems unanswered by diffraction.

In this chapter, we present recent advances in probing and understanding the microscopic structures of charge density wave phases in low dimensional metal chalcogenide compounds obtained using the STM approach. The scope of the chapter is as follows: first, we will describe the STM technique; then we will review applications of STM to elucidating the low temperature CDW phases in the one-dimensional NbSe$_3$ system and further in the 2D 1T-TaS$_2$ system. Next, we will focus on incommensurate CDW structures in chemically impurity doped 1T-TaS$_2$ materials, and the wealth of physics revealed by real-space studies. Finally, we will conclude with a look at future directions such as CDWs in very small 2D nanocrystals.

2. Background: Scanning Tunneling Microscopy

In this section, the instrumentation and theoretical concepts that are essential for understanding STM studies will be discussed. More detailed reviews can be found elsewhere [1,2,6-9,40,41]. A typical microscope is illustrated schematically in Figure 1. The underlying basis for the operation of the microscope is electron tunneling between a sharp metal tip and a conducting sample. When the tip and sample are brought sufficiently close, their wave functions can overlap. If a bias voltage V is then applied to the sample, a tunneling current I will flow between the sample and tip. Electrons will tunnel from filled electronic states in the tip to empty states in the sample when V is positive, and

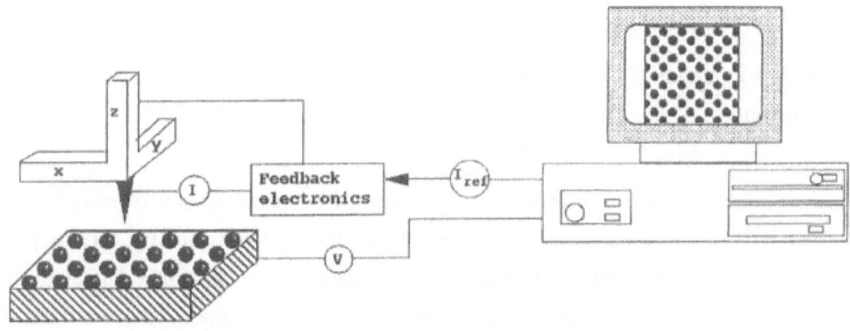

Fig. 1. Schematic view of a tunneling microscope.

conversely, electrons will tunnel from filled sample states to empty tip states when V is negative. The tunneling current that flows when V is applied varies exponentially with the tip-sample separation, and for a typical work function of 4 eV, I decreases 10-fold for an angstrom increase in separation. The strong exponential dependence of the tunneling current on distance enables STM to achieve high vertical resolution. An atomic resolution map of the surface can then be generated by rastering the tip over the sample with angstrom level control using piezoceramic positioners. Experimental images are typically acquired in the constant current mode in which a feedback loop controls the vertical position of the tip above the sample so that I is equal to a reference current (I_{ref}) at all coordinates on the surface. Features in constant current mode images thus correspond to vertical-displacements of the positioner needed to maintain a constant tunneling current.

Essential to the interpretation of such STM data is an understanding of the response of the tunneling current to the barrier properties, applied voltage, etc.; insight into these problems can be obtained from analyses of the tunneling problem [42-46]. As first discussed by Tersoff and Hamann, an expression for I can be readily derived by assuming noninteracting sample and tip wave functions, and then using perturbation theory [42]. In the limit of small bias voltage and low temperature this treatment yields

$$I = (2\pi/\hbar)e^2V\Sigma|M_{st}|^2\delta(E_s-E_f)\delta(E_t-E_f) \qquad (1)$$

where M_{st} is the tunneling matrix element between wave functions on the tip, ψ_t, and sample, ψ_s. As shown by Bardeen [47], the tunneling matrix element can be written as

$$M_{st} = (\hbar^2/2m)\int(\Psi_t^*\nabla\Psi_s - \Psi_s\nabla\Psi_t^*)dS \qquad (2)$$

where the integral corresponds to a surface within the barrier region between the sample and tip. To evaluate M_{st} in a way that the resulting expression for I can be compared quantitatively to STM images in general (i.e., not for one specific choice of sample and tip) requires several approximations. Tersoff and Hamann showed that by assuming the tip forms a locally spherical potential well with only s-wave functions, I could be expressed as

$$I \propto \Sigma_s |\Psi_s(r_0)|^2 \delta(E_s - E_f) \tag{3}$$

By definition the summation is the local density of sample electronic states, $\rho(r_0, E)$, at the center of curvature of the tip

$$\rho(r_0, E) \equiv \Sigma_s |\Psi_s(r_0)|^2 \delta(E_s - E_f) \tag{4}$$

Thus, constant current images correspond to contours of constant density of sample electronic states. Because a CDW causes a variation of the electronic states close to the E_f, STM represents a uniquely sensitive and direct probe of the structure and amplitude of the CDW lattice. In the next section, we review recent STM studies in pure materials and give some background information about CDWs.

3. Charge Density Waves In Pure Materials

3.1 THE ONE-DIMENSIONAL NbSe₃ SYSTEM

A large number of quasi-one-dimensional materials exhibit CDW transitions including NbSe$_3$, TaS$_3$, (TaSe$_4$)$_2$I, K$_{0.3}$MoO$_3$ and organic conductors such as (fluoranthene)$_2$PF$_6$ and (perylene)$_2$Au(maleonitriledithiolate)$_2$ [48-51]. Among them the two most widely studied materials are NbSe$_3$ and K$_{0.3}$MoO$_3$ [51-53]. In this section, we will focus on NbSe$_3$. NbSe$_3$ crystals are built up from a Se trigonal prismatic cage surrounding a metallic Nb chain. The NbSe$_6$ trigonal prisms are stacked on top of each other to form linear chains along the b-axis of a monoclinic unit cell (Figure 2a). In NbSe$_3$, the crystal lattice unit cell ($a_0 = 10.0$ Å, $b_0 = 3.5$ Å, $c_0 = 15.6$ Å) [52] contains three pairs of trigonal chains labeled as types I, II, and III (Figure 2b). These three chains exhibit distinct CDW states [53]. Two independent charge density wave transitions are observed in NbSe$_3$ with transition temperatures of 144 K and 59 K. These transitions were first observed as DC resistivity anomalies in the transport studies by Monceau et al. in the 70's [52]. Subsequent X-ray diffraction and transport studies by Fleming et al. proved the existence of two independent CDWs and showed that these have wave vectors of (0.00, 0.243, 0.00) and (0.5, 0.263, 0.5) at 144 K and 59 K respectively. Hence, both of these CDWs are incommensurate with respect to the atomic lattice; the incommensurability is maintained down to 4.2 K [54].

In the past two decades, NbSe$_3$ has been a model system for experimental condensed matter studies of collective sliding, non-linear phenomena and pinning [49,50,55-59]. The detailed CDW structure has been studied by a variety of techniques including X-ray diffraction [54,60,61], NMR [62] and scanning tunneling microscopy [63]. Synchrotron-radiation X-ray diffraction and NMR studies have found that the 144 K and 59 K CDW modulations are primarily associated with atomic displacements of the type III and type I chains, respectively. These observations are consistent with electronic structure calculations that suggest the type-II chains are not involved in CDW formation [53].

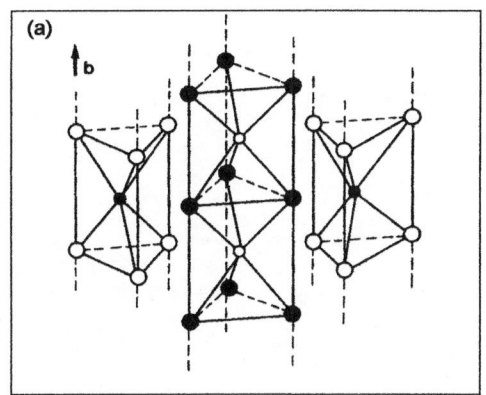

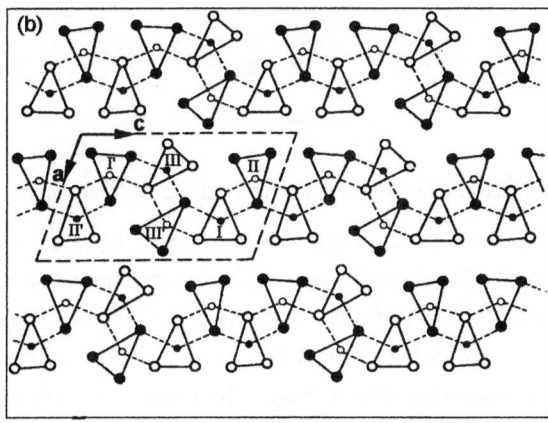

Fig. 2. (a) NbSe₃ chains running along the b axis; (b) The structure of NbSe₃ projected along [0 1 0].

High resolution low-temperature STM studies of NbSe₃ by Coleman and coworkers were also used to probe the complex CDW states in real space [63]. Shown in Figure 3 is an STM image of the b-c plane of NbSe₃ acquired at 4.2 K by Coleman et al. [63]. In contrast to synchrotron X-ray diffraction and NMR studies, CDWs were observed on all three types of chains. Comparison of these 4.2 K images with STM images acquired at 77 K [64-66] led the authors to conclude that the 59 K CDW forms on type I and II chains, while at 144 K a CDW forms on the type III chain. Furthermore, it can be seen that the type I and II chains of adjacent unit cells exhibit CDW modulations that are out of phase by 180°, meaning that the CDW maxima matches the CDW minima on adjacent type I or II chains. The type I and II chains are therefore associated with the low temperature CDW state and have a wavelength of $2c_0$. The CDW on type III chains is found to be in phase in adjacent unit cells, and the repeat distance on each chain is ~$4b_0$; these results are consistent with diffraction data for the high temperature CDW. Besides providing detailed phase information of the complex CDW in NbSe₃, the STM data also showed that the amplitude of CDW modulation on type III chains is large, while the CDW amplitude on type I and II chains is relatively weak.

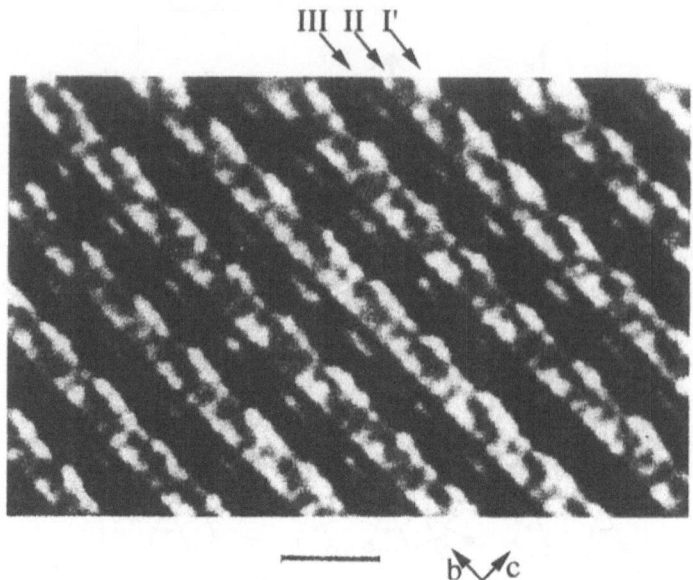

Fig. 3. STM image of charge density waves in NbSe₃ at 4.2 K. Reproduced from Ref. 63.

The STM results of Coleman et al. suggest that all three types of chains in NbSe₃ are conducting and contribute to the Fermi surface sections that are responsible for the CDW formation. These results contradict models [53] and other experiments [60-62] suggesting that the type II chains are nearly insulating and are excluded from CDW formation. As described in another chapter of this book, this conflict was considered by Whangbo [67] and coworkers using electronic band structures calculations and numerical simulation of resulting STM images. This latter work suggests that the character of electrons at the Fermi level is dominated by Nb atoms on type I and III chains. However, due to geometric reasons, STM only detects Se atoms. Their calculations showed that these Se atoms make similar contribution to the LDOS for all type I, II and III chains. Thus, this analysis supports the STM studies of Coleman. These results demonstrate that real space STM imaging and analysis can provide new insight about complex CDW structures that are otherwise difficult to observe and enable a deeper understanding of the electronic structures of one dimensional materials.

3.2 TWO-DIMENSIONAL METAL DICHALCOGENIDES: 1T-TaS₂

The dichalcogenides of the group IV, V, and VI transition metals have layered structures in which the transition metal (M) occupies either octahedral or trigonal prismatic sites between two layers of hexagonally close packed chalcogenide atoms (X) (Figure 4) [68,69]. In general, the MX_2 compounds are quasi-2D materials since the X-M-X layers are held together by strong covalent-ionic bonding, whereas the interactions holding adjacent layers together are much weaker dispersion forces. These materials cleave readily between the weakly bonded layers to yield free surfaces that

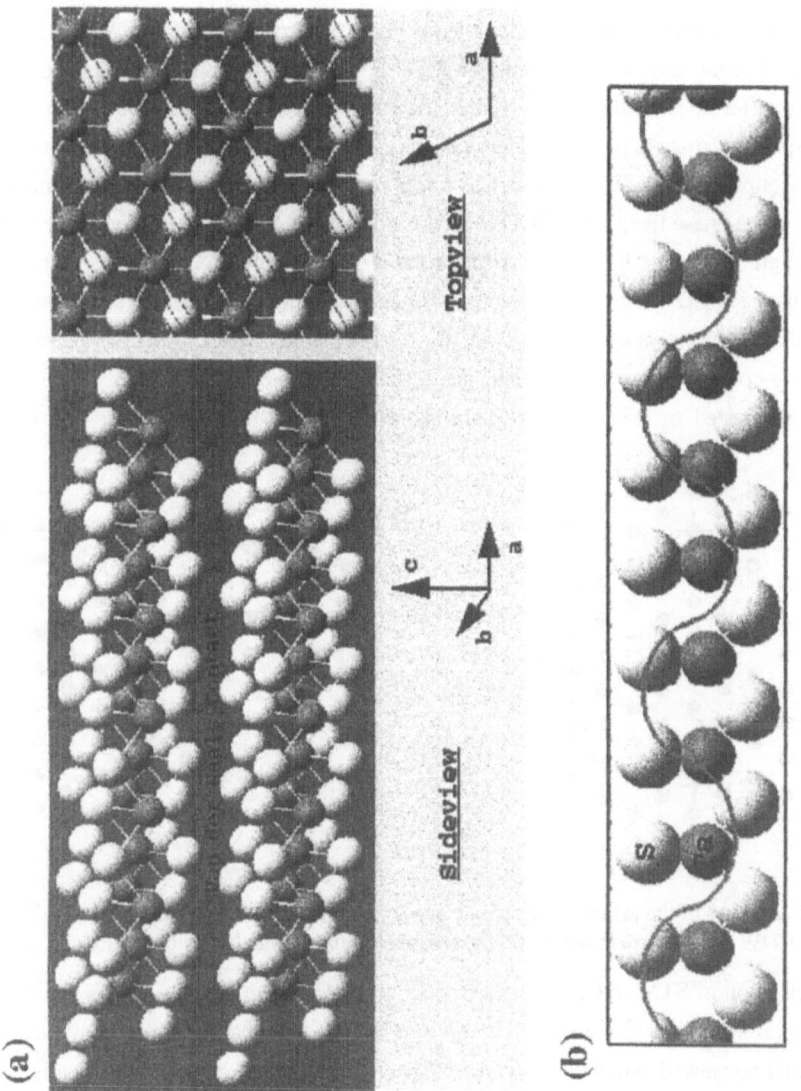

Fig. 4. (a) Schematic view of two M-X-M layers of MX_2 (e.g., M=Ta, X=S) that are bonded primarily via dispersion forces. The sulfur atoms are arranged in hexagonal close packed planes. (b) One-dimensional view of a sinusoidal CDW in a single M-X-M layer.

are structurally and electronically similar to the bulk [68-71].

Among the MX_2 compounds, $1T$-TaS_2 exhibits one of the most complex and interesting CDW phase diagrams. At high temperatures the compound $1T$-TaS_2 is metallic; however, on cooling below 543 K $1T$-TaS_2 exhibits four distinct temperature dependent CDW states [69,70]. A schematic of the CDW and atomic lattice in $1T$-TaS_2 is shown in Figure 4b. Electronic structure calculations have shown that $1T$-TaS_2 exhibits large Fermi surface nesting that leads to CDW formation. Previous X-ray and electron diffraction have showed that the high temperature CDW in $1T$-TaS_2 is incommensurate, while for T < 183 K the CDW rotates 13.9° relative to the atomic lattice to become commensurate [69-71]. The real-space structures of these phases are illustrated schematically in Figure 5. The CDW superlattice of both phases has a regular hexagonal symmetry. In the incommensurate phase the peaks of CDW are located randomly relative to the atomic lattice. In the commensurate phase, the CDW superlattice rotates 13.9° relative to the atomic lattice and expands 2% so that each CDW maxima is oriented over an equivalent atomic lattice site. It was not possible using diffraction techniques, however, to resolve the structures of the intermediate temperature nearly-commensurate and triclinic nearly commensurate CDW phases. These two phase had been suggested to exhibit either uniformly incommensurate structures or domain-like structures [72,73].

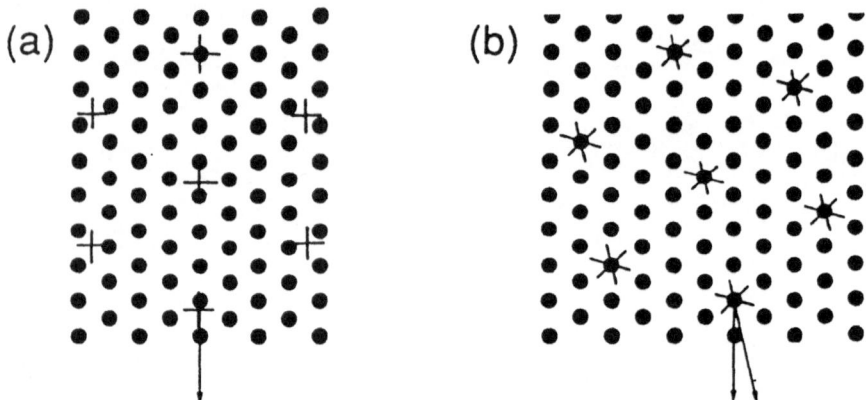

Fig. 5. Schematic topviews of the (a) incommensurate and (b) commensurate CDW phases in $1T$-TaS_2. The CDW maxima are indicated by the crosses and the atomic lattice by the filled.

The first application of STM to CDWs was carried out by Coleman and coworkers in 1985 on $1T$-TaS_2 [74]. This work showed that the charge modulation of a charge density wave and the atomic lattice could be viewed simultaneously in real space. In addition, comparison of early STM data with previous diffraction results demonstrated that the surface sensitive STM experiment truly reflected the bulk properties of $1T$-TaS_2 [10,74]. The ability of STM to determine simultaneously both CDW and atomic lattice positions has subsequently been exploited by our group and others to resolve the complicated structural details of the intermediate temperature CDW phases in TaS_2 that previously could not be determined by diffraction techniques [64,75-78].

STM studies of the real-space structure of 1T-TaS$_2$ at room temperature showed that the nearly-commensurate CDW phase adopts a novel, hexagonal, domain-like structure, in which there is a periodic variation in the CDW amplitude that occurs on a wavelength much larger than the CDW wavelength, λ (Figure 6) [75]. In large area images, the CDW maxima (Figure 6a, filled white circles) define a regular hexagonal superlattice with average wavelength in agreement with

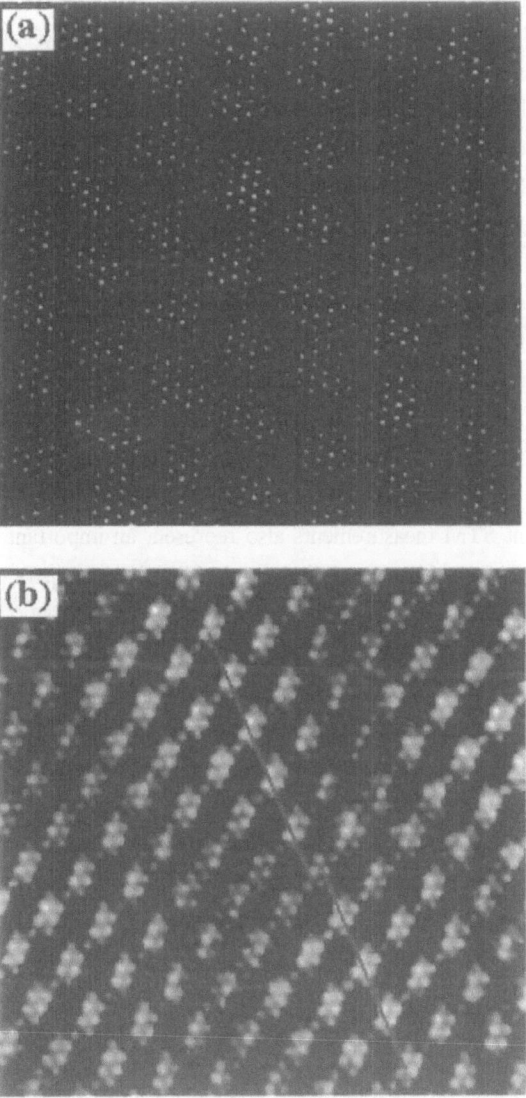

Fig. 6. Room temperature STM images of 1T-TaS$_2$. (a) 50 x 50 nm^2 image showing CDW maxima and domains. (b) Atomic-resolution image of four domains. The domains are highlighted by circular white lines. The white lines indicate the single atom phase shift that occurs between the CDWs in adjacent domains. Reproduced from Ref. 75.

diffraction and STM measurements. The amplitude of the CDW maxima further exhibits a larger periodic modulation that defines domains consisting of relatively high-amplitude CDW maxima separated by lower amplitude domain boundaries. The roughly circular, high amplitude domains are arranged in a hexagonal superstructure with a period of approximately 70 Å at room temperature. Hence, the nearly commensurate CDW exhibits a fascinating hierarchy of structures: the hexagonal atomic lattice with period 3.35 Å, the fundamental hexagonal CDW lattice with period of ~12 Å, and the hexagonal domain superlattice with period 70 Å.

Atomic resolution images of the nearly commensurate phase have provided further insight into the complex structure of the domain phase (Figure 6b). First, these images show that there are well-defined phase shifts of one atomic lattice period between the CDWs in adjacent domains [75,76]. The abrupt changes in CDW phase are clear in real-space images, but are quite difficult to detect in diffraction measurements. Secondly, the high-resolution images show that there is a similar arrangement of atoms at each CDW maxima within a domain, thus indicating that the CDW is approximately commensurate (e.g., see schematic in Figure 5) in the domains. Because the actual angle between the atomic lattice and CDW superlattice can be measured, such images also afford the ability to address quantitatively the issue of commensurability. Significantly, these measurements showed the angle within single domains was that expected for a commensurate CDW, ~13.9°. Taken together, these STM studies unambiguously resolved the long standing controversy about the structure of the nearly commensurate CDW phase in 1T-TaS$_2$.

Temperature-dependent STM measurements also represent an important approach for further probing CDW phases in the transition metal dichalcogenides since temperature variations can be used to assess melting and other phase transitions [10,76,77]. STM studies of the nearly commensurate phase in 1T-TaS$_2$ first demonstrated the power of such measurements [76]. Images recorded between 200 and 350 K show the hexagonal CDW domain structure described above. The period of the domain structure was found to undergo a remarkable and unexpected change with temperature decreasing from 100 Å at 200 K to ~60 Å at 350 K (Figures 7a-c). At still higher temperatures, there is an abrupt loss of the domain structure as the true incommensurate phase is formed (Figure 7d). Significantly, the decrease in domain size appears similar to a second-order melting transition as the sample temperature is raised, however, it is important to point out this corresponds to melting of an electronic (CDW) lattice and not the atomic lattice.

4. Charge Density Wave Pinning

4.1 INTRODUCTION TO CHARGE DENSITY WAVE PINNING

An important goal in condensed matter research is to understand the factors that determine the structure, electronic properties and phase transitions in materials since such knowledge will lead the way to the rational design of new solids with predictable properties [79,80]. Essential to the

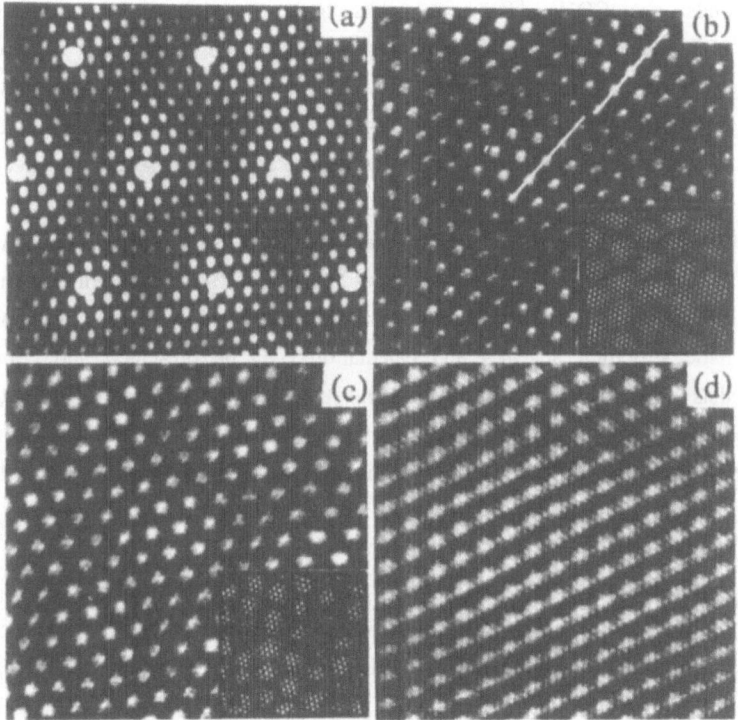

Fig. 7. STM images of 1T-TaS$_2$ recorded at sample temperatures of (a) 242, (b) 298, (c) 349, and (d) 357 K. The images are (a) 30 x 30, (b) 17.5 x 17.5, and (c), (d) 15.5 x15.5 nm^2 in size. The insets in (b) and (c) are 30 x 30 nm^2. The domain center positions for an ideal hexagonal structure are marked with white dots in (a). Reproduced from Ref. 76.

achievement of this goal is a detailed understanding of how material properties vary and can be controlled by atomic level modifications such as substitutional doping. In principle, doping is a straight-forward process with readily predictable effects (e.g., doping semiconductors). However, in many materials, especially those that exhibit cooperative phenomena such as charge density waves and superconductivity, the role of dopants and impurities is not well understood [13,81]. Past difficulties in elucidating the microscopic effects of dopants and impurities on material properties can be traced to the fact that conventional diffraction and spectroscopic techniques provide only an averaged view of a solid [82]. These techniques have not been able to assess unambiguously structural disorder and electronic inhomogeneity in doped solids [13,81,83].

Impurities in charge density wave compounds tend to interact or pin CDWs because their electronic properties differ from the host atoms. It can be envisioned that an impurity in the atomic lattice will attract or repel a CDW maxima, and thus distort the otherwise perfect CDW lattice. However, this qualitative picture encounters much complication due to factors such as the strength of the interaction or pinning, the randomness and concentration of impurities, and the susceptibility (or rigidity) of CDWs to distortion. Previous diffraction studies have found that in metal-doped

TaS$_2$ compounds, the CDW diffraction peaks became progressively broader as dopant concentration is increased [69,71]. These results provided the first structural evidence of CDW pinning by metal dopant atoms, and showed that pinning can distort the overall CDW structure. Nevertheless, a detailed picture of CDW pinning on a microscopic scale was inaccessible from diffraction studies, and thus comparison with theoretical models on structural effects of CDW pinning was not possible.

In the following section, we will focus on STM studies of the Nb-doped 1T-TaS$_2$ system to illustrate the effects of pinning by random impurities in a CDW system. We will first discuss CDW pinning in the framework of a theoretical model, and then present STM images of the pinned CDW state. Systematic structural analyses and calculations will be presented and compared with theoretical models. We will show that a deeper understanding not only of charge density waves, but also of complex yet generic phenomena in two-dimensional systems such as collective pinning, dislocation related hexatic ordering and two-dimensional melting can be obtained using this approach.

4.2 THEORY OF CHARGE DENSITY WAVE PINNING

The one-dimensional Fukuyama-Lee-Rice (FLR) CDW pinning model [84,85] has been the starting point for many studies of CDW-impurity pinning. In the FLR model the electron density distribution $\rho(\mathbf{r})$ is treated as a sinusoidal wave with a constant amplitude ρ_0 modulated over a uniform background electron density ρ_{av}:

$$\rho(\mathbf{r}) = \rho_{av} + \rho_0 \cos (\mathbf{Q} \cdot \mathbf{r} + \phi(\mathbf{r})) \tag{5}$$

The wave vector of the CDW is $\mathbf{Q} = 2\mathbf{k}_f$ and the real-space periodicity of the electron density is π/k_f. The FLR model treats the impurity-CDW interaction within a Ginzburg-Landau framework, where the order parameter is the CDW phase, $\phi(\mathbf{r})$. For a perfect CDW free of impurities, $\phi(\mathbf{r})$ is a constant. Interestingly, if $\phi(\mathbf{r})$ is constant and if the wave vector $\mathbf{Q}$ is incommensurate with the atomic lattice, then the energy of the CDW state is independent of its position relative to the atomic lattice [86].

However, $\phi(\mathbf{r})$ may vary with position when impurities are introduced into the host atomic lattice since impurities can pin the CDW phase. To calculate the expected behavior of the CDW in presence of impurities we consider several limiting cases for a one-dimensional system. In the commensurate state ($\mathbf{Q} = 2\pi/n\mathbf{a}$) the CDW is in registry with the atomic lattice and is thus strongly pinned to the atomic lattice. Impurity pinning does not perturb significantly the commensurate CDW state [87]. On the other hand, an incommensurate CDW ($\mathbf{Q} \neq 2\pi/n\mathbf{a}$) interacts weakly with the underlying lattice. Variations in the lattice potential due to impurity atoms can thus effectively distort or pin an incommensurate CDW. Qualitatively, an attractive pinning potential will distort CDW maxima towards the nearest impurities to minimize the energy. The interaction of the impurities with the incommensurate CDW is examined below.

First, we assume that each impurity has a short-ranged pinning potential and that this potential has the same magnitude for every impurity. The total pinning potential distribution is thus a function of the positions of the randomly quenched impurities:

$$V_{pin}(\mathbf{r}) = V_0 \Sigma_i \delta(\mathbf{r} - \mathbf{R}_i) \tag{6}$$

where R_i is the ith impurity position vector and the sum is over all impurities.

Similar to other solids, the CDW lattice has rigidity that can be quantified by the elastic constant $\kappa \approx hv_f/A_0$, where A_0 is the unit cell area [88]. Because a distortion of the CDW lattice from equilibrium will cost the system elastic energy, it is necessary to account for both pinning and CDW rigidity simultaneously. This is accomplished by computing the potential energy gain associated with pinning and the elastic energy loss due to CDW distortions using the following effective Hamiltonian:

$$H = \int \rho(\mathbf{r}) V_{pin}(\mathbf{r}) d^d\mathbf{r} + \frac{\kappa}{2} \int [\nabla \phi(\mathbf{r})]^2 d^d\mathbf{r} \tag{7}$$

The first term is the pinning energy, the second term is the elastic energy and d is the dimensionality of the system. Combining equations (5), (6) and (7) yields

$$H = \rho_0 V_0 \Sigma_i \cos(\mathbf{Q} \cdot \mathbf{R}_i + \phi(\mathbf{R}_i)) + \frac{\kappa}{2} \int [\nabla \phi(\mathbf{r})]^2 d^d\mathbf{r} \tag{8}$$

The pinning energy depends on the random impurity configuration, and the elastic energy depends on the CDW phase fluctuations caused by pinning.

The phase variation is determined by minimizing the total energy. Two limiting cases, strong and weak pinning, are considered. In the strong pinning regime, the elastic energy is much less than the pinning potential. Hence, the CDW phase is pinned to each impurity site with the phase given by $\mathbf{QR}_i + \phi(\mathbf{R}_i) = -\pi$ (Figure 8). The energy determined for the limiting case of strong pinning is then $-\rho_0|V_0|N_i$, where N_i is the total number of impurities; that is, the energy is simply proportional to the number of impurities. On the other hand, in the weak pinning regime the pinning energy is much smaller than the energy required to deform the CDW. Hence, phase distortions do not occur at each impurity site [$\mathbf{QR}_i + \phi(\mathbf{R}_i) \neq -\pi$], but rather, $\phi(\mathbf{r})$ varies on a length scale much larger than the impurity spacing (Figure 8). Since each term in the sum in equation 8 can be either positive or negative the total pinning energy grows only as $N_i^{1/2}$ (vs. N_i for strong pinning). An interesting consequence of the FLR weak pinning model is that the CDW will be unstable to domain formation for d < 4. These domains should consist of distinct regions containing many impurities within which the CDW phase varies smoothly. The STM studies described below provide a unique test of this prediction.

Quantitatively, a parameter ε is used to characterize the pinning strength and separate the strong and weak pinning regimes:

$$\varepsilon = \frac{\rho_0 V_0}{\kappa n_i^{2/d-1}} \tag{9}$$

where $\rho_0 V_0$ is the pinning potential energy per impurity and $\kappa \cdot n_i^{2/d-1}$ is the dimensionality

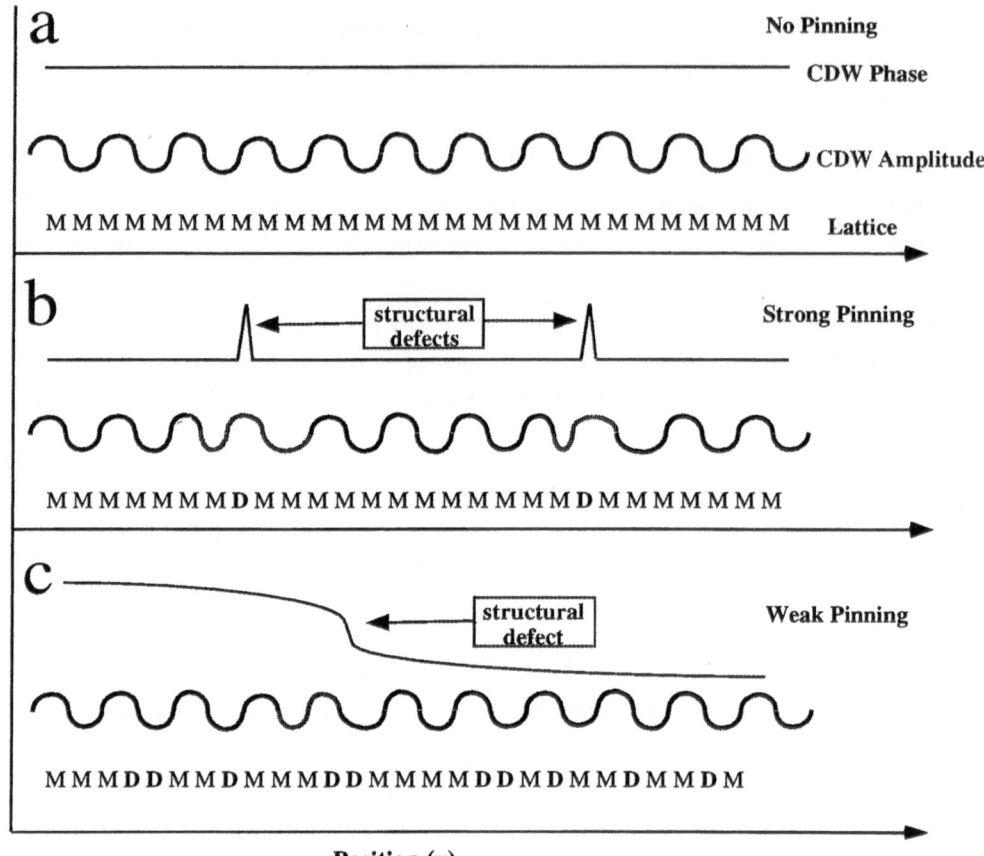

Fig. 8. Plots of the CDW phase, $\phi(r)$, and amplitude versus position for (a) a perfect one-dimensional lattice and (b, c) for lattices that contain impurities, D. The abrupt variation in $\phi(r)$ at each impurity site (b) corresponds to strong pinning. The smooth variation in $\phi(r)$ over a distance containing several impurities (c) corresponds to weak pinning.

dependent elastic energy per impurity [88]. Hence, $\varepsilon \gg 1$ corresponds to the strong pinning regime and $\varepsilon \ll 1$ corresponds to the weak pinning regime.

As indicated above, approximately constant phase domains containing many impurities are expected in the weakly pinning regime. The length scale of these domains can be estimated using scaling arguments of Imry and Ma [89]. Specifically, they showed that the pinning energy scales as $L^{d/2}$ while the elastic energy scales as L^{d-2}. Hence, the pinning energy gain is $\rho_0 V_0 N_i^{1/2} = \rho_0 V_0 n_i^{1/2} L^{d/2}$, and the elastic energy cost is κL^{d-2} (assume $\phi(\mathbf{r})$ varies by $\sim \pi$ over length L). We can thus define a critical length L_c, which defines the average size of constant phase domains, by minimizing the energy with respect to L:

$$L_c = \alpha \cdot \frac{1}{\varepsilon} \cdot d_i, \qquad (10)$$

where α is of order unity and $d_i = n_i^{-1/d}$ is the average impurity spacing.

Specific defects in the CDW lattice arising from CDW-impurity pinning have also been studied theoretically. For example, McMillan showed that the energy cost for a single dislocation diverges logarithmically with the size of the sample [87]; this behavior is similar to that of a dislocation in an ordinary solid. The existence of dislocations can be explained using the scaling argument outlined above. First, a dislocation loop with size L will contain $n_i L^d$ impurities. The pinning energy gain is $\sim \rho_0 V_0 n_i^{1/2} L^{d/2}$ and the elastic energy cost for creating the dislocation is $\sim \kappa$ up to a logarithm [87,90]. Since the total energy is the same as in the treatment of domains we can obtain the critical length given by equation 10. In the case of dislocations, L_c is comparable to the average separation between dislocation.

Lastly, it is important to point out that the FLR model only considers phase fluctuations of the CDW; amplitude fluctuations of the CDW are ignored. A justification for ignoring amplitude fluctuations is that the energy cost for an amplitude fluctuation is significantly higher than that of the long wavelength phase fluctuations treated by FLR. Recent work by Coppersmith shows, however, that amplitude fluctuations do have nontrivial effects on the CDW state [91]. Hence, we believe that the relative importance of phase and amplitude fluctuations is an area that requires additional theoretical and experimental study.

4.3 RESULTS OF CHARGE DENSITY WAVE PINNING IN $Nb_xTa_{1-x}S_2$

4.3.1 STM Images

Here we review studies of niobium-substituted tantalum disulfide, $Nb_xTa_{1-x}S_2$. Substitution of isoelectronic Nb(IV) for Ta(IV) represents the weakest perturbation on the potential that is possible with metal substitution, and thus serves as the best test of the weak pinning regime [15,92,93]. The effect of Nb-substitution on the macroscopic transport properties of these $Nb_xTa_{1-x}S_2$ materials have been assessed by variable-temperature resistivity measurements (Figure 9). In general, it is well-established [13,69,70] that Nb-substitution (and other metal substitutions) decreases the transition temperature from the high-temperature incommensurate (IC) CDW state to the lower temperature nearly-commensurate state. Qualitatively, the suppression in the transition temperature can be rationalized by assuming that the IC state is stabilized by impurity pinning. Macroscopic measurements do not, however, address the fundamental details of pinning that were outlined above. To develop a microscopic understanding of the CDW-impurity interaction we turn to STM studies of the CDW state in these materials.

High resolution STM images of the $Nb_xTa_{1-x}S_2$ materials are shown in Figures 10-12. Images of the IC CDW phase of undoped TaS_2 samples [92,93] exhibit a well-ordered hexagonal structure. In contrast, substitution of Nb causes disorder in the IC CDW lattice (Figures 10-12). The images of the Nb-doped materials exhibit areas in which the CDW lattice has hexagonal order and regions containing defects. These defects introduce disorder into the CDW lattice. The predominant defects observed in the x(Nb) < 0.07 samples are dislocations. Dislocations are formed by the

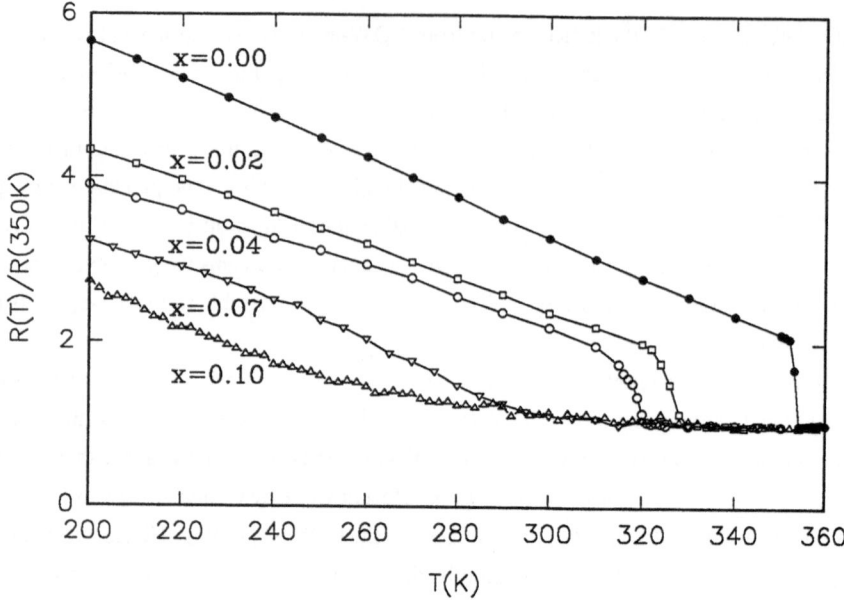

Fig. 9. Normalized resistance versus temperature curves recorded on the $Nb_xTa_{1-x}S_2$ materials. Reproduced from Ref. 15.

insertion of extra half rows of CDW sites in the lattice; black lines in Figure 10a highlight the creation of two dislocations in a x(Nb) = 0.02 sample. Importantly, there is a significant strain field associated with the dislocations [90]. The CDW can relax the strain field by locally deforming through site positional shifts and rotations (Figure 10a), although these deformations introduce disorder into the lattice. We have also found one or more CDW site vacancies close to the dislocation core in many of the images (Figure 10b). This latter observation is interesting since it is completely analogous to the classical behavior of vacancies and dislocations in the atomic lattice of materials [90]. Since STM can be used to study directly such defects we believe that it will be interesting in the future to explore the generality of the $M_xTa_{1-x}S_2$ materials as models for the behavior of crystal defects and the dynamics of vacancy/dislocation formation.

As the impurity concentration increases to x(Nb) = 0.04 a higher density of defects (predominantly dislocations) is observed in the CDW lattice (Figure 11). The CDW rows near the dislocations are distorted as discussed above. However, in areas free of dislocations the CDW lattice is locally ordered. We have highlighted several dislocation cores by constructing Burgers loops (Figure 11). The Burgers loop consists of an equal number of steps along each lattice direction; the loop will remain open if it encloses a single dislocation. The vector pointing from start to end of the loop, the Burgers vector, uniquely defines the dislocation [90]. We find that the Burgers vectors defining the CDW dislocations in the Nb-doped samples occur along each of the three crystallographic axes, and thus it is apparent that impurity-induced dislocations occur randomly in the CDW lattice. In the samples containing higher impurity concentrations, x(Nb) = 0.07 and 0.10 the STM images exhibit extended defects (Figure 12). The CDW lattice in these latter samples

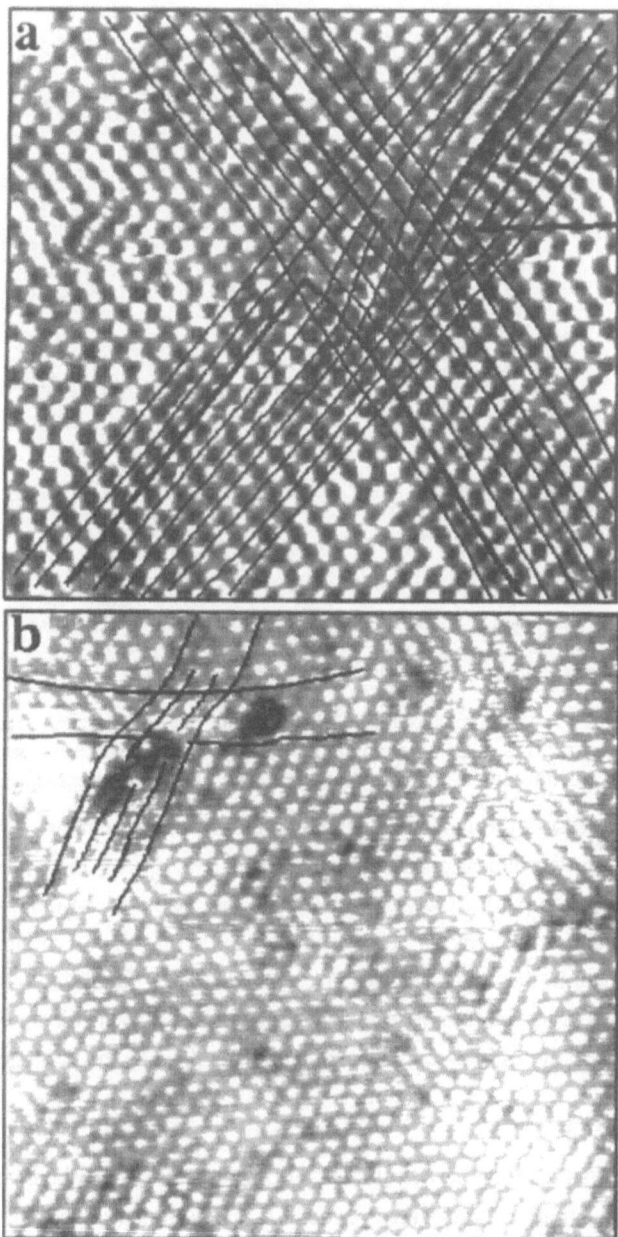

Fig. 10. STM images of the incommensurate CDW state in $Nb_{0.02}Ta_{0.98}S_2$. Black lines highlight the insertion of an extra row of CDW maxima. The extra rows in (a) are highlighted with heavy black lines. Vacancies are also seen in (b). Reproduced from Ref. 92.

exhibits significant disorder with regions of hexagonal order and positional order extending only several lattice constants.

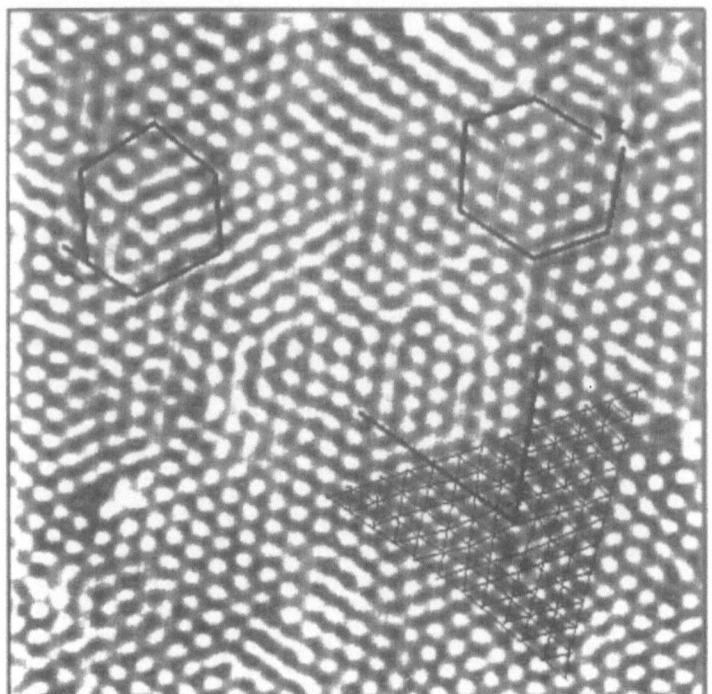

Fig. 11. STM images of the incommensurate CDW state in $Nb_{0.04}Ta_{0.96}S_2$. Black lines highlight the insertion of extra rows of CDW maxima in the lattice. Two distinct Burgers loops and Burgers vectors are drawn to highlight dislocations in the CDW lattice. Reproduced from Ref. 92.

To qualitatively assess the nature of CDW pinning in this system we first calculate the Nb impurity spacing. In two-dimensions the impurity spacing is $d_i(x) = n_i^{-1/2}$, and thus the average impurity spacing for the $x(Nb) = 0.02$, 0.04, 0.07 and 0.10 samples are 0.80, 0.57, 0.43 and 0.36 CDW lattice constants, respectively. This estimate shows that there is more than one impurity atom/CDW maximum in the doped samples investigated. Since the STM images indicate that the CDW lattice exhibits order for at least several lattice constants (which will contain many impurities) these results are suggestive of the weak pinning regime described by the FLR model.

4.3.2 Topological Defects

The topology and density of defects can also be assessed by quantitative image analysis [15,93-96]. The quantitative analysis involves defining the location (x, y coordinates) of each CDW maximum and the unique nearest neighbors in the lattice. Once the lattice points are determined, a sweep-line algorithm [97] is used to construct the Voronoi diagram for the lattice. The Voronoi diagram uniquely defines the nearest neighbors of all lattice points [98]. To illustrate defects in the lattice we triangulate the Voronoi diagram by drawing "bonds" from all CDW lattice points to their nearest neighbors. This resulting plot is termed the Delauney triangulation diagram. In our CDW system fully coordinated lattice sites are indicated by six bonds, while defects contain a smaller or

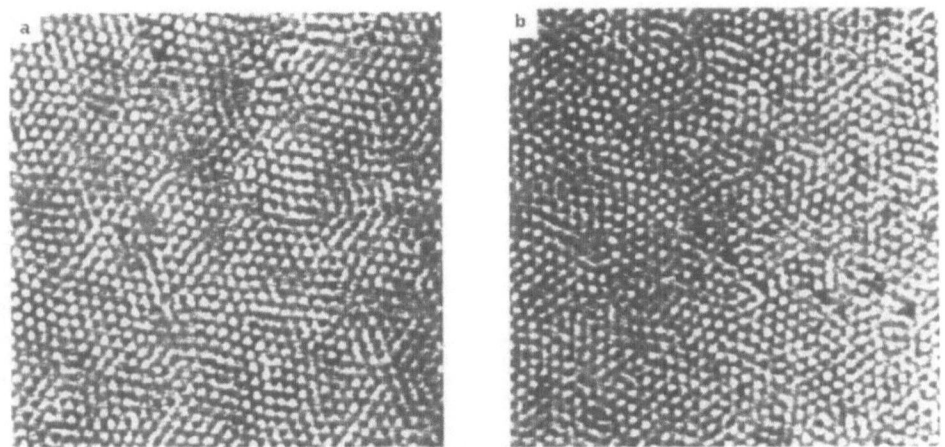

Fig. 12. STM images of the incommensurate CDW phase in (a) $Nb_{0.07}Ta_{0.93}S_2$ and (b) $Nb_{0.1}Ta_{0.9}S_2$.
Reproduced from Ref. 92.

greater number of bonds. Typical results obtained from the analysis of x(Nb) = 0, 0.02, 0.04, 0.07 and 0.10 samples are shown in Figure 13. In these triangulation diagrams we have highlighted the defect sites by shading. Inspection of images recorded on pure TaS_2 samples show that the CDW lattice is free of topological defects; that is, all of the lattice sites are six-fold coordinated (Figure 13a). The triangulations explicitly show, however, that the Nb-doped materials have topological defects in the CDW lattice. At low concentrations of impurities, x(Nb) = 0.02 and 0.04, we find that the dislocations consist of five-fold/seven-fold disclination pairs (Figures 13b, c). Extended defect networks are also obvious in the triangulation data for the x(Nb) = 0.07 and 0.10 samples. These extended topological defects consist of dislocations, vacancies and free disclinations.

This data has also been used to evaluate the average separation between dislocations. We find that the average spacing between dislocations in the x(Nb) = 0.02, 0.04, 0.07 and 0.1 samples is 12, 8, 5, and 3 lattice constants, respectively. As stated above, the average separation between Nb impurity atoms is 0.80, 0.57, 0.43 and 0.26 CDW lattice constants, respectively. The average spacing between dislocations is thus always greater than the average impurity spacing; that is, impurity pinning is a collective effect. Hence, this analysis shows that CDW pinning in the $Nb_xTa_{1-x}S_2$ materials is weak.

The topological defects highlighted in Figure 13 strongly affect positional order. For example, a dislocation displaces lattice points by about a lattice constant, and thus will destroy positional order on a scale of the size of dislocation loop. Areas of the CDW lattice that are free from dislocations should, however, exhibit a smooth variation in the CDW phase; i.e., the domains in the FLR model. In Figure 14 we show a typical large scale image with several phase coherent regions highlighted to demonstrate this important point. This image also shows clearly that the local crystallographic axes of adjacent domains are rotated with respect to one another.

Further analysis of large area images containing different impurity concentrations shows that the

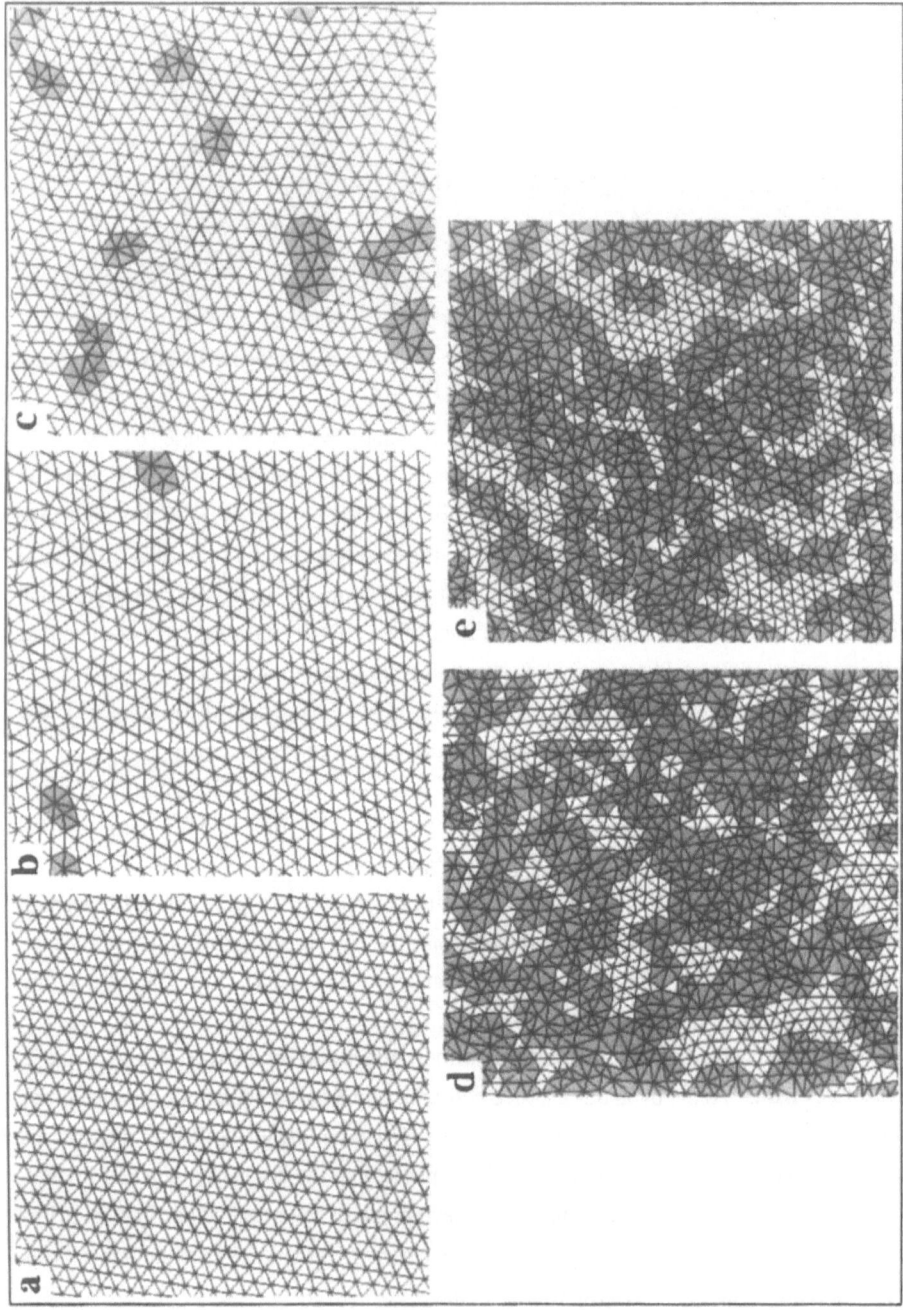

Fig. 13. Delauney triangulations of the STM images recorded on $Nb_xTa_{1-x}S_2$ crystals where (a), (b), (c), (d) and (e) correspond to x = 0, 0.02, 0.04, 0.07 and 0.10, respectively. Lattice sites that do not have six fold coordination are highlighted by shading. Reproduced from Ref. 93.

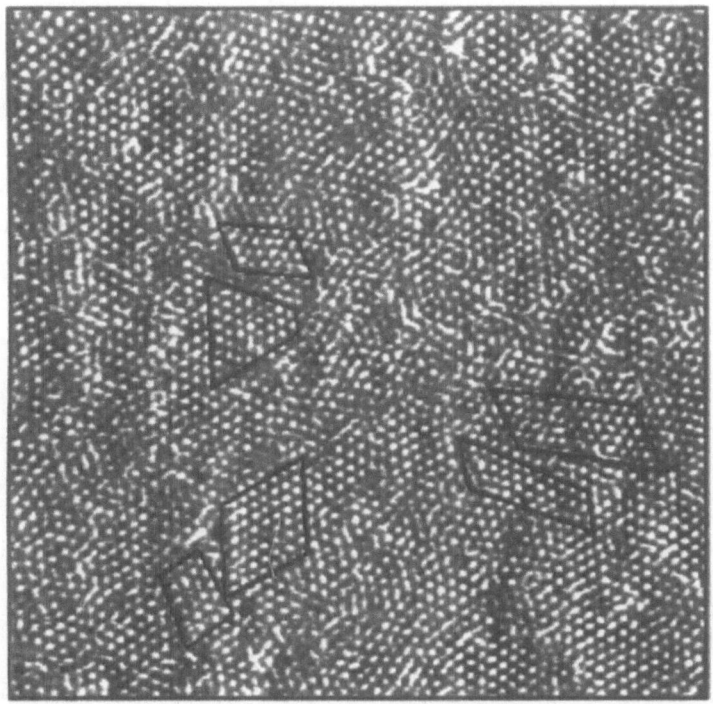

Fig. 14. A large area STM image of the incommensurate CDW lattice of a $Nb_{0.07}Ta_{0.93}S_2$ sample. Several adjacent domains containing positionally ordered lattices are highlighted with black lines. Note that adjacent domains are rotated with respect to each other. Reproduced from Ref. 92.

area of the ordered regions (the domain size) shrinks as the impurity concentration increases. This behavior follows directly from equation 10. Importantly, if we take the dislocation separation as the critical length, L_c, and using the impurity spacing d_i given above, we find that the calculated value of the FLR pinning parameter ε is ~ 0.1. These results quantitatively confirm that the system is in the weakly pinned regime.

4.3.3 Two-Dimensional Order

Above we have characterized the nature of CDW pinning and defects. More generally, it is important to examine the consequences of the disorder produced by weak pinning since this phenomena is important to a number of important physical systems [95,96]. To quantify disorder, we first examine the structure factor $S(\mathbf{k})$.

$$S(\mathbf{k}) = |\rho(\mathbf{k})|^2 \tag{11}$$

where $\rho(\mathbf{k}) = \int \rho(\mathbf{r})e^{i\mathbf{k}\cdot\mathbf{r}}d\mathbf{r}$ is the Fourier transform of the number density of lattice points. $S(\mathbf{k})$ provides an average of the structural effects of pinning. The results for the Nb-substituted materials are shown in Figure 15. $S(\mathbf{k})$ for pure TaS_2 (not shown) exhibit sharp six-fold symmetric peaks (Bragg peaks). These peaks broaden both radially and angularly as the impurity concentration increases to 0.04. For the x(Nb) = 0.07 and 0.10 the first order Bragg peaks have broadened to form a ring whose intensity has a six fold modulation. The angular broadening is due to CDW rotations that were discussed above. This broadening indicates a loss of orientational order; however, it is not possible to provide a quantitative measure of the orientational disorder from this S($\mathbf{k}$) data. In addition, the radial widths of the Bragg peaks indicate that the translational correlation length is short in all of the Nb-doped materials. Notably, our S($\mathbf{k}$) data are qualitatively similar to the results expected for second order melting transition in 2D [92-94,96,99,100].

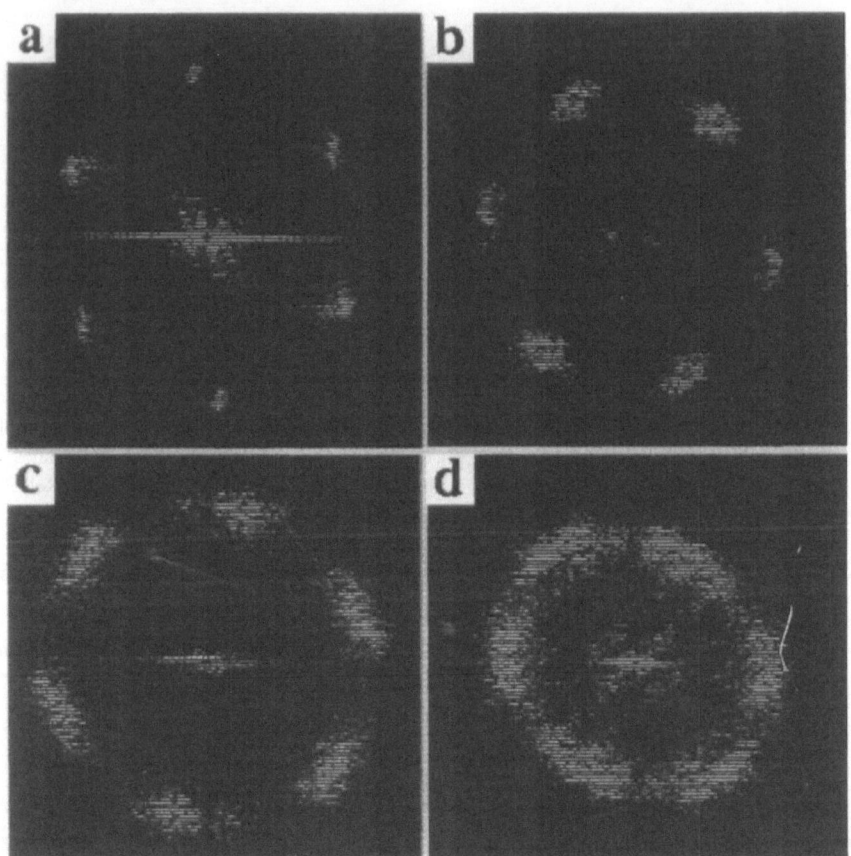

Fig. 15. Structure factors corresponding to the CDW lattices of (a) x = 0.02, (b) x = 0.04, (c) x = 0.07, and (d) x = 0.10 $Nb_xTa_{1-x}S_2$ materials. Reproduced from Ref. 15.

To provide a more quantitative measure of the order in this system and how it is affected by impurity pinning we have investigated systematically the radial distribution, translational correlation and orientational correlation functions. The radial distribution function is defined as g(r) = <n(r)>/n_0, where n(r) is the point density at a distance r from the origin of the structure and n_0 is the average density of points. For a general structure, g(r) tends to 1 as r goes to infinity; that is, the distribution of lattice points at large distances appears uniform. Plots of g(r) determined from

STM images of the $Nb_xTa_{1-x}S_2$ materials are presented in Figure 16. For the pure sample, g(r) oscillates strongly out to more than 10 lattice constants. The persistent oscillation indicates that positional order is long range. When the impurity level increases to 0.02 and 0.04 the amplitude of the oscillations in g(r) decreases and the oscillations die out much more rapidly than in g(r) for the pure material (Figures 16a-c). For the x(Nb) = 0.07 or 0.10 materials the amplitude of the peaks in g(r) is further reduced and those peaks only persist to the second or third nearest neighbors. These results indicate that Nb impurity pinning induced disorder destroys translational order in this CDW system.

The order of the CDW system can be further characterized by calculating the translational and orientational correlation functions. These functions were first introduced by Mermin [101] in his theoretical studies of the positional and orientational order of 2D solids. In two dimensions, the translational correlation function is defined as

$$G_T(\mathbf{r}) = <\psi(\mathbf{0})\psi(\mathbf{r})^*> \tag{12}$$

where
$$\psi(\mathbf{r}) = \frac{1}{3}\sum_{i=1}^{3} e^{i\mathbf{G}_i \cdot \mathbf{r}} \tag{12a}$$

is the translational order parameter for a hexagonal lattice at position r. This order parameter consists of the local Fourier components of the number density of lattice points at reciprocal lattice vectors G. The translational correlation function is equivalent to the Debye-Waller factor and its Fourier transform gives the structure factor S(k). Thus, it measures how lattice points are correlated positionally in real space.

The orientational correlation function for a 2D lattice with hexagonal symmetry is defined as

$$G_6(\mathbf{r}) = <\psi_6(\mathbf{0})\psi_6(\mathbf{r})^*>, \tag{13}$$

where
$$\psi_6(\mathbf{r}) = \sum_{i=1}^{nn} e^{i6\theta_i(\mathbf{r})} \tag{13a}$$

is the orientational order parameter at position $\mathbf{r}$. $\theta(\mathbf{r})$ is the local bond angle with respect to a given direction. Bonds correspond to the lines that connect the nearest neighbors (nn) defined by the Voronoi analysis and Delauney triangulation discussed earlier. The sum is over all nn bonds of a point at position $\mathbf{r}$. This definition for the orientational correlation function is also valid for liquids since the average coordination number of a 2D liquid is also six.

It is instructive to examine the general behavior of these correlation functions before we proceed with the analysis of the CDW system. First, translational and orientational order of a crystal lattice are often coupled to one another. For example, a grain boundary rotates the area (volume) of one part of a crystal lattice relative to an adjacent one; this rotation disrupts both translational and orientational order. However, it is also possible for $G_T(r)$ and $G_6(r)$ to exhibit very different dependencies with r. First, theoretical studies have shown that long wavelength phonon fluctuations destroy only translational order in a 2D solid; orientational order will remain long-range [101,102].

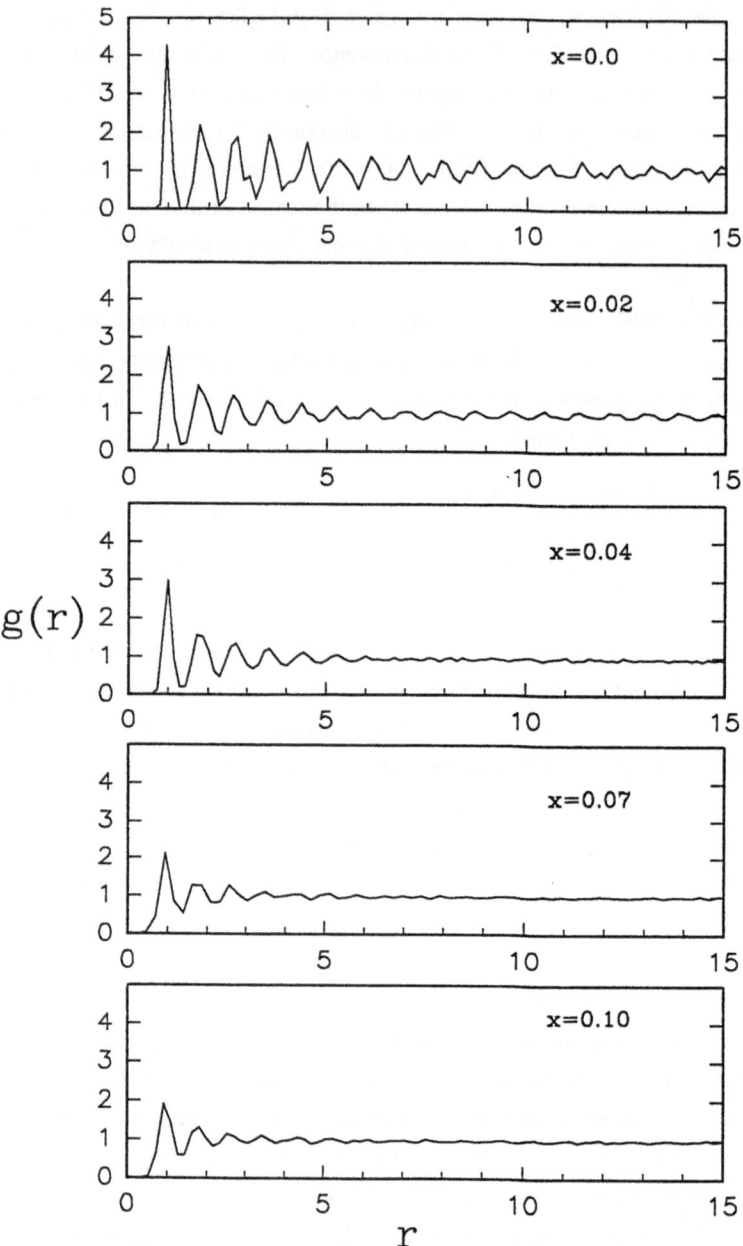

Fig. 16. Radial distribution functions for the $Nb_xTa_{1-x}S_2$ materials. The x-axis corresponds to CDW lattice constants. Reproduced from Ref. 93.

Secondly, Halperin and Nelson [103,104] have shown that dislocations affect translational and orientational order differently in 2D crystals: translational order is destroyed and decays exponentially, while orientational order is quasi long-ranged and decays algebraically.

Calculated results for $G_T(\mathbf{r})$ and $G_6(\mathbf{r})$ for the $Nb_xTa_{1-x}S_2$ materials are shown in Figure 17. $G_T(\mathbf{r})$ and $G_6(\mathbf{r})$ decay very slowly over 20 lattice constants for the incommensurate CDW lattice of the pure sample. This quasi long-range order is indicative of a crystalline state. In contrast, all of the samples that contain Nb impurities exhibit a rapid decay of translational order. This decay in $G_T(\mathbf{r})$ can be fit reasonably well to an exponential of the form $G_T(\mathbf{r}) = \exp(-r/\xi_T)$ where ξ_T is the translational correlation length. The ξ_T's are 7-10, 3-6, 2-3, 1-2 CDW lattice constants for the x(Nb) = 0.02, 0.04, 0.07 and 0.10 samples, respectively. These results demonstrate quantitatively that the dislocations and other defects arising from weak pinning destroy translational order. In contrast, we find that the orientational order decays slowly for the x(Nb) = 0.02 and 0.04 samples, although $G_6(\mathbf{r})$ decays rapidly for x(Nb) = 0.07 and 0.10 (Figure 17). If the x(Nb) = 0.02, 0.04, 0.07, and 0.10 $G_6(\mathbf{r})$ data are fit with an exponential, $G_6(\mathbf{r}) = \exp(-r/\xi_6)$, we obtain orientational correlation lengths (ξ_6) of 200, 100, 11 and 5 lattice constants, respectively. A better fit to $G_6(\mathbf{r})$ for x(Nb) = 0.02 and 0.04 is obtained, however, using an algebraic decay, $G_6(\mathbf{r}) = r^{-\eta}$, with η = 0.03 and 0.12, respectively.

The simultaneous observation of long-range orientational order and short range translational order in the x(Nb) = 0.02 and 0.04 samples is strongly suggestive of the hexatic state that was proposed on the basis of theoretical studies of 2D melting driven by topological defects [103,104]. This theory predicts that 2D melting involves two continuous phase transitions. First, a crystalline solid phase that has quasi long range translational order and long range orientational order ($G_T(\mathbf{r}) = r^{-\xi}$; $G_6(\mathbf{r}) \sim 1$) undergoes a second order transition to the hexatic phase that has short range translational order and quasi long range orientational order: $G_T(\mathbf{r}) = \exp(-r/\xi_T)$; $G_6(\mathbf{r}) = r^{-\eta}$. At higher temperature, the hexatic undergoes a second continuous transition to a liquid phase that has short range translational and orientational order: $G_T(\mathbf{r}) = \exp(-r/\xi_T)$; $G_6(\mathbf{r}) = \exp(-r/\xi_6)$. The results for the CDW phase in $Nb_xTa_{1-x}S_2$ show similar behavior, and thus we suggest that the CDW lattice evolves from crystalline (x = 0) through hexatic glass (x = 0.02 - 0.04) to liquid-like (x > 0.07) states. We assign the intermediate state to a hexatic glass (versus equilibrium hexatic) since the impurity distribution in these materials is quenched. The fundamental difference between the 2D melting theory and our study is that the topological defects arise from thermal fluctuations in the former and from pinning to a quenched impurity distribution in our work. Since statistical averaging differs for equilibrium versus quenched disorder [105], we have compared further the CDW system to equilibrium melting.

First, we can estimate the power law exponent η using the relationship $\eta = 9c/\pi$, where c is a fractional area of dislocation cores, derived for dislocations in thermal equilibrium [94]. The values of η calculated in this way, 0.02 and 0.13, are in excellent agreement with the values obtained from fits to $G_6(\mathbf{r})$ for x(Nb) = 0.02 and 0.04. This agreement suggests that dislocations arising from impurity pinning are responsible for the decay in orientational correlation. Although the average dislocation spacing (ξ_D) is similar to ξ_T, ξ_D is always larger than ξ_T at the smallest impurity concentration (Table I). The fact that $\xi_T < \xi_D$ in the x(Nb)=0.02 samples indicates that factors in

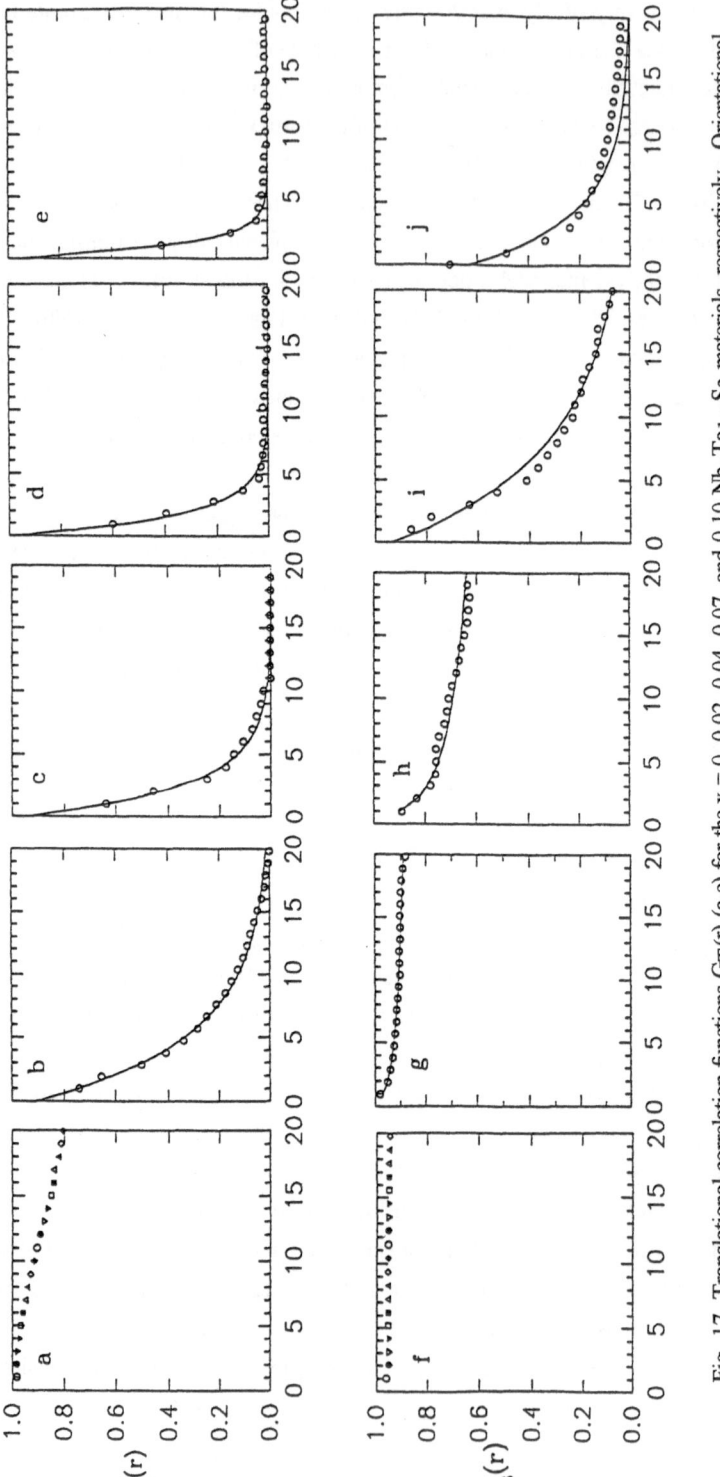

Fig. 17. Translational correlation functions $G_T(r)$ (a-e) for the x = 0, 0.02, 0.04, 0.07, and 0.10 $Nb_xTa_{1-x}S_2$ materials, respectively. Orientational correlation functions $G_6(r)$ (f-j) for x = 0, 0.02, 0.04, 0.07 and 0.10, respectively. The x-axes in these figures correspond to CDW lattice constants. The points represent the experimental data and the lines correspond to exponential fits, $\exp(-r/\xi)$, (a-e) and (i, j) or power law fits, $r^{-\eta}$, (g, h). Reproduced from Ref. 93.

addition to dislocations decrease translational order; this observation contrasts the prediction of equilibrium theory. However, we can explain this result (i.e., $\xi_T < \xi_D$) within the context of weak pinning theory. As discussed above, the CDW phase varies smoothly from $-\pi$ to π in regions separated by dislocations (i.e., within domains). The smooth variation in phase causes small distortions of the CDW lattice positions that reduce the translational order. Indeed, Chudnovsky has explicitly shown how elastic distortions (i.e., smooth variations in the phase vs. a singular phase variation at a dislocation) destroy the positional order of weakly pinned lattices [106]. Hence, a combination of the equilibrium melting and weak pinning theories provide a better description of the changes in translational correlation for this metal-substituted CDW system.

TABLE I.
Important length scales in the $Nb_xTa_{1-x}S_2$ CDW system.
(the units for d_i, ε_D, ε_T, ε_6 are CDW lattice constants)

$Nb_xTa_{1-x}S_2$	d_i	ε_D	ε_T	ε_6	η
x = 0.02	0.80	12	7 - 10	200	0.02
x = 0.04	0.57	8	3 - 6	100	0.13
x = 0.07	0.43	5	2 - 3	11	—
x = 0.10	0.36	3	1 - 2	5	—

4.3.4 Other Comments: Pinning and Dimensionality

The analysis discussed above indicates that the CDW state in $Nb_xTa_{1-x}S_2$ can be treated as a two-dimensional system. Because dimensionality plays an essential role in determining the physical properties of materials (e.g., the physics of critical phenomena), we examine further the consequences of the dimensionality of a system below.

The CDW pinning theory developed earlier shows that different effects should arise from pinning in systems of different dimensionality. First, domain formation should occur for a weakly pinned CDW for d < 4 [84,85,89]. This prediction for domain formation does not, however, lead to a distinguishable criteria in our system. Secondly, the impurity concentration dependence for pinning also varies significantly for systems of different dimensionality. This variation is readily apparent upon examination of the parameter ε in equation 9. For a 1D CDW chain, $\varepsilon \approx \rho_0 V_0 / n_i \kappa \propto n_i^{-1}$. Hence, in a one-dimensional system, the pinning should be strong if the impurities are dilute and pinning should be weak if the impurities are dense. In 2D, however, $\varepsilon \approx \rho_0 V_0 / \kappa$. This indicates that the pinning strength (weak or strong) should be independent of impurity concentration. Our studies are consistent with this prediction since the results for the $Nb_xTa_{1-x}S_2$ materials indicate that pinning is weak for $0 \le x \le 0.1$.

As indicated above, melting in 2D also differs fundamentally from melting in 3D. The solid to liquid phase transition in a 3D solid is an abrupt first order phase transition with discontinuities in the thermodynamic parameters. However, 2D melting is characterized by two continuous phase transitions that are mediated by topological defects. The first transition is from a crystalline solid to a hexatic phase and is caused by paired dislocations unbinding into isolated dislocations. The second transition is from the hexatic phase to a liquid phase and is caused by dislocation unbinding into isolated disclinations. Our investigations of the Nb-doped materials bear a strong resemblance to the theoretical predictions for 2D melting, and do not exhibit an abrupt phase transition based on our analysis of the structural order. However, these systems are clearly not ideal 2D systems and it is expected there is at least small 3D couplings (i.e., between layers). The effect of interlayer coupling has been considered theoretically by McMillan [87]. He argued that weak coupling of the CDW phase between layers will provide 3D stiffness to the CDW, and hence, it should not undergo a 2D melting transition. To clarify this issue, it will be necessary to assess experimentally the strength of the interlayer interactions and the 3D stiffness of the CDW. In the future these parameters could be probed by determining thermodynamic quantities (e.g., heat capacity) and the structure as a function of temperature.

A central result presented above is the direct elucidation of the structural manifestations of quenched impurity pinning. We have identified three different phases of the pinned IC CDW lattice, these are: (1) crystalline solid; (2) hexatic glass; and a (3) liquid-like state. It is interesting to ask whether there are observable consequences of these novel structural phases. Indeed, we find that these structurally distinct CDW states can be mapped to distinct macroscopic transport behavior. Resistivity data from the pure material exhibit a sharp first-order phase transition. However, in the hexatic glass regime ($x(Nb)$ = 0.02 and 0.04) resistance versus temperature plots show a broadened phase transitions. Finally, in the liquid-like state ($x(Nb)$ 0.07) no distinct transition is observed in the resistivity data. Thus, the three structurally distinct CDW phases may correspond to true thermodynamic phases. Other physical measurements, such as heat capacity and magnetic susceptibility, should also be carried out to confirm this suggestion.

5. New Directions: CDWs In Nanocrystals

It is widely recognized that Fermi surface nesting plays a vital role in CDW formation [56,75,77,86-88,107-112]. The Fermi surface of a nanocrystal can be modified by varying the size and shape of the crystal, and furthermore, for sufficiently small sizes the very concept of a Fermi surface will be ill-defined. As mentioned above, STM is an exciting tool for research on single nanoscale structures since it can in principle both create and probe the properties of very small structures. For example, STM has been used to manipulate individual atoms and molecules into structures [27,29,113-115], to probe quantum behavior in nanostructures 27,117], and to lithographically pattern surfaces [117-125].

Recently, we discovered a new approach to fabricate well defined nanocrystals on the surface of layered transition metal dichalcogenide [126]. We can create T phase $TaSe_2$ nanocrystals in the surface layer of $2H\text{-}TaSe_2$ crystals by STM tip induced local solid-solid phase transition. Furthermore, it is possible to control the size of the T-phase $TaSe_2$ nanocrystals by varying the applied bias voltage during the tip-induced modification as shown in Figure 18. The three T-phase nanocrystals in these images have dimensions of 70, 35 and 7 nm. The two large nanocrystals (Figure 18a, b) exhibit relatively uniform commensurate CDW states. The CDW superlattice in both nanocrystals also show different domains that likely result from pinning to atomic lattice defects or the T/H-phase interface. In the smallest nanocrystal (Figure 18c), both the intensity and wavelength of the CDW are distorted relative to the uniform state observed in single crystals or the large nanocrystals. The CDW amplitude is larger at the center of the triangular nanocrystal, and furthermore, the wavelength appears to decrease from 12.5 Å at the center to 10.0 Å at the edge. We believe these results demonstrate that it is indeed possible to probe effects of size on the properties of a CDW state, although future systematic investigation will be needed to define the CDW physics in this new size regime.

6. Concluding Remarks

In this chapter, we have reviewed some of the recent advances in elucidating charge density wave structural and electronic properties in low dimensional metal chalcogenide compounds by scanning tunneling microscopy. Probing and understanding charge density waves in both pure and doped materials has been presented. In the one-dimensional pure $NbSe_3$ system, STM research has shown that all three atomic chains in this material are involved in forming the CDW phases at low temperature. In the two-dimensional $1T\text{-}TaS_2$ compound, our STM studies have unambiguously identified the structure of the nearly-commensurate CDW state and thus resolved a long-standing problem. We have also shown how STM can be used to elucidate systematically CDW pinning and disorder with studies of $Nb_xTa_{1-x}S_2$ solids. Analyses of STM images demonstrate that Nb impurities introduce topological defects into the CDW lattice. Quantitative analysis of the density of dislocations and comparisons of this data with theoretical scaling arguments demonstrates that the pinning of the CDW by Nb impurities is in the weak-pinning regime. In addition, we have shown how pinning affects the translational and orientational order of the 2D CDW lattice. Calculations of the translational and orientational correlation functions suggest that the CDW lattice evolves continuously from a crystalline solid, $x(Nb)$ $= 0$, to a hexatic glass, $0 < x(Nb) \leq 0.04$, and finally to an amorphous state, $x(Nb) \geq 0.07$. Finally, we have shown that the structural evolution of the CDW lattice with increasing impurity concentration has many analogous features to equilibrium melting in 2D, although there are differences that likely arise from the quenched disorder and pinning in $Nb_xTa_{1-x}S_2$. We believe that $M_xTa_{1-x}S_2$ materials will be ideal systems for future studies since

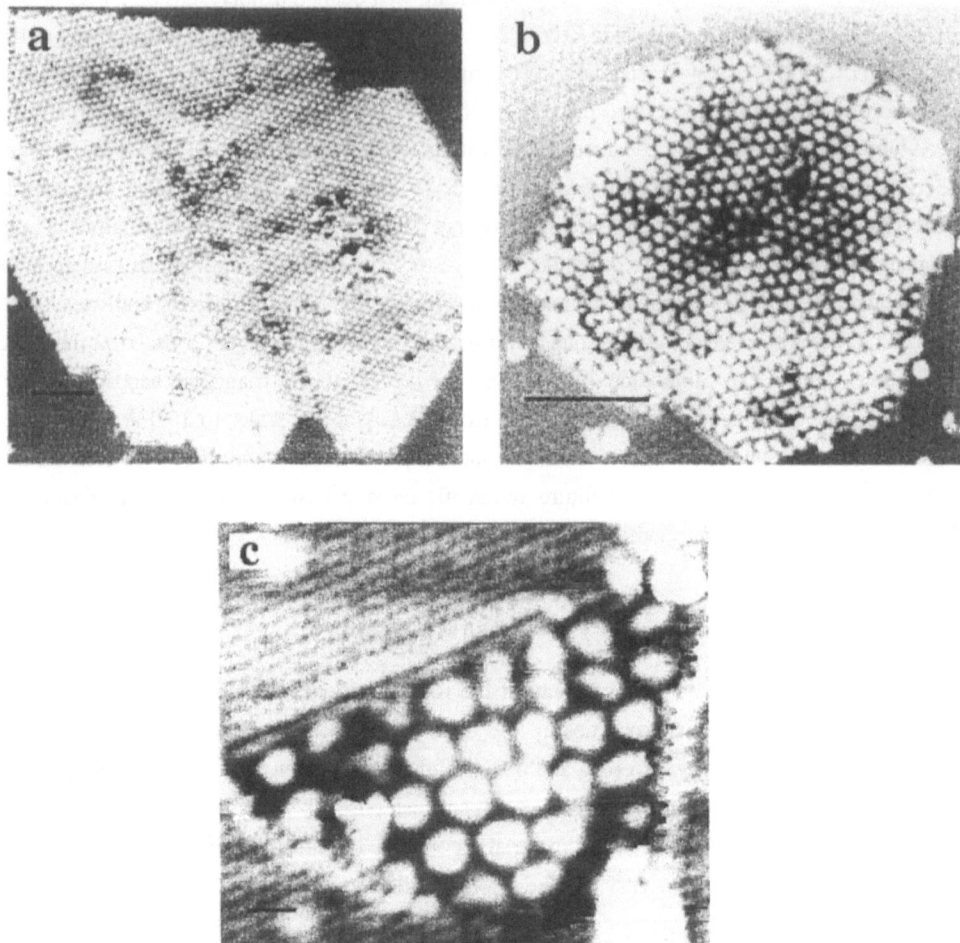

Fig. 18. STM images of T-phase nanocrystals that were created using different applied voltage. The T-phase nanocrystals in these images are all located near the center of the image frame. The nanocrystals in (a), (b) and (c) were made using voltages of -1400 mV, -1300 mV and -1200 mV respectively. The scale bar in (a), (b) and (c) are 10, 10 and 1 nm, respectively. Reproduced from Ref. 127.

the impurity concentration and potential can be varied systematically, and since the CDW-lattice coupling can be changed through variations in the temperature. Hence, experimental and theoretical investigations of this system will lead to a much deeper understanding of pinning and non-equilibrium disorder in 2D systems. In a more general sense, it is clear from this work that STM can provide key insight into understanding the microscopic properties of other complex low-dimensional solids [4]. Future investigations of such materials will undoubtedly lead to significant advances in our understanding of materials.

Acknowledgment

C.M.L. acknowledges support from the National Science Foundation (DMR-9306684), the Air Force Office of Scientific Research (AFOSR 94-1-0010) and the Materials Science and Engineering Center of the National Science Foundation (DMR-9400396).

References

1. G. Binnig, H. Rohrer, C. Gerber, and E. Weibel, *Phys. Rev. Lett.*, **49**, 57 (1982).
2. G. Binnig and H. Rohrer, *Sci. Am.*, *August*, 50 (1985).
3. C.F. Quate, *Phys. Today*, **39,** 26 (1986).
4. H.J. Guntherodt and R. Wiesendanger, Eds., *Scanning Tunneling Microscopy*, Springer-Verlag, Berlin, 1992.
5. C.J. Chen, *Introduction to Scanning Tunneling Microscopy*, Oxford University Press, New York, 1993.
6. P. K. Hansma and J. Tersoff, *J. Appl. Phys*, **61**, R1 (1987).
7. Y. Kuk and P. J. Silverman, *Rev. Sci. Instru.*, **60**, 165 (1989).
8. J. E. Griffith and G. P. Kochanski, *Annu. Rev. Mater. Sci.*, **20**, 194 (1990).
9. P. Avouris, *J. Phys. Chem.*, **94**, 2246 (1990).
10. R.V. Coleman, B. Giambattista, P.K. Hansma, A. Johnson, W.W. McNairy, and C.G. Slough, *Adv. Phys.*, **37**, 559 (1988).
11. C.M. Lieber, J. Liu, and P.E. Sheehan, *Angew. Chem. Int. Ed. Engl.*, **35**, 687 (1996).
12. C.M. Lieber, *C&EN News,* April 18, 28(1994).
13. X.L. Wu and C.M. Lieber, *Progress in Inorganic Chemistry*, S. J. Lippard, Ed., John Wiley & Sons, Inc, New York, 1991.
14. C.M. Lieber and X.L. Wu, *Acc. Chem. Res.*, **24**, 170 (1991).
15. H. Dai and C.M. Lieber, *Annu. Rev. Phys. Chem.*, **44**, 237 (1993).
16. Q. Zhong, D. Inniss, K. Kjoller, and V.B. Elings, *Surf. Sci. Lett.*, **290**, L688 (1993).
17. M.T. Beal-Monod, C. Bourbonnais, and V.J. Emery, *Phys. Rev. B*, **34**, 7716 (1986).
18. G. Binnig, C.F. Quate, and C. Gerber, *Phy. Rev. Lett.*, **56**, 930 (1986).
19. R.J. Colton and J.S. Murday, *Naval Res. Rev.*, **3**, 2 (1988).
20. P.K. Hansma, V.B. Elings, O. Marti, and C.E. Braker, *Science*, **242**, 209 (1988).
21. C.F. Quate, *Surface Science*, **299/300**, 980 (1994).
22. J.P. Spatz, S. Sheiko, M. Moller, R.G. Winkler, P. Reineker, and O. Marti, *Nanotechnology*, **6**, 40 (1995).
23. S. Yoon, H. Dai, J. Liu, and C.M. Lieber, *Science*, **265**, 215 (1994).
24. J. Frommer, *Angew. Chem. Int. Ed. Engl.*, **31**, 1298 (1992).
25. D. Rugar and P. Hansma, *Physics Today*, October, 23 (1990).
26. P.E. Sheehan and C.M. Lieber, *Science*, **272**, 1158 (1996).
27. M.F. Crommie, C.P. Lutz, and D.M. Eigler, *Science*, **262**, 218 (1993).
28. P. Avouris, *Acc. Chem. Res.*, **28**, 95 (1995).
29. J.A. Stroscio and D.M. Eigler, *Science*, **254**, 1319 (1991).
30. P. Avouris, *Acc. Chem. Res.*, **27**, 159 (1994).
31. J.C. Patrin and J.H. Weaver, *Phys. Rev. B*, **49**, 17913 (1993).
32. D. Rioux, R.J. Pechman, M. Chander, and J.H. Weaver, *Phys. Rev. B*, **50**, 4430 (1994).
33. K. Koguchi, T. Matsumoto, and T. Kawai, *Science*, **267**, 71 (1995).
34. M. Kanai, T. Kanai, K. Motui, X.D. Wang, T. Hashizume, and T. Sakurai, *Surface Science*, **326**, L619 (1995).
35. F.J. DiSalvo and T.M. Rice, *Physics Today,* April, 23 (1979).
36. D.M. Ginsburg, *Physical Properties of High Temperature Superconductors*, World Scientific, Singapore, 1989.
37. H.F. Hess, *Methods of Experimental Physics*, Academic Press, Inc., New York, 1993.
38. H.F. Hess, R.B. Robinson, and J.V. Waszczak, *Phys. Rev. Lett.*, **64**, 2711 (1990).
39. H.F. Hess, R.B. Robinson, R.C. Dynes, J.M. Valles, and J.V. Waszczak, *Phys. Rev. Lett.*, **62**, 214 (1989).
40. G. Binnig and H. Rohrer, *Angew. Chem. Int. Ed. Engl.*, **26**, 606 (1987).
41. R.J. Hamers, *Annu. Rev. Phys. Chem.*, **40**, 351 (1989).

42. J. Tersoff and D.R. Hamann, *Phys. Rev. B.*, **31**, 805 (1985).
43. J. Tersoff, *Phys. Rev. B*, **41**, 1235 (1990).
44. N.D. Lang, *Phys. Rev. Lett.*, **56**, 1164 (1986).
45. A. Selloni, P. Caenevalli, P.E. Tosatti, and C.D. Chen, *Phys. Rev. B*, **33**, 5770 (1986).
46. C.J. Chen, *Phys. Rev. Lett.*, **65**, 448 (1990).
47. J. Bardeen, *Phys. Rev. Lett.*, **6**, 57 (1963).
48. S. Kagoshima, H. Nagasawa, and T. Sambongi, *One-Dimensional Conductors*, Springer-Verlag, New York, 1985.
49. R.E. Thorne, *Physics Today*, May, 42 (1996).
50. *Electronic Properties of Inorganic Quasi-One-Dimensional Compounds*, Ed. P. Monceau, Reidel, Boston, 1985.
51. *Low-Dimensional Electronic Properties of Molybdenum Bronzes and Oxides*, Ed. C. Schlenker, Kluwer, Dordrecht, 1989.
52. P. Monceau, N.P. Ong, and A.M. Portis, *Phys. Rev. Lett.*, **37**, 602 (1976).
53. N. Shima and H. Kamimura, *Theoretical Aspects of Band Structure and Electronic Properties of Pseudo-One-Dimensional Solids*, Reidel, Boston, 1985.
54. R.M. Fleming, D.E. Moncton, and D.B. McWhan, *Phys. Rev. B*, **18**, 5560 (1978).
55. G. Gruner, *Rev. Mod. Phys*, **60**, 1129 (1988).
56. J. McCarten, D.A. Dicarlo, M.P. Maher, T.L. Adelman, and R.E. Thorne, *Phys. Rev. B*, **46**, 4456 (1992).
57. G. Gruner and A. Zettl, *Phys. Rep.*, **119**, 117 (1985).
58. L.P. Gorkov and G. Gruner, *Charge-Density Waves in Solids*, North-Holland, Amsterdam, 1989.
59. D. Jerome, *Low-Dimensional Conductors and Superconductors*, Vol. 155, Plenum, 1987.
60. S. van Smaalen, J.L. DeBoer, P. Coppen, and H. Graafsma, *Phys. Rev. Lett.*, **67**, 1471 (1991).
61. S. van Smaalen, J.L. DeBoer, A. Meersma, H. Graafsma, H.S. Sheu, A. Darovskikh, and P. Coppens, *Phys. Rev. B*, **45**, 3103 (1992).
62. J.H. Ross, Z. Wang, and C.P. Schlichter, *Phys. Rev. Lett.*, **56**, 633 (1986).
63. Z. Dai, C.G. Slough, and R.V. Coleman, *Phys. Rev. Lett.*, **67**, 1472 (1991).
64. C.G. Slough, W.W. McNairy, R.V. Coleman, J. Garnaes, C.B. Prater, and P.K. Hansma, *Phys. Rev. B*, **42**, 9255 (1990).
65. C.G. Slough and R.V. Coleman, *Phys. Rev. B*, **40**, 8042 (1989).
66. C.G. Slough, B. Giambattista, A. Johnson, W.W. McNairy, and R.V. Coleman, *Phys. Rev. B*, **39**, 5496 (1989).
67. J. Ren and M. H. Whangbo, *Phys. Rev. B*, **46**, 4917 (1992).
68. R.M.A. Leith and J.C. Terhell, *Preparation and Crystal Growth of Materials with Layered Structures*, Reidel, Dordretch, 1977.
69. J.A. Wilson, F.J. DiSalvo, and S. Mahajan, *Adv. Phys.*, **24**, 117 (1975).
70. F.J. DiSalvo, J.A. Wilson, B.G. Bagley, and J.V. Waszczak, *Phys. Rev. B.*, **12**, 2220 (1975).
71. R.L. Withers and J.A. Wilson, *J. Phys. C: Solid State Phys.*, **19**, 4809 (1986).
72. K. Nakanishi, H. Takatera, Y. Yamada, and H. Shiba, *J. Phys. Soc. Jpn.*, **43**, 1509 (1977).
73. K. Nakanishi and H. Shiba, *J. Phys. Soc. Jpn.*, **43**, 1839 (1977).
74. R.V. Coleman, B. Drake, P.K. Hansma, and C.G. Slough, *Phys. Rev. Lett.*, **55**, 394 (1985).
75. X.L. Wu and C.M. Lieber, *Science*, **243**, 1703 (1989).
76. X.L. Wu and C.M. Lieber, *Phys. Rev. Lett.*, **64**, 1150 (1990).
77. R.E. Thomson, U. Walter, E. Ganz, J. Clark, and A. Zettle, *Phys. Rev. B.*, **38**, 10734 (1988).
78. B. Giambattista, C.G. Slough, W.W. McNairy, and R.V. Coleman, *Phys. Rev. B*, **41**, 10082 (1990).
79. F. J. DiSalvo, *Science*, **247**, 649 (1992).
80. *Physics Through 1990's: Condensed Matter Physics*, Natl. Acad. Press,, Washington, 1986.
81. Z. Zhang, , Harvard University, 1993.
82. A.K. Cheetham and P. Day, *Solid State Chemistry: Techniques*, Clarendon, Oxford, 1987.
83. X.L. Wu, Y. Wang, Z. Zhang, J. Huang, and C.M. Lieber, *Science*, **248**, 1211 (1990).
84. H. Fukuyama and P.A. Lee, *Phys. Rev. B*, **17**, 535 (1978).
85. P.A. Lee and T.M. Rice, *Phys. Rev. B*, **19**, 3970 (1979).
86. H. Frohlich, *Proc. Roy. Soc. A.*, **223**, 296 (1954).
87. W.L. McMillan, *Phys. Rev. B.*, **12**, 1187 (1975).
88. H. Matsukawa and H. Takayama, *J. Phys. Soc. Japan*, **56**, 1507 (1987).
89. Y. Imry and S.-K. Ma, *Phys. Rev. Lett.*, **35**, 1399 (1975).
90. F.R.N. Nabarro, *Theory of Dislocations*, Dover, New York, 1967.

91. S.N. Coppersmith, *Phys. Rev. Lett.*, **65**, 1044 (1990).
92. H. Dai, H.-F. Chen, and C.M. Lieber, *Phys. Rev. Lett.*, **66**, 3183 (1991).
93. H. Dai and C.M. Lieber, *Phys. Rev. Lett.*, **69**, 1576 (1992).
94. D.R. Nelson, M. Rubinstein, and F. Spapen, *Philos. Mag. A.*, **46**, 105 (1982).
95. C.A. Murray, P.L. Gammel, D.J. Bishop, D. J. Mitzi, and A. Kapitulnik, *Phys. Rev. Lett.*, **64**, 2312 (1990).
96. R. Seshadri and R.M. Westervelt, *Phys. Rev. Lett.*, **66**, 2774 (1991).
97. S. Fortune, *Algorithmica*, **2**, 153 (1987).
98. G.F. Voronoi, *J. Reine Agnew. Math.*, **134**, 198 (1908).
99. C.A. Murray, W.O. Sprenger, and R.A. Wenk, *Phys. Rev. B.*, **42**, 688 (1990).
100. D.G. Grier, C.A. Murray, C.A. Bolle, P.L. Gammel, D.J. Bishop, D.B. Mitzi, and A. Kalpitulnik, *Phys. Rev. Lett.*, **66**, 2270 (1991).
101. N.D. Mermin, *Phys. Rev.*, **176**, 250 (1968).
102. R.E. Peierls, *Helv. Phys. Acta.*, **7**, 81 (1923).
103. B.I. Halperin and D.R. Nelson, *Phys. Rev. Lett.*, **4**, 121 (1978).
104. D.R. Nelson and B.I. Halperin, *Phys. Rev. B.*, **19**, 2457 (1979).
105. S.-K. Ma, *Modern Theory of Critical Phenomena*, Benjamin/Cummings, Reading, 1976.
106. E.M. Chudnovsky, *Phys. Rev. B.*, **43**, 7873 (1991).
107. W.L. McMillan, *Phys. Rev. B.*, **14**, 1976 (1976).
108. L. F. Mattheiss, *Phys. Rev. B.*, **8**, 3719 (1973).
109. H.W. Myron and A.J. Freeman, *Phys. Rev. B*, **15**, 885 (1977).
110. A.W. Overhauser, *Phys. Rev.*, **128**, 1437 (1962).
111. A. W. Overhauser, *Phys. Rev.*, **167** (1968).
112. M.-H. Whangbo, R.J. Cava, F.J. DiSalvo, and R.M. Fleming, *Solid State Communications*, **43**, 277 (1982).
113. L.J. Whitman, J.A. Stroscio, R.A. Dragoset, and R.J. Celotta, *Science*, **251**, 1206 (1991).
114. I.-W. Lyo and P. Avouris, *Science*, **253**, 173 (1991).
115. T.A. Jung, R.R. Schlitter, J.K. Gimzewski, H. Tang, and C. Joachim, *Science*, **271**, 181 (1996).
116. P. Avouris and I.-W. Lyo, *Science*, **264**, 942 (1994).
117. S. Rubel, M. Trochet, E.E. Ehrichs, W.F. Smith, and A.L. d. Lozanne, *J. Vac. Sci. Technol. B*, **12**, 1894 (1994).
118. J. A. Dagata, *Science*, **270**, 1625 (1995).
119. E.A. Dobisz and C.R.K. Marrian, *Appl. Phys. Lett.*, **58**, 2526 (1991).
120. A.D. Kent, T.M. Shaw, S.V. Molar, and D.D. Awschalom, *Science*, **262**, 1249 (1993).
121. A. Zaluska, L. Zaluski, and A. Witek, *Mat. Sci. Eng.*, *A*, **122**, 251 (1989).
122. U. Staufer, R. Wiesendanger, L. Eng, L. Rosenthaler, H.-R. Hidber, H.-J. Guntherodt, and N. Garcia, *J. Vac. Sci. Technol. A*, **6**, 537 (1988).
123. A. Sato and Y. Tsukamoto, *Nature*, **363**, 431 (1993).
124. A. Sato and Y. Tsukamoto, *Advanced Materials*, **6**, 79 (1994).
125. A. Sato, S. Momose, and Y. Tsukamoto, *J. Vac. Sci. Technol. B, Microelectron. Nanometer Struct.*, **13**, 2832 (1995).
126. J. Zhang, J. Liu, J. Huang, P. Kim and C.M. Lieber, *Science*, **274**, 757 (1996).

INDEX OF SUBJECTS